AF411174

Responses of Forest Ecosystems to Nitrogen Deposition

Responses of Forest Ecosystems to Nitrogen Deposition

Editor

Frank S. Gilliam

MDPI • Basel • Beijing • Wuhan • Barcelona • Belgrade • Manchester • Tokyo • Cluj • Tianjin

Editor
Frank S. Gilliam
University of West Florida
USA

Editorial Office
MDPI
St. Alban-Anlage 66
4052 Basel, Switzerland

This is a reprint of articles from the Special Issue published online in the open access journal *Forests* (ISSN 1999-4907) (available at: https://www.mdpi.com/journal/forests/special_issues/Nitrogen).

For citation purposes, cite each article independently as indicated on the article page online and as indicated below:

LastName, A.A.; LastName, B.B.; LastName, C.C. Article Title. *Journal Name* **Year**, *Volume Number*, Page Range.

ISBN 978-3-0365-2047-6 (Hbk)
ISBN 978-3-0365-2048-3 (PDF)

Cover image courtesy of Frank S. Gilliam.

Contents

About the Editor vii

Frank S. Gilliam
Responses of Forest Ecosystems to Nitrogen Deposition
Reprinted from: *Forests* **2021**, *12*, 1190, doi:10.3390/f12091190 1

Jin Lee, Masahiro Nakamura and Tsutom Hiura
Effects of Large-Scale Nitrogen Fertilization on Insect–Plant Interactions in the Canopy of Tall
Alder Trees with N_2-Fixing Traits in a Cool Temperate Forest
Reprinted from: *Forests* **2021**, *12*, 210, doi:10.3390/f12020210 5

**Xiankai Lu, Qinggong Mao, Zhuohang Wang, Taiki Mori, Jiangming Mo, Fanglong Su and
Zongqing Pang**
Long-Term Nitrogen Addition Decreases Soil Carbon Mineralization in an N-Rich Primary
Tropical Forest
Reprinted from: *Forests* **2021**, *12*, 734, doi:10.3390/f12060734 15

Qinggong Mao, Hao Chen, Cong Wang, Zongqing Pang, Jiangming Mo and Xiankai Lu
Effect of Long-Term Nitrogen and Phosphorus Additions on Understory Plant Nutrients in a
Primary Tropical Forest
Reprinted from: *Forests* **2021**, *12*, 803, doi:10.3390/f12060803 25

**Thomas Dirnböck, Heike Brielmann, Ika Djukic, Sarah Geiger, Andreas Hartmann,
Franko Humer, Johannes Kobler, Martin Kralik, Yan Liu, Michael Mirtl and Gisela Pröll**
Long- and Short-Term Inorganic Nitrogen Runoff from a Karst Catchment in Austria
Reprinted from: *Forests* **2020**, *11*, 1112, doi:10.3390/f11101112 37

**Florian Achilles, Alexander Tischer, Markus Bernhardt-Römermann, Ines Chmara,
Mareike Achilles and Beate Michalzik**
Effects of Moderate Nitrate and Low Sulphate Depositions on the Status of Soil Base Cation
Pools and Recent Mineral Soil Acidification at Forest Conversion Sites with European Beech
("Green Eyes") Embedded in Norway Spruce and Scots Pine Stands
Reprinted from: *Forests* **2021**, *12*, 573, doi:10.3390/f12050573 57

**Christopher A. Walter, Zachariah K. Fowler, Mary Beth Adams, Mark B. Burnham,
Brenden E. McNeil and William T. Peterjohn**
Nitrogen Fertilization Increases Windstorm Damage in an Aggrading Forest
Reprinted from: *Forests* **2021**, *12*, 443, doi:10.3390/f12040443 83

Lacey J. Smith and Kirsten Stephan
Nitrogen Fertilization, Stand Age, and Overstory Tree Species Impact the Herbaceous Layer in
a Central Appalachian Hardwood Forest
Reprinted from: *Forests* **2021**, *12*, 829, doi:10.3390/f12070829 95

Alexander Storm, Mary Beth Adams and Jamie Schuler
Long-Term Projection of Species-Specific Responses to Chronic Additions of Nitrogen, Sulfur,
and Lime
Reprinted from: *Forests* **2021**, *12*, 1069, doi:10.3390/f12081069 119

Mark B. Burnham, Martin J. Christ, Mary Beth Adams and William T. Peterjohn
Assessing the Linkages between Tree Species Composition and Stream Water Nitrate in a
Reference Watershed in Central Appalachia
Reprinted from: *Forests* **2021**, *12*, 1116, doi:10.3390/f12081116 . . . . **137**

Frank S. Gilliam
Response of Temperate Forest Ecosystems under Decreased Nitrogen Deposition: Research
Challenges and Opportunities
Reprinted from: *Forests* **2021**, *12*, 509, doi:10.3390/f12040509 . . . . **151**

About the Editor

Frank S. Gilliam is a professor of biology at the University of West Florida and a professor emeritus at Marshall University. He completed his B.S. in biology at Vanderbilt University and received a Ph.D. in plant ecology at Duke University, studying the fire ecology of southeastern coastal plain pine forests. Following post-doctoral appointments at Kansas State University to study the fire ecology of tallgrass prairie and at the University of Virginia to study hardwood forest canopy–atmosphere interactions, Dr. Gilliam began his 28-year tenure with the faculty of Marshall University in 1990. His research lies primarily at the conceptual boundary between terrestrial plant communities and ecosystems, including the movement and cycling of plant nutrients, especially nitrogen (N). These interests extend to fire ecology and the effects of fire on nutrient cycling, plants, and soils in fire-prone ecosystems. Additionally, related to his ecosystem approach to ecological research is an interest in atmospheric deposition and precipitation chemistry, leading to the study of pollutant conditions (acid deposition, excess N, and ozone) in forested areas. Other work includes secondary succession and the species dynamics of the herbaceous layer of forests, as well as the variety of biotic and abiotic factors that influence species composition and change within this vegetation stratum. Ongoing work includes vegetation dynamics in forest ecosystems, N cycling in forest ecosystems, and species composition and stand structure in longleaf pine forests. Dr. Gilliam currently serves as Associate Editor for Journal of Ecology, Journal of Plant Ecology, and Forests. He has authored or co-authored more than 100 peer-reviewed articles, in addition to book chapters and reviews of books, current scientific articles, and software, and has authored/co-authored the books Terrestrial Plant Ecology, 3rd Edition (1999), The Herbaceous Layer in Forests of Eastern North America (2003; 2nd Edition 2014), and Causes and Consequences of Species Diversity in Forest Ecosystems (2019). He is the grateful husband of Laura P. Gilliam and father of Rachel M. Gilliam, M.Div., and Ian S. Gilliam LT USN.

Editorial

Responses of Forest Ecosystems to Nitrogen Deposition

Frank S. Gilliam

Department of Biology, University of West Florida, Pensacola, FL 32514, USA; fgilliam@uwf.edu

Citation: Gilliam, F.S. Responses of Forest Ecosystems to Nitrogen Deposition. *Forests* **2021**, *12*, 1190. https://doi.org/10.3390/f12091190

Received: 24 August 2021
Accepted: 31 August 2021
Published: 2 September 2021

Publisher's Note: MDPI stays neutral with regard to jurisdictional claims in published maps and institutional affiliations.

Environmental legislation in countries around the world has led to notable recent declines in the atmospheric deposition of nitrogen (N), although most decreases relate to oxidized N, with reduced N increasing in many areas [1]. Still, deposition of N remains high in many regions globally. For areas where chronic atmospheric deposition of N has led to N saturation, excess N chronically threatens the structure and function of ecosystems. Indeed, critical loads for N remain widely exceeded for many forests, leading to a variety of deleterious effects, including loss of biodiversity and altered biogeochemical cycles, all of which threaten the sustainability of impacted forests [2]. It is likely that the recovery of N-impacted sites might require extended periods of time, especially in locations where base cations, such as Ca^{++}, have been depleted by accelerated NO_3^- leaching. Thus, understanding the potential responses of forest ecosystems to N deposition remains essential.

This Special Issue of *Forests* explores the multifaceted responses of forest ecosystems to both increases and decreases in N deposition on a global scale. This includes effects on plants and plant assemblages, as well as effects of N on forest biogeochemistry, and comprises studies from Asia, Europe, and North America.

Certainly, the fundamental paradigms of N cycling in terrestrial ecosystems have shifted greatly over the past several decades [3], and N has been the focus of extensive research as both basic and applied science. More recently and more specifically this has been carried out by biogeochemists, extending from the discovery of N as an element in 1772, to its central place in von Liebig's Law of Minimum for plant growth articulated in 1827, to the discovery of symbiotic N fixation in 1888, to the development of the Haber-Bosch process in 1913 (initiating of its use as fertilizer in crop production), and finally to the present awareness that excess N in the environment can alter the structure and function of ecosystems [4].

Chronic N deposition has been an environmental threat throughout most of Asia. In Japan, research has demonstrated that effects of excess N can include trophic interactions involving tree foliar tissue and defoliating insects, which can be particularly notable in N_2-fixing trees [5]. Excess N also influences the biogeochemical cycles of tropical Asian forests. Experimental additions of N significantly inhibited mineralization of soil carbon (C) in tropical forest soils of southern China (the Dinghushan Biosphere Reserve—DHSBR), increasing concentrations of dissolved organic carbon [6]. Deposition of N can interact with that of other nutrients, e.g., phosphorus (P), to alter plant tissue nutrient stoichiometry, especially in humid tropical forests that are typically N-rich, but limited by P. Another factorial field experiment at DHSBR demonstrated that additions of P, rather than N, greatly altered foliar N and P concentrations in understory plants [7].

Europe has long experienced excess N deposition from increased emissions of gaseous N from industrial, domestic, and agricultural sources. Long-term (~30 yr) data from the Australian Alps has shown that, although N deposition approximated or exceeded critical loads, >80% of this was retained in the ecosystem, maintaining lower hydrologic N loss; tree growth was identified as the main sink for N [8]. Indeed, forest dynamics often mitigate effects of high N deposition, including the response of soil nutrient cations, i.e., Ca^{++}, Mg^{++}, and K^+. Work in hardwood and conifer stands of central Germany found significant differences between these stand types regarding the effects of excess N on the biogeochemical cycling of these essential plant nutrients [9].

Most research on effects of excess N on forest ecosystems has been carried out via plot-based field designs which have the advantage of being fully factorial and involving replications, but the disadvantage of lacking realistic simulations of increased deposition of N from the atmosphere. Accordingly, whole-watershed manipulations, wherein N is applied from rotary or fixed-wing aircraft, are rare. Research at the Fernow Experimental Forest (FEF), Tucker County, West Virginia, comprises both plot-based and whole-watershed approaches, with aerial additions of $(NH_4)_2SO_4$ at 35 kg N ha^{-1} yr^{-1} to the treatment watershed beginning in 1989. Although this site is an eastern deciduous forest, results from over a quarter century have relevant implications for other forest types, as well [10,11].

Past and ongoing work at FEF is highly varied with respect to forest response variables. An unexpected outcome has been the discovery that excess N can make some hardwood species more susceptible to damage from storm-related winds, which has implications for climate change and the future of impacted forests [12]. Excess N can also greatly alter the species composition and diversity of forest herbaceous communities, which is where up to 90% of plant diversity is found in forest ecosystems [10,13,14]. Recent efforts have demonstrated that this response can vary greatly with stand age and dominant tree species [15]. Not surprisingly, the response of forest trees to excess N can be highly species specific, and work at FEF has shown that several hardwood species, such as sweet birch (*Betula lenta* L.), black cherry (*Prunus serotina* Ehrh.), and red maple (*Acer rubrum* L.) responded positively to N additions, whereas yellow poplar (*Liriodendron tulipifera* L.) responded negatively [16]. These species-specific responses have implications for ecosystem-level effects of N on forests, as some hardwood species, especially sugar maple (*Acer saccharum* L.), can facilitate greater loss of N, via leaching of NO_3^-, by enhancing rates of net nitrification [17].

Recent studies have speculated about the future of N-impacted forest ecosystems under conditions of decreased N deposition [1,18–20]. In 2019, aerial applications of NS ceased after 30 yr of treatment at FEF. This new phase of research at FEF now allows researchers the opportunity to test these predictions empirically [21], further advancing our knowledge and understanding of the complexity of N biogeochemistry in forest ecosystems.

Funding: This research received no external funding.

Acknowledgments: I gratefully acknowledge the work and contributions of the authors and co-authors of the papers featured in this Special Issue. I also thank Special Issue Editor for invaluable assistance throughout the publishing process.

Conflicts of Interest: The author declares no conflict of interest.

References

1. Gilliam, F.S.; Burns, D.A.; Driscoll, C.T.; Frey, S.D.; Lovett, G.M.; Watmough, S.A. Decreased atmospheric nitrogen deposition in eastern North America: Predicted responses of forest ecosystems. *Environ. Pollut.* **2019**, *244*, 560–574. [CrossRef] [PubMed]
2. Pardo, L.H.; Fenn, M.E.; Goodale, C.L.; Geiser, L.H.; Driscoll, C.T.; Allen, E.B.; Baron, J.S.; Bobbink, R.; Bowman, W.D.; Clark, C.M.; et al. Effects of nitrogen deposition and empirical nitrogen critical loads for ecoregions of the United States. *Ecol. Appl.* **2011**, *21*, 3049–3082. [CrossRef]
3. Bobbink, R.; Hicks, K.; Galloway, J.; Spranger, T.; Alkemade, R.; Ashmore, M.; Bustamante, M.; Cinderby, S.; Davidson, E.; Dentener, F.; et al. Global assessment of nitrogen deposition effects on terrestrial plant diversity effects of terrestrial ecosystems: A synthesis. *Ecol. Appl.* **2010**, *20*, 30–59. [CrossRef] [PubMed]
4. Galloway, J.N.; Dentener, F.J.; Capone, D.G.; Boyer, E.W.; Howarth, R.W.; Seitzinger, S.P.; Asner, G.P.; Cleveland, C.C.; Green, P.A.; Holland, E.A.; et al. Nitrogen cycles: Past, present, and future. *Biogeochemistry* **2004**, *70*, 153–226. [CrossRef]
5. Lee, J.; Nakamura, M.; Hiura, T. Effects of large-scale nitrogen fertilization on insect–plant interactions in the canopy of tall alder trees with N_2-fixing traits in a cool temperate forest. *Forests* **2021**, *12*, 210. [CrossRef]
6. Lu, X.; Mao, Q.; Wang, Z.; Mori, T.; Mo, J.; Su, F.; Pang, Z. Long-term nitrogen addition decreases soil carbon mineralization in an N-rich primary tropical forest. *Forests* **2021**, *12*, 734. [CrossRef]
7. Mao, Q.; Chen, H.; Wang, C.; Pang, Z.; Mo, J.; Lu, X. Effect of long-term nitrogen and phosphorus additions on understory plant nutrients in a primary tropical forest. *Forests* **2021**, *12*, 803. [CrossRef]

8. Dirnböck, T.; Brielmann, H.; Djukic, I.; Geiger, S.; Hartmann, A.; Humer, F.; Kobler, J.; Kralik, M.; Liu, Y.; Mirtl, M.; et al. Long- and short-term inorganic nitrogen runoff from a karst catchment in Austria. *Forests* **2020**, *11*, 1112. [CrossRef]
9. Achilles, F.; Tischer, A.; Bernhardt-Römermann, M.; Chmara, I.; Achilles, M.; Michalzik, B. Effects of moderate nitrate and low sulphate depositions on the status of soil base cation pools and recent mineral soil acidification at forest conversion sites with European beech ("Green Eyes") embedded in Norway spruce and Scots pine stands. *Forests* **2021**, *12*, 573. [CrossRef]
10. Gilliam, F.S.; Welch, N.T.; Phillips, A.H.; Billmyer, J.H.; Peterjohn, W.T.; Fowler, Z.K.; Walter, C.A.; Burnham, M.B.; May, J.D.; Adams, M.B. Twenty-five year response of the herbaceous layer of a temperate hardwood forest to elevated nitrogen deposition. *Ecosphere* **2016**, *7*, e01250. [CrossRef]
11. Gilliam, F.S.; Walter, C.A.; Adams, M.B.; Peterjohn, W.T. Nitrogen (N) dynamics in the mineral soil of a Central Appalachian hardwood forest during a quarter century of whole-watershed N additions. *Ecosystems* **2018**, *21*, 1489–1504. [CrossRef]
12. Walter, C.A.; Fowler, Z.K.; Adams, M.B.; Burnham, M.B.; McNeil, B.E.; Peterjohn, W.T. Nitrogen Fertilization increases Windstorm Damage in an Aggrading Forest. *Forests* **2021**, *12*, 443. [CrossRef]
13. Gilliam, F.S. Response of the herbaceous layer of forest ecosystems to excess nitrogen deposition. *J. Ecol.* **2006**, *94*, 1176–1191. [CrossRef]
14. Gilliam, F.S. *The Herbaceous Layer in Forests of Eastern North America*, 2nd ed.; Oxford University Press, Inc.: New York, NY, USA, 2014.
15. Smith, L.J.; Stephan, K. Nitrogen fertilization, stand age, and overstory tree species impact the herbaceous layer in a Central Appalachian hardwood forest. *Forests* **2021**, *12*, 829. [CrossRef]
16. Storm, A.; Adams, M.B.; Schuler, J. Long-term projection of species-specific responses to chronic additions of nitrogen, sulfur, and lime. *Forests* **2021**, *12*, 1069. [CrossRef]
17. Burnham, M.B.; Christ, M.J.; Adams, M.B.; Peterjohn, W.T. Assessing the linkages between tree species composition and stream water nitrate in a reference watershed in Central Appalachia. *Forests* **2021**, *12*, 1116. [CrossRef]
18. Schmitz, A.; Sanders, T.G.M.; Bolte, A.; Bussotti, F.; Dirnböck, T.; Johnson, J.; Peñuelas, J.; Pollastrini, M.; Prescher, A.-K.; Sardans, J.; et al. Responses of forest ecosystems in Europe to decreasing nitrogen deposition. *Environ. Pollut.* **2019**, *244*, 980–994. [CrossRef] [PubMed]
19. Chiwa, M. Long-term changes in atmospheric nitrogen deposition and stream water nitrate leaching from forested watersheds in western Japan. *Environ. Pollut.* **2021**, *287*, 117634. [CrossRef] [PubMed]
20. Reid, H.; Aherne, J. Staggering reductions in atmospheric nitrogen dioxide across Canada in response to legislated transportation emissions reductions. *Atmos. Environ.* **2016**, *146*, 252–260. [CrossRef]
21. Gilliam, F.S. Response of temperate forest ecosystems under decreased nitrogen deposition: Research challenges and opportunities. *Forests* **2021**, *12*, 509. [CrossRef]

 *forests*

MDPI

Article

Effects of Large-Scale Nitrogen Fertilization on Insect–Plant Interactions in the Canopy of Tall Alder Trees with N$_2$-Fixing Traits in a Cool Temperate Forest

Jin Lee [1,*], Masahiro Nakamura [2] and Tsutom Hiura [3]

[1] Division of Restoration Research, Research Center for Endangered Species, National Institute of Ecology, Yeongyang 36531, Korea

[2] Wakayama Experimental Forest, Field Science Center for Northern Biosphere, Hokkaido University, Higashimuro, Wakayama 649-4563, Japan; masahiro@fsc.hokudai.ac.jp

[3] Graduate School of Agriculture and Life Science, The University of Tokyo, Tokyo 113-8657, Japan; hiura@g.ecc.u-tokyo.ac.jp

* Correspondence: jinlee@nie.re.kr; Tel.: +82-54-680-7324

Abstract: Nitrogen (N) deposition is expected to influence forests. The effects of large-scale N fertilization on canopy layer insect–plant interactions in stands of tall, atmospheric nitrogen (N$_2$)-fixing tree species have never been assessed. We conducted a large-scale fertilization experiment (100 kg N ha^{-1} year^{-1} applied to approximately 9 ha) over three years (2012–2014) in a cool temperate forest in northern Japan. Our goal was to evaluate relational responses between alder (*Alnus hirsuta* [Turcz.]) and their insect herbivores to N deposition. Specifically, we assessed leaf traits (N concentration, C:N ratio, condensed tannin concentration, and leaf mass per unit area (LMA)) and herbivory by three feeding guilds (leaf damage by chewers and the densities of gallers and miners) between the fertilized site and an unfertilized control. Fertilization led to increased galler density in spring 2013 and increased leaf damage by chewers in late summer 2014. For leaf traits, the LMA decreased in spring 2013 and late summer 2014, and the C:N ratio decreased in late summer 2013. The N and condensed tannin concentrations remained unchanged throughout the study period. There was a negative correlation between LMA and leaf damage by chewers, but LMA was not correlated with galler density. These results show that large-scale N fertilization had a positive plant-mediated (i.e., indirect) effect on leaf damage by chewers via a decrease in LMA in the canopy layer. Changes in physical defenses in canopy leaves may be a mechanism by which N fertilization affects the herbivory in tall N$_2$-fixing trees.

Keywords: *Alnus hirsuta*; forest communities; insect herbivory; nitrogen deposition; leaf traits; physical defenses

Citation: Lee, J.; Nakamura, M.; Hiura, T. Effects of Large-Scale Nitrogen Fertilization on Insect–Plant Interactions in the Canopy of Tall Alder Trees with N$_2$-Fixing Traits in a Cool Temperate Forest. *Forests* **2021**, *12*, 210. https://doi.org/10.3390/f12020210

Academic Editors: Frank S. Gilliam and Mariangela Fotelli

Received: 24 December 2020

Accepted: 8 February 2021

Published: 10 February 2021

Publisher's Note: MDPI stays neutral with regard to jurisdictional claims in published maps and institutional affiliations.

1. Introduction

The effects of enhanced atmospheric nitrogen (N$_2$) deposition, a result of increased anthropogenic activities, on terrestrial ecosystem processes is a growing environmental problem [1,2]. Enhanced deposition leads to changes in plant community composition, decreases in plant diversity, and altered nitrogen (N) cycling [3–6]. There are several potential mechanisms by which deposition affects terrestrial ecosystem processes. Deposited N can act as a fertilizer, directly stimulating plant growth and increasing the N content in plant tissue; many studies have focused on this phenomenon [7]. On the other hand, N deposition may indirectly affect plant communities and terrestrial ecosystems by modifying plant–insect interactions [7,8]. Because increased N deposition can affect ecosystem processes directly (e.g., nutrient cycling [9]) or indirectly (e.g., causing shifts in herbivory resulting from altered host plant traits), more consideration should be given to changes in plant–insect interactions following increased N deposition, in order to fully understand N deposition effects on terrestrial ecosystem processes.

Many studies have shown that shifts in leaf traits, such as an increase in foliar N content and/or a decrease in carbon (C)-based chemical defenses, which can be caused by increased N deposition, positively affect the performance and/or abundance of insect herbivores, including Lepidoptera, Hymenoptera, and Coleoptera [10–14]. However, these studies have been based on small-scale ecosystem manipulations [10,12,15]. Experiments employing open-top chambers and N fertilization in <10 m × 10 m field plots [10,12,15] have important limitations, such as the lamp effect, wherein insects that prefer the conditions of a treated site (e.g., moths) congregate at an artificially high density, while others avoid the site due to those same conditions [16]. Thus, extrapolating data obtained from small-scale experiments to predicting the effects of these changes at large scales will not yield robust results. To avoid these limitations, N deposition studies should focus on large-scale manipulations and examine how N fertilization affects the performance and/or abundance of insect herbivores [17,18].

Plant tissues often have much lower N levels than herbivorous insect tissues [19]. N-containing nutrients in plant tissues are often a limiting factor for insect herbivores, either because of limited resource utilization, a high population density, or both [13,20]. According to the C-nutrient balance (CNB) hypothesis, when N is abundant, more C is allocated to growth than to C-based chemical defenses (e.g., condensed tannins and total phenolics) in plants [21,22]. Thus, the majority of N deposition studies have focused solely on changes in nutrients and C-based chemical defenses. However, N fertilization could also affect the hardness and thickness of foliage (i.e., leaf mass per unit area (LMA)) because elevated N enhances photosynthesis more than growth; thus, LMA may decrease [23]. It is probable that N fertilization could lead to shifts in leaf traits to the advantage of insect herbivores, not only by decreasing chemical defenses, but also by reducing physical defenses—specifically, the toughness and thickness of leaves. LMA has been shown to be a significant component of leaf toughness. Leaf toughness is widely recognized as one of the most effective forms of physical defense in leaves [24,25].

Numerous N fertilization experiments were conducted using N-limited species, which indicated that shifts in foliar chemical and nutritional status could affect plant–insect interactions, specifically with insect herbivores [7]. Alder (*Alnus*) species have a greater potential to accumulate N (mg/g) in their tissues than other species, owing to their N_2-fixing traits, in association with symbiotic *Frankia* species (alder, >25 mg/g; oak, 15–25 mg/g; pine, 12–17 mg/g) [26]. Alder species are typically the largest natural pathway for N input in certain forest types [27,28]. Given the research focus on N-limited species, it remains unclear how the leaf traits of N_2-fixing plants may be affected by increased N deposition, and how, in turn, plant–insect interactions may be affected, even when N fertilization promotes plant growth [29,30].

Classifying herbivorous insects into ecologically and evolutionarily related feeding guilds [31] allows for comparisons and generalizations that cannot be made using taxonomic groups alone [9,32]. In an insect herbivore community, chewers are external herbivores and gallers and miners are internal herbivores. These guilds (i.e., chewers, gallers, and miners) may respond differently to N fertilization. For example, N fertilization was shown to affect chewer performance (body size, development time, and survival) as a result of altered plant traits (increased N content and decreased C-based chemical defenses) [7]. By contrast, other studies showed no effect of N fertilization on the density and growth rate of the galler *Eurosta solidaginis* [33], and the abundance and diversity of miners [34]. A thorough understanding of insect herbivore responses to N fertilization requires assessing the responses of major feeding guilds on the same host plant.

Here, we report the first manipulative test of large-scale N fertilization effects on alder, an N_2-fixing species, and its insect herbivores. We aimed to determine how fertilization affects leaf traits in canopy-sized trees and their interactions with insect herbivores. We asked how large-scale fertilization influences leaf traits in mature alder, how fertilization affects the three major insect feeding guilds, and if certain interactions between leaf traits and feed guilds are affected by N fertilization.

2. Materials and Methods

2.1. Site Description and Study Species

This study was conducted in Tomakomai Experimental Forest, Hokkaido University, Japan (42°40′ N, 141°36′ E). The forest is formed on a 2 m deep volcanic ash layer that accumulated following the eruptions of Mt. Tarumae in 1699 and 1739; soils are very shallow [35]. The study area was divided into control (~10 ha) and treatment (fertilized) (~9 ha) compartments (sites). Tree basal area was investigated in four plots (20 m × 20 m) at each site; we found no difference between the control (27.5 ± 4.6 cm/m^2) and the fertilized sites (24.8 ± 4.6 cm/m^2) using a *t*-test (*t* = 0.827; *p* = 0.440). The dominant tree species at the study site were *Quercus crispula, Acer mono, Sorbus alnifolia*, and *Tilia japonica* [36]. Mature tall alder (*Alnus hirsuta* [Turcz.]) trees were the focus of our study. The alder trees were not suppressed by surrounding trees, and the distances between the trees were more than 10 m. The buds typically open in early spring (late April or early May) and the leaves emerge twice per year—in spring (early leaves) and in late summer (late leaves) [37].

2.2. Large-Scale N Fertilization Experiment

The fertilized site (~9 ha) was fertilized with a urea solution (100 kg N ha^{-1} year^{-1}; CO[NH$_2$]$_2$) designed for dry application; it was applied to simulate a pulse of N input as rain melts the urea into the soil. The fertilizer was distributed by helicopter in 2013 (April 15 and 16) and manually in 2014 (April 21 and 22). Akata et al. [38] found that the N fertilization experiment had a larger diameter and high nitrogen concentrations in fine roots of oak species and dissolved organic carbon in soil. We did not have site replications, because individual trees could be considered as independent within and between the treatments. When conducting field experiments, the best way to do this is to create replications of sites, but if this is difficult to conduct such as in large-scale experiments, and individuals can be considered as replications, this can be another option [39,40]. Moreover, we had a limited number of alder trees in the study site, although it was worth-studying them due to their specific N$_2$-fixing traits.

2.3. Estimation of Leaf Traits

We assessed 10 mature (>10 m in height) alder trees (diameter at breast height (DBH) = 30.7 ± 7.5 cm in the control site (*n* = 5) and 24.0 ± 4.6 cm in the fertilized site (*n* = 5) growing along roads through the control and fertilized compartments. One tree from the fertilized site died in 2014 and was thus excluded. During each sampling period, two branches (1 cm in diameter) were randomly sampled from the canopy of each tree using a crane car. All branches were located on the top of the canopy and were fully exposed to sunlight. Samples were collected from the selected trees from 2012–2014, including a visit on June 28–29, 2012, prior to N fertilization, which was used as a preliminary assessment period. Otherwise, sampling was conducted during early and late periods in each year, on June 17–18 and September 17–18, 2013, and June 19–20 and August 27–28, 2014, respectively, because the seasonal fluctuation of moth communities including chewers and gallers showed two large peaks in mid-June and early to mid-August in our study site [41].

Leaf traits (N concentration, C:N ratio, condensed tannin concentration (a measure of C-based defense), and LMA) were determined from at least five leaves randomly collected from the two sampled branches from each tree. LMA was calculated using:

$$\text{LMA} = W_{dry} / A_{area}$$

where W_{dry} is the mass of leaves dried at 60 °C for 48 h, and A_{area} is the leaf area, calculated from leaf disks obtained by leaf punching (1.5 cm in diameter). Dried samples were maintained in an oven at 40 °C for another week, then ground to a fine powder using an analytical mill with a micro-smasher to assess leaf traits. C:N ratio was used as an indicator of C-based chemical defenses [21]. The leaf N concentration and C:N ratio were determined using an NC analyzer (Sumigraph NC-900; Sumika Chemical Analysis Service, Osaka,

Japan). The condensed tannin concentration was determined using the method described by Julkunen–Tiitto [42].

2.4. Insect Herbivore Feeding Guilds

We collected 364 individuals accounting for 33 species of Coleoptera, Hemiptera, Hymenoptera, and Lepidoptera (Table S1).

Three major feeding guilds (chewers, gallers, and miners) were used to assess the community-level response to N fertilization. Feeding guilds are closely related to host specificity: chewers are generalist herbivores, whereas gallers and miners are specialists [43]. The three feeding guilds were assessed in 10 randomly selected current-year shoots obtained from the two canopy branches (diameter 1 cm) taken from each tree. First, leaf damage by chewers was defined as the percentage of leaf area chewed with respect to the total leaf area and was scored visually using six categories: 0% = 0, 1–10% = 1, 11–25% = 2, 26–50% = 3, 51–75% = 4, and 76–100% = 5 [44,45]. The median of each group was used in the statistical analysis of leaf damage by chewers (i.e., 0, 5, 18, 38, 63, and 88%, respectively). We then counted the number of gallers and miners on each shoot to determine their respective densities.

2.5. Statistical Analyses

Differences in leaf traits (N concentration, C:N ratio, condensed tannin concentration, and LMA) and guild-level insect herbivory (leaf damage by chewers and the respective densities of gallers and miners) were compared between the control and fertilized sites before and after N fertilization using *t*-tests. Pearson's correlation coefficient was used to assess correlations between leaf traits and feeding guilds that were significantly altered by N fertilization. All variables were tested for normality using the Shapiro–Wilk test, and non-normal variables were log-transformed. All statistical analyses were conducted using R version 2.9.2 [46].

3. Results

3.1. Effect of N Fertilization on Leaf Traits

Prior to the first application of N in early 2012, there were no significant differences in any of the leaf traits between the control and fertilized treatment ($p > 0.050$, *t*-test; Figure 1). Following N fertilization, there was a significant decrease in C:N ratio in late 2013 ($t = 3.537$, $p = 0.007$, Figure 1B), and in LMA in early 2013 ($t = 2.610$, $p = 0.003$, Figure 1C) and marginally in late 2014 ($t = 1.982$, $p = 0.080$, Figure 1C). There were no significant differences in N concentration ($p > 0.050$, Figure 1A) and condensed tannin concentration ($p > 0.050$, Figure 1D) between the control and fertilization treatment throughout the study period, which included two years following N fertilization.

3.2. Insect Herbivore Responses

As with leaf traits, no significant differences in leaf damage and galler or miner density were detected between the control and fertilization site prior to N fertilization ($p > 0.050$, Figure 2). Following N fertilization, there was a significant increase in leaf damage by chewers in late 2014 ($t = -2.571$, $p = 0.037$, Figure 2A). Galler density decreased marginally in early 2013 ($t = 2.221$, $p = 0.057$, Figure 2B). There was no difference in miner density between the control and fertilized sites throughout the study period ($p > 0.050$, *t*-test; Figure 2C).

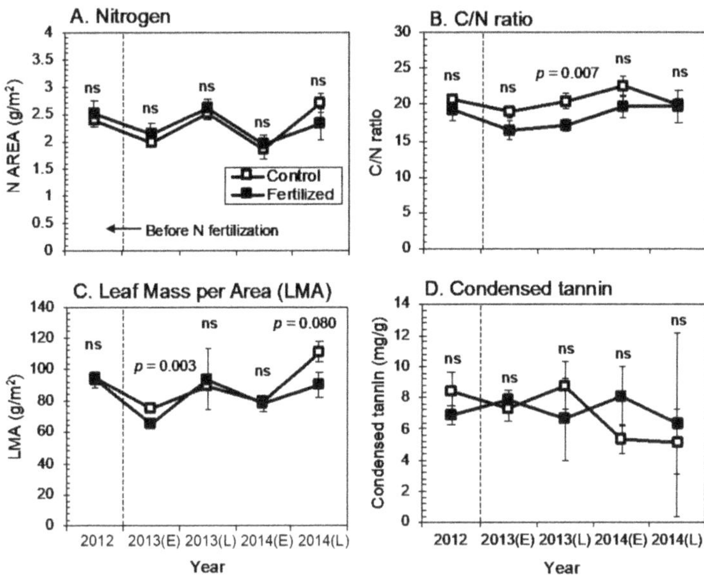

Figure 1. Changes in leaf nutrients of alder trees. (**A**) Nitrogen content, (**B**) C/N ratio, (**C**) leaf mass per area (LMA), and (**D**) condensed tannin in the leaves of alder trees at the control (empty boxes, $n = 5$) and fertilized (filed boxes, $n = 5$) sites from 2012 (before N fertilization) to 2014. E and L indicate early and late seasons. Each point is mean ± SE and the p-values are the results of t-tests (ns, $p > 0.05$).

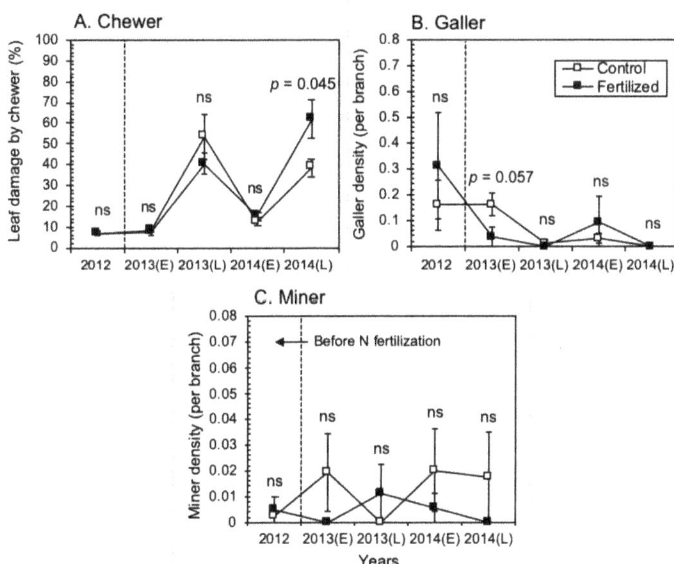

Figure 2. Changes in the three major guilds of insect herbivores categorized based on feeding type. (**A**) Leaf damage by chewers, (**B**) galler density, and (**C**) miner density on trees in control (empty boxes, $n = 5$) and fertilized (filed boxes, $n = 5$) sites from 2012 (before N fertilization) to 2014. E and L indicate early and late seasons. Each point is mean ± SE and the p-values are the results of t-tests (ns, $p > 0.05$).

3.3. Correlation Analysis between Leaf Traits and Insect Herbivory

We assessed correlations in the variables (LMA in early 2013 and late 2014, C:N ratio in late 2013, leaf damage by chewers in late 2014, and galler density in early 2013) that were significantly or marginally altered following N fertilization. There was a negative correlation between LMA and leaf damage by chewers in late 2014 ($r = -0.887$, $p = 0.001$, Figure 3). However, LMA was not correlated with galler density in early 2013 ($p > 0.05$). The timing of changes in C:N ratio and herbivory by any feeding guild did not align in any sampling period.

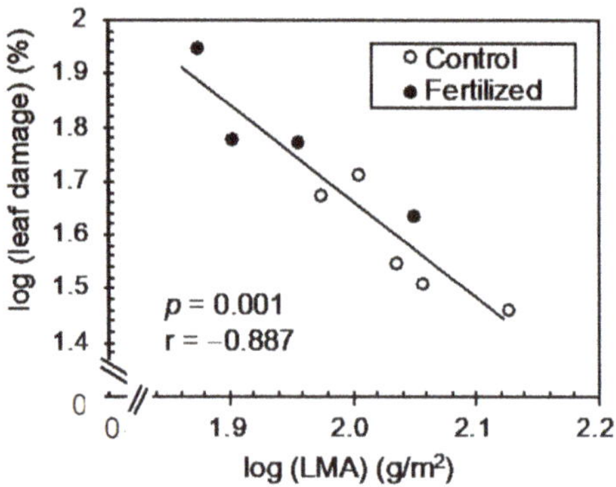

Figure 3. Pearson's correlation between the LMA and leaf damage by chewers in late 2014. Circles in white and black indicate control and fertilized trees, respectively. Regression results and the line of best fit are indicated.

4. Discussion

Large-scale N fertilization led to a decrease in the C:N ratio and LMA in mature *A. hirsuta*, an N_2-fixing species, and an increase in leaf damage by chewers and a decrease in the density of gallers, at least once during the study period; although other leaf traits (i.e., N and condensed tannin concentrations) remained unchanged throughout the study period. We found a negative relationship between LMA and leaf damage by chewers, but LMA was not correlated with the density of gallers. Our results suggest that large-scale N fertilization may have a positive, indirect (plant-mediated) effect on leaf damage by chewers, via a decrease in physical defense (LMA) in the canopy leaves of mature alder trees.

Interestingly, although we found no effect on N and condensed tannin concentrations, Koike et al. [15] reported that N addition led to a decrease in C:N ratio and chemical defense compounds, such as total phenolics and condensed tannin, in alder saplings. This discrepancy may be related to maturity and plant size. The responses of woody plants to shifting site conditions may vary depending on their stage of development [47–50]. In *Q. crispula*, N fertilization was shown to influence leaf traits (e.g., N, C:N, and condensed tannins) in two-year-old seedlings [15], but not in mature trees [51]. Given that our study employed a large-scale manipulation, which avoided issues such as the lamp effect [16], our findings suggest the actual effects of large-scale global change in N deposition [52]. However, no clear conclusions on the responses of secondary defense compounds can be drawn by the present study, because only tannins were measured, and future research should include a broad range of secondary metabolites.

LMA has been shown to decrease following N fertilization [13,52], especially when plant nutrients are limited [53,54]. In our study, N fertilization decreased the LMA of *A. hirsuta* (N_2-fixing species) (Figure 1). Ruess et al. [55] also reported that N-fertilization significantly decreased leaf thickness of thin-leaf alder (*A. tenuifolia*), due to the down-regulation of N-fixation. The decrease in LMA may be driven primarily by a shift from

N to C limitation in the fertilized site—carbon allocated in leaf growth is partitioned to increase leaf area though the production of thinner leaves.

Lee et al. [56] showed that large-scale N fertilization enhanced leaf quality (e.g., increase in N concentration and decrease in C:N ratio) in mature oak trees >10 m in height. These results are contrary to our findings with *A. hirsuta* (only C:N ratio), and may be reflective of differences in the severity of N deficiency between N-limited and N_2-fixing species. N-limited species may be more sensitive to N fertilization than N-fixing species because of more N deficiency. Furthermore, leaf density and/or thickness reflect variation in anatomical structure arising from environmental conditions and species traits [57], although the role of physiological regulation in determining leaf traits remains unclear [23].

Major feeding guilds of insect herbivores may respond differently to N deposition-induced changes in the quality of their host plants [20]. Most previous studies have suggested that N fertilization affects chewer insects by altering chemical and nutritional qualities [10–14], whereas gallers and miners have been more poorly studied [13]. To assess insect herbivore responses to N fertilization, we assessed how each of the three major feeding guilds were affected on the same host plants and found that large-scale N fertilization significantly increased leaf damage by chewers (Figure 2). Leaf damage was negatively associated with LMA (Figure 3). As a physical defense, LMA has been reported to influence the performance of chewers [13,54,58,59]. These results suggest a mechanism by which large-scale N fertilization affects leaf damage by chewers on mature alder—through changes in physical defenses. Additionally, our results indicate that N fertilization marginally decreased galler density (Figure 2), as opposed to Hartley and Lawton's findings [33], which suggested that leaf N concentration was not correlated with the density of the gall midge *Eurosta solidaginis*. It appears that N deposition-related changes in plant quality did not affect galler density because density was not associated with any of the assessed leaf traits. Several review papers have suggested the possibility of N deposition-mediated changes in herbivore susceptibility to predators as an important mechanism [7,8], although only a few studies have explored this. The decrease in galler density after N deposition may be related to changes in plant volatiles that are essential for successful foraging and attacks by predators. Our understanding of this mechanism remains limited, and we strongly encourage future work on the effect of N deposition on the susceptibility of insect herbivores to predation.

5. Conclusions

Our large-scale N fertilization experiment is the first manipulative study on tall N_2-fixing trees and their insect herbivores. Fertilization has a positive, plant-mediated effect on leaf damage by chewers via a decrease in physical defense (LMA). Our study provides valuable information on large-scale N fertilization (i.e., N deposition) effects on insect–plant interactions in N_2-fixing plants, which may be helpful in predicting future climate change impacts.

Supplementary Materials: The following are available online at https://www.mdpi.com/1999-4907/12/2/210/s1, Table S1: List of herbivory species collected in this study.

Author Contributions: Writing—original draft preparation, J.L.; writing—review and editing, M.N. and T.H. All authors have read and agreed to the published version of the manuscript.

Funding: This study was partly supported by JSPS [grant number 2529207903 to TH] and by the National Institute of Ecology [NIE-C-2021-46] in Korea.

Institutional Review Board Statement: Not applicable.

Informed Consent Statement: Not applicable.

Data Availability Statement: The data presented in this study are available on request from the corresponding author.

Acknowledgments: We thank the members of Tomakomai Experimental Forest (TOEF) for help in N fertilization experiments and sampling.

Conflicts of Interest: The authors declare no conflict of interest.

References

1. Bobbink, R.; Hornung, M.; Roelofs, J.G. The effects of air-borne nitrogen pollutants on species diversity in natural and semi-natural European vegetation. *J. Ecol.* **1998**, *86*, 717–738. [CrossRef]
2. Rockström, J.; Steffen, W.; Noone, K.; Persson, A.; Chapin, F.S., III; Lambin, E.F.; Lenton, T.M.; Scheffer, M.; Folke, C.; Schellnhuber, H.J.; et al. A safe operating space for humanity. *Nature* **2009**, *461*, 472–475.
3. Aber, J.D.; Magill, A.; Mcnulty, S.G.; Boone, R.D.; Nadelhoffer, K.J.; Downs, M.; Hallett, R. Forest biogeochemistry and primary production altered by nitrogen saturation. *Water Air Soil Pollut.* **1995**, *85*, 1665–1670. [CrossRef]
4. Fenn, M.E.; Baron, J.S.; Allen, E.B.; Rueth, H.M.; Nydick, K.R.; Geiser, L.; Bowman, W.D.; Sickman, J.O.; Meixner, T.; Johnson, D.W.; et al. Ecological effects of nitrogen deposition in the western United States. *BioScience* **2003**, *53*, 404–420. [CrossRef]
5. Jones, M.E.; Paine, T.D.; Fenn, M.E. The effect of nitrogen additions on oak foliage and herbivore communities at sites with high and low atmospheric pollution. *Envion. Pollut.* **2008**, *151*, 434–442. [CrossRef] [PubMed]
6. Tilman, D. Secondary succession and the pattern of plant dominance along experimental nitrogen gradients. *Ecol. Monogr.* **1987**, *57*, 189–214. [CrossRef]
7. Throop, H.L.; Lerdau, M.T. Effects of nitrogen deposition on insect herbivory: Implications for community and ecosystem processes. *Ecosystems* **2004**, *7*, 109–133. [CrossRef]
8. Chen, Y.; Olson, D.M.; Ruberson, J.R. Effects of nitrogen fertilization on tritrophic interactions. *Arthropod Plant Interact.* **2010**, *4*, 81–94. [CrossRef]
9. Andrew, N.R.; Hughes, L. Herbivore damage along a latitudinal gradient: Relative impacts of different feeding guilds. *Oikos* **2005**, *108*, 176–182. [CrossRef]
10. Dittrich, A.D.; Helden, A.J. Experimental sward islets: The effect of dung and fertilisation on Hemiptera and Araneae. *Insect Conserv. Divers.* **2012**, *5*, 46–56. [CrossRef]
11. Hargrove, W.W.; Crossley, D.A., Jr.; Seastedt, T.R. Shifts in insect herbivory in the canopy of black locust, *Robinia pseudacacia*, after fertilization. *Oikos* **1984**, *43*, 322–328. [CrossRef]
12. Jones, M.E.; Fenn, M.E.; Paine, T.D. The effect of nitrogen additions on bracken fern and its insect herbivores at sites with high and low atmospheric pollution. *Arthropod Plant Interact.* **2011**, *5*, 163–173. [CrossRef]
13. Kytö, M.; Niemelä, P. Larsson S Insects on trees: Population and individual response to fertilization. *Oikos* **1996**, *75*, 148–159. [CrossRef]
14. Scriber, J.M.; Slansky, F., Jr. The nutritional ecology of immature insects. *Annu. Rev. Entomol.* **1981**, *26*, 183–211. [CrossRef]
15. Koike, T.; Tobita, H.; Shibata, T.; Matsuki, S.; Konno, K.; Kitao, M.; Yamashita, N.; Maruyama, Y. Defense characteristics of several deciduous broad-leaved tree seedlings grown under differing levels of CO_2 and nitrogen. *Popul. Ecol.* **2006**, *48*, 23–29. [CrossRef]
16. Moise, E.R.; Henry, H.A. Like moths to a street lamp: Exaggerated animal densities in plot−level global change field experiments. *Oikos* **2010**, *119*, 791–795. [CrossRef]
17. Englund, G.; Cooper, S.D. Scale effects and extrapolation in ecological experiments. *Adv. Ecol. Res.* **2003**, *33*, 161–213.
18. Fay, T.M.; Turner, E.C.; Basset, Y.; Ewers, R.M.; Glen, R.; Novotny, V. Whole-ecosystem experimental manipulations of tropical forests. *Trends Ecol. Evol.* **2015**, *30*, 334–346.
19. Southwood, T. Insect/plant relationship—An evolutionary perspective. In *Insect-Plant Relationships*; Blackwell: Oxford, UK, 1972; pp. 3–30.
20. Awmack, C.S.; Leather, S.R. Host plant quality and fecundity in herbivorous insects. *Annu. Rev. Entomol.* **2002**, *47*, 817–844. [CrossRef]
21. Bryant, J.P.; Chapin, F.S., III; Klein, D.R. Carbon/nutrient balance of boreal plants in relation to vertebrate herbivory. *Oikos* **1983**, *40*, 357–368. [CrossRef]
22. Coley, P.D.; Bryant, J.P.; Chapin, F.S. Resource availability and plant antiherbivore defense. *Science* **1985**, *230*, 895–899. [CrossRef] [PubMed]
23. Poorter, H.; Niinemets, Ü.; Poorter, L.; Wright, I.J. Villar R Causes and consequences of variation in leaf mass per area (LMA): A meta-analysis. *New Phytol.* **2009**, *182*, 565–588. [CrossRef]
24. Webber, B.L.; Woodrow, I.E. In Intra-plant variation in cyanogenesis and the continuum of foliar plant defense traits in the rainforest tree *Ryparosa kurrangii* (Achariaceae). *Tree Physiol.* **2008**, *28*, 977–984. [CrossRef] [PubMed]
25. Webber, B.L.; Woodrow, I.E. Chemical and physical plant defence across multiple ontogenetic stages in a tropical rain forest understorey tree. *J Ecol.* **2009**, *97*, 761–771. [CrossRef]
26. Stefan, K.; Fürst, A.; Hacker, R.; Bartels, U. *Forest Foliar Condition in Europe. Results of Large-Scale Foliar Chemistry Surveys 1995*; EC-UN/ECE-FBVA: Brussels, Belgium; Geneva, Switzerland, 1997.
27. Perakis, S.S.; Matkins, J.J.; Hibbs, D.E. N_2-fixing red alder indirectly accelerates ecosystem nitrogen cycling. *Ecosystems* **2012**, *15*, 1182–1193. [CrossRef]
28. Rytter, L.; Arveby, A.S.; Granhall, U. Dinitrogen (C_2H_2) fixation in relation to nitrogen fertilization of grey alder [*Alnus incana* (L.) Moench.] plantations in a peat bog. *Biol. Fertil. Soils* **1991**, *10*, 233–240. [CrossRef]

29. Gaulke, L.S.; Henry, C.L.; Brown, S.L. Nitrogen fixation and growth response of *Alnus rubra* following fertilization with urea or biosolids. *Sci. Agric.* **2006**, *63*, 361–369. [CrossRef]
30. Troelstra, S.R.; Wagenaar, R.; Smant, W. Growth of actinorhizal plants as influenced by the form of nitrogen with special reference to *Myrica gale* and *Alnus glutinosa*. *J. Exp. Bot.* **1992**, *43*, 1349–1359. [CrossRef]
31. Simberloff, D.; Dayan, T. The Guild Concept and the Structure of Ecological Communities. *Annu. Rev. Eco. Syst.* **1991**, *22*, 115–143. [CrossRef]
32. Landsberg, J.; Smith, M.S. A functional scheme for predicting the outbreak potential of herbivorous insects under global atmospheric change. *Aust. J. Bot.* **1992**, *40*, 565–577. [CrossRef]
33. Hartley, S.E.; Lawton, J.H. Host-plant manipulation by gall-insects: A test of the nutrition hypothesis. *J. Anim. Ecol.* **1992**, *61*, 113–119. [CrossRef]
34. Faeth, S.H.; Mopper, S.; Simberloff, D. Abundances and diversity of leaf-mining insects on three oak host species: Effects of host-plant phenology and nitrogen content of leaves. *Oikos* **1981**, *37*, 238–251. [CrossRef]
35. Shibata, H.; Kirikae, M.; Tanaka, Y.; Sakuma, T.; Hatano, R. Proton budgets of forest ecosystems on volcanogenous regosols in Hokkaido, Northern Japan. *Water Air Soil Pollut.* **1998**, *105*, 63–72. [CrossRef]
36. Hiura, T. Stochasticity of species assemblage of canopy trees and understory plants in a temperate secondary forest created by major disturbances. *Ecol. Res.* **2001**, *16*, 887–893. [CrossRef]
37. Kikuzawa, K. Leaf survival and evolution in Betulaceae. *Ann. Bot.* **1982**, *50*, 345–353. [CrossRef]
38. Ataka, M.; Sun, L.; Nakaji, T.; Katayama, A.; Hiura, T. Five-year nitrogen addition affects fine root exudation and its correlation with root respiration in a dominant species, *Quercus crispula*, of a cool temperate forest, Japan. *Tree Physiol.* **2020**, *40*, 367–376. [CrossRef]
39. Oksanen, L. Logic of experiments in ecology: Is pseudoreplication a pseudoissue? *Oikos* **2001**, *94*, 27–38. [CrossRef]
40. Colegrave, N.; Ruxton, G.D. Using biological insight and pragmatism when thinking about pseudoreplication. *Trends Ecol Evol.* **2018**, *33*, 28–35. [CrossRef]
41. Yoshida, K. Seasonal fluctuation of moth community in Tomakomai Experiment Forest of Hokkaido University. *Res. Bull. Hokkaido Univ. For.* **1980**, *37*, 675–685.
42. Julkunen-Tiitto, R. Phenolic constituents in the leaves of northern willows: Methods for the analysis of certain phenolics. *J. Agric. Food Chem.* **1985**, *33*, 213–217. [CrossRef]
43. Barton, K.E.; Valkama, E.; Vehviläinen, H.; Ruohomäki, K.; Knight, T.M.; Koricheva, J. Additive and non-additive effects of birch genotypic diversity on arthropod herbivory in a long-term field experiment. *Oikos* **2015**, *124*, 697–706. [CrossRef]
44. Kudo, G. Herbivory pattern and induced responses to simulated herbivory in *Quercus mongolica* var. *grosseserrata*. *Ecol. Res.* **1996**, *11*, 283–289. [CrossRef]
45. Nakamura, M.; Hina, T.; Nabeshima, E.; Hiura, T. Do spatial variation in leaf traits and herbivory within a canopy respond to selective cutting and fertilization? *Can. J. For. Res.* **2008**, *38*, 1603–1610. [CrossRef]
46. R Development Core Team R. *A Language and Environment for Statistical Computing*; R Foundation for Statistical Computing: Vienna, Austria, 2011.
47. Augspurger, C.K.; Bartlett, E.A. Differences in leaf phenology between juvenile and adult trees in a temperate deciduous forest. *Tree Physiol.* **2003**, *23*, 517–525. [CrossRef]
48. Delagrange, S.; Messier, C.; Lechowicz, M.J.; Dizengremel, P. Physiological, morphological and allocational plasticity in understory deciduous trees: Importance of plant size and light availability. *Tree Physiol.* **2004**, *24*, 775–784. [CrossRef] [PubMed]
49. Vitasse, Y. Ontogenic changes rather than difference in temperature cause understory trees to leaf out earlier. *New Phytol.* **2013**, *198*, 149–155. [CrossRef]
50. Weiner, J.; Thomas, S.C. The nature of tree growth and the "age-related decline in forest productivity". *Oikos* **2001**, *94*, 374–376. [CrossRef]
51. Struve, D.K. A review of shade tree nitrogen fertilization research in the United States. *J. Arboric.* **2002**, *28*, 252–263.
52. Coley, P.D. Herbivory and defensive characteristics of tree species in a lowland tropical forest. *Ecol. Monogr.* **1983**, *53*, 209–234. [CrossRef]
53. Bussotti, F.; Borghini, F.; Celesti, C.; Leonzio, C.; Bruschi, P. Leaf morphology and macronutrients in broadleaved trees in central Italy. *Trees* **2000**, *14*, 361–368. [CrossRef]
54. Sardans, J.; Peñuelas, J.; Rodà, F. Plasticity of leaf morphological traits, leaf nutrient content, and water capture in the Mediterranean evergreen oak *Quercus ilex* subsp. ballota in response to fertilization and changes in competitive conditions. *Ecoscience* **2006**, *13*, 258–270.
55. Ruess, R.W.; Anderson, M.D.; McFarland, J.M.; Kielland, K.; Olson, K.; Taylor, D.L. Ecosystem−level consequences of symbiont partnerships in an N-fixing shrub from interior Alaskan floodplains. *Ecol Monogr.* **2013**, *83*, 177–194. [CrossRef]
56. Lee, J.; Nakumara, M.; Hiura, T. Does large-scale N fertilization have time-delayed effect on insects community structure by changing oak quantity and quality. *Arthropod Plant Interact.* **2017**, *11*, 515–523. [CrossRef]
57. Witkowski, E.T.F.; Lamont, B.B. Leaf specific mass confounds leaf density and thickness. *Oecologia* **1991**, *88*, 486–493. [CrossRef] [PubMed]
58. Clissold, F.J.; Sanson, G.D.; Read, J. The paradoxical effects of nutrient ratios and supply rates on an outbreaking insect herbivore, the Australian plague locust. *J. Anim. Ecol.* **2006**, *75*, 1000–1013. [CrossRef]
59. Clissold, F.J.; Sanson, G.D.; Read, J.; Simpson, S.J. Gross vs. net income: How plant toughness affects performance of an insect herbivore. *Ecology* **2009**, *90*, 3393–3405. [CrossRef] [PubMed]

Article

Long-Term Nitrogen Addition Decreases Soil Carbon Mineralization in an N-Rich Primary Tropical Forest

Xiankai Lu [1,2,*], Qinggong Mao [1,2], Zhuohang Wang [1], Taiki Mori [1], Jiangming Mo [1,2], Fanglong Su [1] and Zongqing Pang [1]

[1] Key Laboratory of Vegetation Restoration and Management of Degraded Ecosystems, South China Botanical Garden, Chinese Academy of Sciences, Guangzhou 510650, China; maoqinggong@scbg.ac.cn (Q.M.); wangzhuohang347@berrygenomics.com (Z.W.); taikimori7@gmail.com (T.M.); mojm@scbg.ac.cn (J.M.); sufanglong@scbg.ac.cn (F.S.); pangzongqing@scbg.ac.cn (Z.P.)

[2] Center for Plant Ecology, Core Botanical Gardens, Chinese Academy of Sciences, Guangzhou 510650, China

[*] Correspondence: luxiankai@scbg.ac.cn; Tel./Fax: +86-20-3708-4353

Abstract: Anthropogenic elevated nitrogen (N) deposition has an accelerated terrestrial N cycle, shaping soil carbon dynamics and storage through altering soil organic carbon mineralization processes. However, it remains unclear how long-term high N deposition affects soil carbon mineralization in tropical forests. To address this question, we established a long-term N deposition experiment in an N-rich lowland tropical forest of Southern China with N additions such as NH_4NO_3 of 0 (Control), 50 (Low-N), 100 (Medium-N) and 150 (High-N) kg N ha^{-1} yr^{-1}, and laboratory incubation experiment, used to explore the response of soil carbon mineralization to the N additions therein. The results showed that 15 years of N additions significantly decreased soil carbon mineralization rates. During the incubation period from the 14th day to 56th day, the average decreases in soil CO_2 emission rates were 18%, 33% and 47% in the low-N, medium-N and high-N treatments, respectively, compared with the Control. These negative effects were primarily aroused by the reduced soil microbial biomass and modified microbial functions (e.g., a decrease in bacteria relative abundance), which could be attributed to N-addition-induced soil acidification and potential phosphorus limitation in this forest. We further found that N additions greatly increased soil-dissolved organic carbon (DOC), and there were significantly negative relationships between microbial biomass and soil DOC, indicating that microbial consumption on soil-soluble carbon pool may decrease. These results suggests that long-term N deposition can increase soil carbon stability and benefit carbon sequestration through decreased carbon mineralization in N-rich tropical forests. This study can help us understand how microbes control soil carbon cycling and carbon sink in the tropics under both elevated N deposition and carbon dioxide in the future.

Keywords: nitrogen deposition; soil carbon mineralization; carbon sequestration; soil heterotrophic respiration; microbial activity; tropical forests

Citation: Lu, X.; Mao, Q.; Wang, Z.; Mori, T.; Mo, J.; Su, F.; Pang, Z. Long-Term Nitrogen Addition Decreases Soil Carbon Mineralization in an N-Rich Primary Tropical Forest. *Forests* **2021**, *12*, 734. https://doi.org/10.3390/f12060734

Academic Editor: Frank S. Gilliam

Received: 6 April 2021
Accepted: 2 June 2021
Published: 4 June 2021

Publisher's Note: MDPI stays neutral with regard to jurisdictional claims in published maps and institutional affiliations.

1. Introduction

Anthropogenic-elevated nitrogen (N) deposition, significantly accelerating N cycle on Earth, has become an important driver of global change, especially in the tropics, in the coming decades [1,2]. Due to the coupled relationships between carbon (C) and N cycles, elevated N deposition will inevitably affect the stability of soil organic C (SOC) in terrestrial ecosystems [3], where soils are the largest C reservoir [4]. Soil C stability and sequent C sequestration are tightly related to SOC mineralization, one of the most important processes in the terrestrial ecosystem C cycle. A better understanding of the responses of SOC mineralization to elevated N deposition is essential for better C management under global changes.

Elevated N deposition can increase the terrestrial C sink through increasing net primary productivity in N-limited ecosystems, and even can stimulate soil C sequestration through impeding organic matter decomposition in temperate forests, where N is not limiting microbial growth [5–8]. In tropical forests, where ecosystems are more likely to be N-rich [9], soil C cycling has not been well-assessed. Tropical forests, with more than half of global forest C stocks, are a globally significant terrestrial C sink, and have a disproportionately large influence on the global C cycle [10–12]. Studies in subtropical successional forests showed that the contribution of heterotrophic soil respiration to total soil respiration reached 64% [13], indicating that microbial-driven soil heterotrophic respiration is a critical CO_2 flux in the atmosphere in tropical ecosystems [14]. Many studies showed that high N inputs decreased soil respiration, and lower soil pH was suggested to be a key driver due to its negative priming effects [7,15–17], pointing to a further understanding of the role of soil C mineralization. In tropical forests, a few studies addressed the responses of organic C mineralization, using soils with one-time N addition [14,18,19] or soils from field N fertilization plots [20]. However, there are still large uncertainties regarding how and to what extent continuing N deposition affects the soil C mineralization process over a longer time-scale (e.g., >10 years), especially in regions with high background N deposition [1]. Understanding of N accumulation effect is urgent in the prediction of soil C stability and sequestration with the globalization of N deposition.

Here, we aim to explore how long-term N deposition (N accumulation effect) with varied N input rates affects soil C mineralization in N-rich tropical ecosystems. In 2002, we chose an N-rich primary tropical forest at the Dinghushan Biosphere Reserve in southern China (23°09′21″ N to 23°11′30″ N and 112°30′39″ E to 112°33′41″ E). The study site was naturally high in N status, as are many lowland tropical forests, and it was already affected by anthropogenic N deposition [21]. In 2009–2010, total atmospheric N deposition was about 48.6 kg N ha^{-1} y^{-1} [21]. Soils in the study site are lateritic red earths formed from sandstone, and are highly weathered, with poor soil-buffering capacity [22]. We hypothesize that long-term N additions decrease soil C mineralization in this N-rich forest, considering N-addition-induced soil acidification and alteration in microbial community composition in the studied site [22–24] and the global negative effects of nitrogen deposition on soil microbes [25].

2. Materials and Methods

2.1. Study Site

This study was conducted at the Dinghushan Biosphere Reserve (DHSBR), which is an UNESCO's Man and the Biosphere (MAB) Programme site in the middle of Guangdong Province, southern China (112°10′ E, 23°10′ N). The DHSBR has a monsoon climate, and is in a subtropical/tropical moist forest zone. The annual average precipitation is 1748 mm, from 2002 to 2012, mainly concentrated from April to September [21]. Annual mean relative humidity is 80%. Mean annual temperature is 21.0 °C. The reserve has been experiencing high atmospheric N deposition in precipitation (e.g., commonly >30 kg N ha^{-1} y^{-1}) since the 1990s. In 2009–2010, total atmospheric N deposition was about 48.6 kg N ha^{-1}·y^{-1}, with a wet N deposition of 34.4 kg N ha^{-1} y^{-1} (see Lu et al., 2018 for further references) [21].

We established our research site in 2002 in a mature monsoon evergreen broadleaf forest (primary forest), between 250 and 300 m above sea level. The forest was protected from disturbances related to land-use for >400 years [26], and supports a rich assemblage of plant species, most of which are natives of tropical and subtropical China, including *Castanopsis chinensis* Hance, *Schima superba* Chardn. and Champ., *Cryptocarya chinensis* (Hance) Hemsl., *Machilus chinensis* (Champ. Ex Benth.) Hemsl., *Syzygium rehderianum* Merr. & Perry, and *Acmena acuminatissima* (Blume) Merr. et Perry. Canopy closure is typically above 95% [27]. The topography is heterogeneous, with slopes ranging from 25° to 35°. Soils in the study site are lateritic red earths (Oxisols) formed from sandstone, and are highly weathered, with poor soil-buffering capacity [22]. The mature forest is a typically N-rich ecosystem, as are many lowland tropical forests [21].

2.2. Experimental Treatments

Nitrogen amendments were initiated in July 2003, with four rates: Control (0 N added), Low-N (50 kg N ha^{-1} y^{-1}, close to the total background atmospheric N deposition), Medium-N (100 kg N ha^{-1} y^{-1}) and High-N (150 kg N ha^{-1} y^{-1}). These were based on the atmospheric N deposition rate of the 1990s [21], and the increase expected in the future due to the rapid development of agricultural and industrial activity [1]. In total, there were twelve 10-m $\times$ 20-m plots surrounded by buffer strips of at least 20 m in width, with treatments replicated in triplicate and randomly assigned. Monthly applications of NH_4NO_3 solution were administered by hand to the forest floor of these plots, in the form of 12 equal applications over the whole year. During each application, fertilizer was weighed, mixed with 20 L of deionized water, and applied to each plot, using a backpack sprayer below the canopy. Two passes were made across each plot to ensure an even distribution of fertilizer. The control plots received equal volumes (20 L) of water without N fertilizer.

2.3. Field Sampling and Laboratory Analysis

In December 2017, we took soil samples (upper 0–10 cm depth) from these plots, after removing litter layers. In each plot, soil samples were taken from six randomly selected points. In the laboratory, soils were sieved (2 mm) to remove roots and stones, and mixed thoroughly by hand for subsequent chemical analysis. Soil pH was measured using a soil to water ratio of 1:2.5 (10 g soil and 25 mL deionized water; pH meter: FE28-Standard FiveEasyPlus™, Mettler Toledo instruments Co., Ltd., Greifensee, Switzerland). Dissolved organic C was extracted by the addition of deionized water using fresh soils (soil to water ratio of 1:5). Soil inorganic N was extracted with 2M KCl (soil to water ratio of 1:5) and the filtrates were analyzed for NH_4^+-N and NO_3^--N by colorimetric method using a flow-injection auto-analyzer (Lachat Quik-Chem 8000; Lachat Instrument, Mequon, WI, USA). The method for NH_4^+-N analysis is based on the Berthelot reaction (Lachat QuikChem Method: 10-107-06-1-C) and the method for NO_3^--N analysis uses a copperized cadmium column to reduce nitrate to nitrite (Lachat QuikChem Method: 10-107-04-1-C). Soil microbial biomass C (MBC) and N (MBN) were determined by chloroform fumigation and extraction in 0.5 M K_2SO_4 by standard procedures [28,29]. The difference in organic C and total N between fumigated and unfumigated samples was assumed to originate from microbial biomass. Both MBC and MBN concentrations were corrected for unrecovered biomass using a k factor of 0.45 [30]. The amount of organic C and total N in the extracts were simultaneously measured with a Total Organic Carbon Analyzer (TOC-5000, Shimadzu Co. Ltd., Kyoto, Japan).

For laboratory incubation, soil subsamples with 20 g oven-dry equivalent weight were placed in each incubation unit (250 mL erlenmeyer flask). The soil water content was adjusted to 60% of field holding capacity, and soils were pre-incubated at 20 °C for 2 days prior to the start of the incubation. Soil CO_2 fluxes were measured on days 1, 3, 5, 7, 14, 21, 28, 35, 44, and 58 days of the incubation. Incubation units were covered with white polyethylene plastic wrap to prevent water evaporation. After the polyethylene plastic wrap was removed, the upper air of the flasks was mixed so that the accumulated CO_2 was removed. The CO_2 concentrations, after the closure of a lid equipped with a syringe, were determined immediately using gas chromatography (Agilent 7890A; Agilent Technologies Inc., Santa Clara, CA, USA). The CO_2 flux rate was calculated as the accumulation of CO_2 over the given time period (20 min). We calculated microbial metabolic quotient (MMQ), microbial respiration per unit of biomass, which is a critically important parameter in the understanding of microbial controls for carbon cycling, particularly heterotrophic respiration [31].

2.4. Statistical Analyses

Repeated measures analysis of variance (ANOVA) was performed to examine the effects of N treatments on soil CO_2 fluxes. One-way ANOVA with Fisher LSD test was employed to identify N-treatment effects on soil microbial and soil parameters. Linear regression analysis was also used to examine the relationships between these parameters. All analyses were conducted using SPSS 16.0 for Windows (SPSS, Chicago, IL, USA) at $p < 0.05$.

3. Results and Discussion

Repeated measures ANOVA further showed that long-term N additions significantly decreased soil C mineralization rates and cumulative CO_2 emissions over the incubation period, with the lowest values found in the High-N treatments (Figure 1a,b). Compared to the control plots, soil CO_2 emission rates at the beginning of incubation were significantly decreased by 14%–23% after 15 years of N additions (Figure 1a). The CO_2 emission rates decreased rapidly in the first week and leveled off over subsequent days. During the period from 14th day to 56th day, the average decreases were 18%, 33% and 47% in the Low-N, Medium-N and High-N treatments, respectively. These findings supported our hypothesis, and gave a plausible explanation of our previous finding that N additions reduced soil respiration in the field monitoring of this N-rich site [15,24] (Table 1).

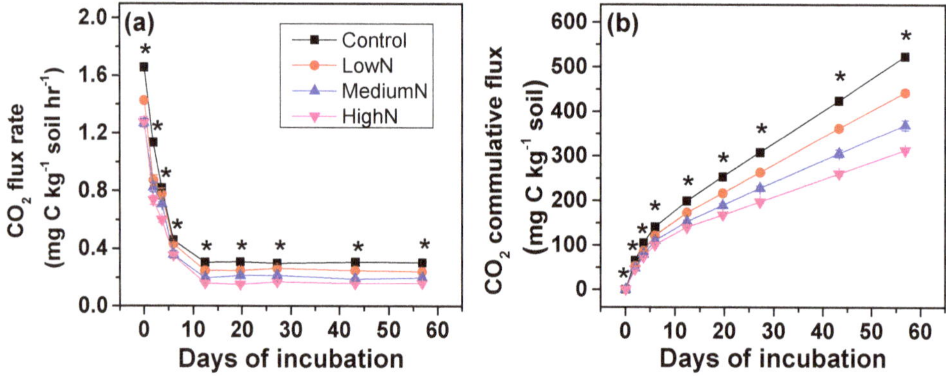

Figure 1. Effects of long-term N additions on soil organic carbon mineralization rate (**a**) and CO2 cumulative flux (**b**) in the N-rich mature tropical forest at Dinghushan reserve. Note: Asterisks (*) means significant differences between Controls and N treatments using planned-contrast analysis; values are shown as means ± standard error (n = 3); h means hour.

The soil C mineralization process is largely mediated by soil microbial activity, which depends on both microbial traits and soil environmental conditions. We found that long-term N addition did not alter specific microbial activity (MMQ; Table 1), indicating no changes in microbial C use efficiency [31]. Furthermore, microbial C:N ratios showed no response to N additions, which is consistent with the results of soil C:N ratios [23]. Carbon and N stoichiometric requirements for biomass production force microbial communities to adapt their foraging strategies to the available substrates, which affects the rate of microbial growth and respiration [32]. Hence, we could exclude the potential effect of soil C and N stoichiometry on soil microbes. Our previous study of the same site showed that thirteen years of N additions significantly decreased microbial biomass, with distinct shifts in microbial community composition leading to reductions in the relative abundance of bacteria as well as the genes responsible for cellulose and chitin degradation [23,24]. Hence, we suggest that an N-induced change in microbial biomass and community composition can dominate the soil C mineralization process in N-rich forests (Table 1).

Table 1. Responses of soil carbon mineralization, soil microbial biomass, and soil properties at 0–10 cm layer to long-term N additions in the N-rich mature tropical forest of Southern China.

	Control	Low-N	Medium-N	High-N
MBC (mg C kg^{-1})	596.18 ± 9.75 a	552.02 ± 20.44 a	535.34 ± 42.51 ab	459.48 ± 18.42 b
MBN (mg N kg^{-1})	77.73 ± 0.64 a	70.71 ± 2.06 ab	68.36 ± 3.44 ab	64.23 ± 5.88 b
MBC/MBN	7.67 ± 0.07 a	7.82 ± 0.33 a	7.90 ± 0.88 a	7.23 ± 0.48 a
DOC (mg C kg^{-1})	231.46 ± 6.21 a	300.25 ± 27.37 ab	346.80 ± 22.06 b	338.89 ± 38.09 b
C mineralization rate (mg C kg^{-1} hr^{-1})	1.66 ± 0.01 a	1.43 ± 0.02 b	1.27 ± 0.03 c	1.28 ± 0.04 c
MMQ(µg mg^{-1} hr^{-1})	2.78 ± 0.04 a	2.59 ± 0.07 a	2.41 ± 0.19 a	2.79 ± 0.15 a
NO$_3^-$-N (mg N kg^{-1})	7.90 ± 0.42 a	10.98 ± 0.33 b	12.53 ± 0.49 b	12.33 ± 1.00 b
NH$_4^+$-N (mg N kg^{-1})	8.24 ± 0.68 a	12.90 ± 0.42 b	14.90 ± 1.09 b	15.55 ± 1.25 b
pH (H$_2$O)	3.91 ± 0.01 a	3.86 ± 0.01 b	3.84 ± 0.01 b	3.71 ± 0.01 b
Soil organic carbon * (g C kg^{-1})	25.4 ± 3.44 a	29.6 ± 2.54 ab	30.2 ± 0.46 b	31.9 ± 0.77 b
Field CO$_2$ emission * (mg CO$_2$ m^{-2} hr^{-1})	91.8 ± 4.05 a	88.2 ± 2.27 ab	85.4 ± 1.4 ab	61.2 ± 0.86 b
Bacterial abundance * (mole%)	52.4 ± 0.24 a	52.0 ± 0.31 a	51.4 ± 0.24 ab	50.0 ± 0.24 b
Fungal abundance * (mole %)	10 ± 0.51 a	9.3 ± 0.16 a	9.9 ± 0.14 a	9.7 ± 0.39 a
Fungal:Bacterial ratio *	0.19 ± 0.01 a	0.18 ± 0.00 a	0.19 ± 0.00 a	0.19 ± 0.01 a

Notes: MBC, microbial biomass carbon; MBN, microbial biomass nitrogen; DOC, dissolved organic carbon; qCO$_2$, microbial metabolic quotient. Different lowercase letters within the same horizontal line indicate significant differences between treatments at $p < 0.05$ level; Values are means ± standard error (n = 3). "*" means data are cited from Tian et al. 2019 [24].

We found that long-term N additions significantly increased soil available N and accelerated soil acidification, and there were significant negative relationships between N treatment rates and soil pH (Table 2). Ecosystem N saturation is a key reason to drive soil acidification under excess N inputs [22,33]. In general, a lower soil pH is known to restrict microbial growth [34,35] and to change the microbial community [23,24,36]. There were significantly positive relationships between microbial biomass and soil pH (Table 2), indicating that accelerated soil acidification contributed to the decline in MBC and MBN under long-term N inputs, which confirmed the results of our recent study [23]. Another reason for the decreased microbial biomass may be phosphorus (P) limitation, which is common for soil microbial processes in moist tropical forests [37–39]. Studying P leaching dynamics in this forest, we found that high N inputs enlarged the imbalance between N and P as influxes into soils [40]. Furthermore, experimental P additions significantly increased soil microbial biomass and soil respiration, suggesting that P availability is an important limiting factor for microbial growth in N-rich forests [41]. The decrease in microbial P utilization genes after N amendment suggested that N amendments exacerbated soil P deficiency in this study [24]. Meanwhile, bacterial relative abundance had the largest total effect on soil respiration in this site, and high N treatments decreased the abundances of genes responsible for labile and recalcitrant C [24]. A decrease in the bacteria's relative abundance could inhibit the production of C degradation enzymes by soil bacterial biomass, providing a straightforward explanation for the measured decrease in CO$_2$ emission under both field and incubation conditions (Table 1).

Interestingly, we found that long-term N additions significantly increased soil DOC contents (Table 1), which was a primary source of energy and cellular C for the soil microbial community. Considering that there were no changes in specific microbial activity (MMQ) among treatments (Table 1) and negative relationships between microbial biomass and soil DOC (Table 2), we suggested that decreases in the microbial consumption of soil-soluble C pool or retardation in organic C decomposition should be an important reason for increased soil DOC. DOC is a significant source of organic C in the mineral soil, and contributes to soil-forming processes though its downward movement and adsorption in mineral soil [42,43], which supports the increased SOC used under N treatments in this study (Table 1). Based on our findings and previous study, we developed a conceptual model hypothesis of how long-term N deposition affects soil C soil mineralization and storage through alterations in soil microbial activity, which will determine the fate of soil C and soil C storage potential (Figure 2). In this model, if N deposition increases soil microbial activity, such as microbial biomass and community functions related to C cycling, soil mineralization rates will increase, and more CO_2 will be emitted; thus, more C sources, such as DOC, will be consumed by soil microbes, and finally, soil C storage will decrease. Alternatively, if N deposition decreases soil microbial activity ("N"), soil mineralization rates will decrease and CO_2 emission will be inhibited, so that soil DOC and SOC will increase compared to the background conditions. Soil microorganisms play a key role in determining the longevity and stability of soil C, and C efflux from soils.

Table 2. Pearson correlation coefficients between N treatment, soil carbon mineralization, soil microbial biomass, and soil properties across all the plots.

	Treatment	MBC	MBN	DOC	C-min	MMQ	NO_3^--N	NH_4^+-N	pH
Treatment	1.00								
MBC	−0.78 **	1.00							
MBN	−0.68 *	0.57 *	1.00						
DOC	0.70 *	−0.54	−0.38	1.00					
C-min	−0.90 **	0.66 *	0.72 **	−0.83 **	1.00				
MMQ	−0.07	−0.47	0.10	−0.35	0.36	1.00			
NO_3^--N	0.81 **	−0.66 *	−0.41	0.64 *	−0.84 **	−0.16	1.00		
NH_4^+-N	0.85 **	−0.50	−0.68 *	0.77 **	−0.94 **	−0.47	0.72 **	1.00	
pH	−0.92 **	0.82 **	0.72 **	−0.54	0.74 **	−0.17	−0.62 *	−0.72 **	1.00

Notes: (1) *, ** indicate correlation is significant at the 0.05 and 0.01 level (2-tailed), respectively; (2) MBC, microbial biomass carbon; MBN, microbial biomass nitrogen; DOC, dissolved organic carbon; C-min, C mineralization rate; MNQ, microbial metabolic quotient; pH, soil pH value.

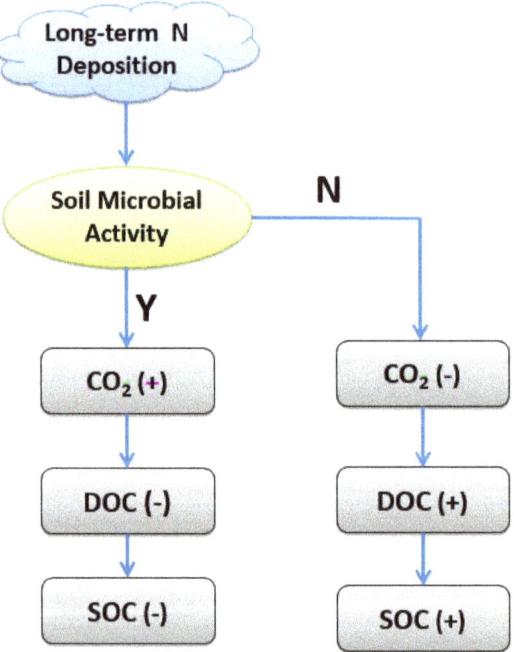

Figure 2. A conceptual model on how long-term N deposition affects soil C soil mineralization and storage through altering soil microbial activity in both microbial biomass and community functions. Notes: "Y" means positive effects while "N" means negative effects; "+" and "−" mean increasing and decreasing, respectively: DOC, dissolved organic carbon; SOC, soil organic carbon.

4. Conclusions

In summary, we conclude that continuous high N deposition can inhibit soil C mineralization process at more than a decade scale in N-rich forests, which is primarily attributed to the decreased microbial activity in microbial biomass and community functions. This inhibiting effect is beneficial to soil C stabilization and accumulation in forest ecosystems. It is noteworthy that the inhibiting effect increased with elevated N addition, indicating that higher soil C storage may occur with increased N addition, as shown in Table 1. This study can help us to understand how microbes control soil carbon cycling and carbon sink in the tropics under both elevated N deposition and carbon dioxide. However, there are several limitations which need to be overcome in future studies. (1) We mainly focus on soils in the present study, but their interactions with plants and the resources they produce above- and belowground should be considered in the field conditions. (2) The DOC can be recalcitrant and not easily accessible to microbes, and a more detailed analysis of DOC chemistry is required in the future. (3) To confirm whether higher DOC under N fertilization leads to greater C retention, it is necessary to obtain more detailed measurements along the soil profile, over a longer period of time, to understand the fate of DOC. (4) It is worth exploring temporal and spatial (deeper soil layers) dynamics, because N addition may change plant phenology and there could be seasonal differences in resource availability and microbial activity, which could eventually lead to reversed patterns in other parts of the year. Finally, our findings require further verification in different forest ecosystems in the tropics, where intensifying industrial and agricultural activities will enhance atmospheric N deposition to much higher levels in the coming decades [1].

Author Contributions: Conceptualization, X.L.; formal analysis, X.L.; investigation, X.L., Q.M. and Z.W.; data curation, X.L.; writing—original draft preparation, X.L.; writing—review and editing, X.L., Q.M., Z.W., T.M., J.M., F.S. and Z.P.; visualization, X.L.; project administration, X.L. and Z.W.; funding acquisition, X.L. and J.M. All authors have read and agreed to the published version of the manuscript.

Funding: This study was funded by the National Natural Science Foundation of China (No. 41922056, 41731176), Key Research and Development Program of Guangdong Province (2020B1111530004), and Youth Innovation Promotion Association CAS (No. Y201965).

Data Availability Statement: Data is available for use upon request.

Acknowledgments: We appreciate Shaoming Cai and Hui Mo, and Xiaoping Pan for their skillful assistance in laboratory and field work, and Dinghushan Forest Ecosystem Research Station for the support in the field work.

Conflicts of Interest: The authors declare no conflict of interest.

References

1. Galloway, J.N.; Dentener, F.J.; Capone, D.G.; Boyer, E.W.; Howarth, R.W.; Seitzinger, S.P.; Asner, G.P.; Cleveland, C.C.; Green, P.A.; Holland, E.A.; et al. Nitrogen Cycles: Past, Present, and Future. *Biogeochemistry* **2004**, *70*, 153–226. [CrossRef]
2. Hietz, P.; Turner, B.L.; Wanek, W.; Richter, A.; Nock, C.A.; Wright, S.J. Long-Term Change in the Nitrogen Cycle of Tropical Forests. *Science* **2011**, *334*, 664–666. [CrossRef] [PubMed]
3. Lu, X.; Hou, E.; Guo, J.; Gilliam, F.S.; Li, J.; Tang, S.; Kuang, Y. Nitrogen addition stimulates soil aggregation and enhances carbon storage in terrestrial ecosystems of China: A meta-analysis. *Glob. Chang. Biol.* **2021**, *27*, 2780–2792. [CrossRef]
4. Lal, R. Forest soils and carbon sequestration. *For. Ecol. Manag.* **2005**, *220*, 242–258. [CrossRef]
5. Reay, D.S.; Dentener, F.; Smith, P.; Grace, J.; Feely, R.A. Global nitrogen deposition and carbon sinks. *Nat. Geosci.* **2008**, *1*, 430–437. [CrossRef]
6. Gruber, N.; Galloway, J.N. An Earth-system perspective of the global nitrogen cycle. *Nature* **2008**, *451*, 293–296. [CrossRef]
7. Janssens, I.A.; Dieleman, W.; Luyssaert, S.; Subke, J.A.; Reichstein, M.; Ceulemans, R.; Ciais, P.; Dolman, A.J.; Grace, J.; Matteucci, G.; et al. Reduction of forest soil respiration in response to nitrogen deposition. *Nat. Geosci.* **2010**, *3*, 315–322. [CrossRef]
8. Schulte-Uebbing, L.; de Vries, W. Global-scale impacts of nitrogen deposition on tree carbon sequestration in tropical, temperate, and boreal forests: A meta-analysis. *Glob. Chang. Biol.* **2018**, *24*, E416–E431. [CrossRef]
9. Matson, P.A.; McDowell, W.H.; Townsend, A.R.; Vitousek, P.M. The globalization of N deposition: Ecosystem consequences in tropical environments. *Biogeochemistry* **1999**, *46*, 67–83. [CrossRef]

10. Phillips, O.L.; Malhi, Y.; Higuchi, N.; Laurance, W.F.; Nunez, P.V.; Vasquez, R.M.; Laurance, S.G.; Ferreira, L.V.; Stern, M.; Brown, S.; et al. Changes in the carbon balance of tropical forests: Evidence from long-term plots. *Science* **1998**, *282*, 439–442. [CrossRef]

11. Pan, Y.; Birdsey, R.A.; Fang, J.; Houghton, R.; Kauppi, P.E.; Kurz, W.A.; Phillips, O.L.; Shvidenko, A.; Lewis, S.L.; Canadell, J.G.; et al. A Large and Persistent Carbon Sink in the World's Forests. *Science* **2011**, *333*, 988–993. [CrossRef]

12. Baccini, A.; Walker, W.; Carvalho, L.; Farina, M.; Sulla-Menashe, D.; Houghton, R.A. Tropical forests are a net carbon source based on aboveground measurements of gain and loss. *Science* **2017**, *358*, 230–233. [CrossRef] [PubMed]

13. Huang, W.; Han, T.; Liu, J.; Wang, G.; Zhou, G. Changes in soil respiration components and their specific respiration along three successional forests in the subtropics. *Func. Ecol.* **2016**, *30*, 1466–1474. [CrossRef]

14. Fanin, N.; Barantal, S.; Fromin, N.; Schimann, H.; Schevin, P.; Haettenschwiler, S. Distinct Microbial Limitations in Litter and Underlying Soil Revealed by Carbon and Nutrient Fertilization in a Tropical Rainforest. *PLoS ONE* **2012**, *7*, e49990. [CrossRef]

15. Mo, J.; Zhang, W.; Zhu, W.; Gundersen, P.; Fang, Y.; Li, D.; Wang, H. Nitrogen addition reduces soil respiration in a mature tropical forest in southern China. *Glob. Chang. Biol.* **2008**, *14*, 403–412. [CrossRef]

16. Liu, L.; Greaver, T.L. A global perspective on belowground carbon dynamics under nitrogen enrichment. *Ecol. Lett.* **2010**, *13*, 819–828. [CrossRef]

17. Frey, S.D.; Ollinger, S.; Nadelhoffer, K.; Bowden, R.; Brzostek, E.; Burton, A.; Caldwell, B.A.; Crow, S.; Goodale, C.L.; Grandy, A.S.; et al. Chronic nitrogen additions suppress decomposition and sequester soil carbon in temperate forests. *Biogeochemistry* **2014**, *121*, 305–316. [CrossRef]

18. Cleveland, C.C.; Townsend, A.R. Nutrient additions to a tropical rain forest drive substantial soil carbon dioxide losses to the atmosphere. *Proc. Natl. Acad. Sci. USA* **2006**, *103*, 10316–10321. [CrossRef]

19. Soong, J.L.; Maranon-Jimenez, S.; Cotrufo, M.F.; Boeckx, P.; Bode, S.; Guenet, B.; Penuelas, J.; Richter, A.; Stahl, C.; Verbruggen, E.; et al. Soil microbial CNP and respiration responses to organic matter and nutrient additions: Evidence from a tropical soil incubation. *Soil Biol. Biochem.* **2018**, *122*, 141–149. [CrossRef]

20. Cusack, D.F.; Torn, M.S.; McDowell, W.H.; Silver, W.L. The response of heterotrophic activity and carbon cycling to nitrogen additions and warming in two tropical soils. *Glob. Chang. Biol.* **2010**, *16*, 2555–2572. [CrossRef]

21. Lu, X.; Vitousek, P.M.; Mao, Q.; Gilliam, F.S.; Luo, Y.; Zhou, G.; Zou, X.; Bai, E.; Scanlon, T.M.; Hou, E.; et al. Plant acclimation to long-term high nitrogen deposition in an N-rich tropical forest. *Proc. Natl. Acad. Sci. USA* **2018**, *115*, 5187–5192. [CrossRef]

22. Lu, X.; Mao, Q.; Gilliam, F.S.; Luo, Y.; Mo, J. Nitrogen deposition contributes to soil acidification in tropical ecosystems. *Glob. Chang. Biol.* **2014**, *20*, 3790–3801. [CrossRef]

23. Wang, C.; Lu, X.; Mori, T.; Mao, Q.; Zhou, K.; Zhou, G.; Nie, Y.; Mo, J. Responses of soil microbial community to continuous experimental nitrogen additions for 13 years in a nitrogen-rich tropical forest. *Soil Biol. Biochem.* **2018**, *121*, 103–112. [CrossRef]

24. Tian, J.; Dungait, J.A.J.; Lu, X.; Yang, Y.; Hartley, I.P.; Zhang, W.; Mo, J.; Yu, G.; Zhou, J.; Kuzyakov, Y. Long-term nitrogen addition modifies microbial composition and functions for slow carbon cycling and increased sequestration in tropical forest soil. *Glob. Chang. Biol.* **2019**, *25*, 3267–3281. [CrossRef]

25. Zhang, T.; Chen, H.Y.H.; Ruan, H. Global negative effects of nitrogen deposition on soil microbes. *ISME J.* **2018**, *12*, 1817–1825. [CrossRef]

26. Shen, C.D.; Liu, D.S.; Peng, S.L.; Sun, Y.M.; Jiang, M.T.; Yi, W.X.; Xing, C.P.; Gao, Q.Z.; Li, Z.; Zhou, G.Y. C-14 measurement of forest soils in Dinghushan Biosphere Reserve. *Chin. Sci. Bull.* **1999**, *44*, 251–256. [CrossRef]

27. Lu, X.; Mo, J.; Gilliam, F.S.; Zhou, G.; Fang, Y. Effects of experimental nitrogen additions on plant diversity in an old-growth tropical forest. *Glob. Chang. Biol.* **2010**, *16*, 2688–2700. [CrossRef]

28. Brookes, P.C.; Landman, A.; Pruden, G.; Jenkinson, D.S. Chloroform fumigation and the release of soil-nitrogen-a rapid direct extraction method to measure microbial biomass nitrogen in soil. *Soil Biol. Biochem.* **1985**, *17*, 837–842. [CrossRef]

29. Vance, E.D.; Brookes, P.C.; Jenkinson, D.S. An extraction method for measuring soil microbial biomass-c. *Soil Biol. Biochem.* **1987**, *19*, 703–707. [CrossRef]

30. Jenkinson, D.S.; Brookes, P.C.; Powlson, D.S. Measuring soil microbial biomass. *Soil Biol. Biochem.* **2004**, *36*, 5–7. [CrossRef]

31. Xu, X.; Schimel, J.P.; Janssens, I.A.; Song, X.; Song, C.; Yu, G.; Sinsabaugh, R.L.; Tang, D.; Zhang, X.; Thornton, P.E. Global pattern and controls of soil microbial metabolic quotient. *Ecol. Monogr.* **2017**, *87*, 429–441. [CrossRef]

32. Sinsabaugh, R.L.; Manzoni, S.; Moorhead, D.L.; Richter, A. Carbon use efficiency of microbial communities: Stoichiometry, methodology and modelling. *Ecol. Lett.* **2013**, *16*, 930–939. [CrossRef] [PubMed]

33. Aber, J.; Mcdowell, W.; Nadelhoffer, K.; Magill, A.; Berntson, G.; Kamakea, M.; Mcnulty, S.; Currie, W.; Rustad, L.; Fernandez, I. Nitrogen Saturation in Temperate Forest Ecosystems. *Bioscience* **1998**, *48*, 921–934. [CrossRef]

34. Nilsson, L.O.; Baath, E.; Falkengren-Grerup, U.; Wallander, H. Growth of ectomycorrhizal mycelia and composition of soil microbial communities in oak forest soils along a nitrogen deposition gradient. *Oecologia* **2007**, *153*, 375–384. [CrossRef]

35. Lauber, C.L.; Hamady, M.; Knight, R.; Fierer, N. Pyrosequencing-Based Assessment of Soil pH as a Predictor of Soil Bacterial Community Structure at the Continental Scale. *Appl. Environ. Microbiol.* **2009**, *75*, 5111–5120. [CrossRef] [PubMed]

36. Kaspari, M.; Bujan, J.; Weiser, M.D.; Ning, D.; Michaletz, S.T.; He, Z.; Enquist, B.J.; Waide, R.B.; Zhou, J.; Turner, B.L.; et al. Biogeochemistry drives diversity in the prokaryotes, fungi, and invertebrates of a Panama forest. *Ecology* **2017**, *98*, 2019–2028. [CrossRef]

37. Cleveland, C.C.; Townsend, A.R.; Schmidt, S.K. Phosphorus limitation of microbial processes in moist tropical forests: Evidence from short-term laboratory incubations and field studies. *Ecosystems* **2002**, *5*, 680–691. [CrossRef]
38. Vitousek, P.M.; Porder, S.; Chadwick, H.O.A. Terrestrial phosphorus limitation: Mechanisms, implications, and nitrogen—Phosphorus interactions. *Ecol. Appl.* **2010**, *20*, 5–15. [CrossRef]
39. Nottingham, A.T.; Turner, B.L.; Stott, A.W.; Tanner, E.V.J. Nitrogen and phosphorus constrain labile and stable carbon turnover in lowland tropical forest soils. *Soil Biol. Biochem.* **2015**, *80*, 26–33. [CrossRef]
40. Zhou, K.; Lu, X.; Mori, T.; Mao, Q.; Wang, C.; Zheng, M.; Mo, H.; Hou, E.; Mo, J. Effects of long-term nitrogen deposition on phosphorus leaching dynamics in a mature tropical forest. *Biogeochemistry* **2018**, *138*, 215–224. [CrossRef]
41. Liu, L.; Gundersen, P.; Zhang, T.; Mo, J. Effects of phosphorus addition on soil microbial biomass and community composition in three forest types in tropical China. *Soil Biol. Biochem.* **2012**, *44*, 31–38. [CrossRef]
42. Neff, J.C.; Asner, G.P. Dissolved organic carbon in terrestrial ecosystems: Synthesis and a model. *Ecosystems* **2001**, *4*, 29–48. [CrossRef]
43. Kalbitz, K.; Schwesig, D.; Rethemeyer, J.; Matzner, E. Stabilization of dissolved organic matter by sorption to the mineral soil. *Soil Biol. Biochem.* **2005**, *37*, 1319–1331. [CrossRef]

forests

Article

Effect of Long-Term Nitrogen and Phosphorus Additions on Understory Plant Nutrients in a Primary Tropical Forest

Qinggong Mao [1,2], Hao Chen [3], Cong Wang [4], Zongqing Pang [1,2], Jiangming Mo [1,2] and Xiankai Lu [1,2,*]

[1] Key Laboratory of Vegetation Restoration and Management of Degraded Ecosystems, Guangdong Provincial Key Laboratory of Applied Botany, South China Botanical Garden, Chinese Academy of Sciences, Guangzhou 510650, China; maoqinggong@scbg.ac.cn (Q.M.); pangzongqing@scbg.ac.cn (Z.P.); mojm@scbg.ac.cn (J.M.)

[2] Center of Plant Ecology, Core Botanical Gardens, Chinese Academy of Sciences, Guangzhou 510650, China

[3] School of Ecology, Sun Yat-sen University, Guangzhou 510006, China; chenhao27@mail.sysu.edu.cn

[4] State Key Laboratory of Mycology, Institute of Microbiology, Chinese Academy of Sciences, Beijing 100101, China; wangc@im.ac.cn

* Correspondence: luxiankai@scbg.ac.cn; Tel.: +86-020-37252712

Abstract: Humid tropical forests are commonly characterized as N-rich but P-deficient. Increased N deposition may drive N saturation and aggravate P limitation in tropical forests. Thus, P addition is proposed to mitigate the negative effects of N deposition by stimulating N cycling. However, little is known regarding the effect of altered N and P supply on the nutrient status of understory plants in tropical forests, which is critical for predicting the consequences of disturbed nutrient cycles. We assessed the responses of N concentration, P concentration, and N:P ratios of seven understory species to N and P addition in an 8-year fertilization experiment in a primary forest in south China. The results showed that N addition had no effect on plant N concentration, P concentration, and N:P ratios for most species. In contrast, P addition significantly increased P concentration, and decreased N:P ratios but had no effect on plant N concentration. The magnitude of P concentration responses to P addition largely depended on the types of organs and species. The increased P was more concentrated in the fine roots and branches than in the leaves. The gymnospermous liana *Gnetum montanum* Markgr. had particularly lower foliar N: P (~9.8) and was much more responsive to P addition than the other species studied. These results indicate that most plants are saturated in N but have great potential to restore P in primary tropical forests. N deposition does not necessarily aggravate plant P deficiency, and P addition does not increase the retention of deposited N by increasing the N concentration. In the long term, P inputs may alter the community composition in tropical forests owing to species-specific responses.

Keywords: nitrogen deposition; phosphorus; nutrient use strategy; understory plants; tropical forests; *Gnetum montanum*; south China

Citation: Mao, Q.; Chen, H.; Wang, C.; Pang, Z.; Mo, J.; Lu, X. Effect of Long-Term Nitrogen and Phosphorus Additions on Understory Plant Nutrients in a Primary Tropical Forest. *Forests* **2021**, *12*, 803. https://doi.org/10.3390/f12060803

Academic Editor: Mariangela Fotelli

Received: 7 May 2021
Accepted: 15 June 2021
Published: 18 June 2021

Publisher's Note: MDPI stays neutral with regard to jurisdictional claims in published maps and institutional affiliations.

1. Introduction

Humid tropical forests support the main organic carbon pool and harbor diverse species in terrestrial ecosystems [1]. These tropical forests are commonly rich in N but are relatively poor in P availability because of the substantial biological N fixation and P leaching loss during pedogenesis [2]. However, global N and P cycling has changed because of the increased N deposition and use of P fertilizers in the last several decades [3,4]. Because nutrient supply controls plant nutrients, nutrient dynamics can influence the structure and function of ecosystems by changing community composition and productivity [5,6]. Considering that N and P are essential nutrients for plants, studying plant N and P status under disturbed nutrient cycling is critically important.

The effects of N input on plant N status have been widely studied. Generally, it is believed that N inputs increase plant N concentration ([N]) in N-limited ecosystems but have a minor effect on plant [N] in N-rich ecosystems [7]. Although N deposition rate has been reduced in many areas, it is projected to increase in the tropics with increasing population and economy [3]. In particular, a large proportion of biodiversity hotspots in humid tropical forest regions have been included in areas affected by N deposition [8]. In N-rich tropical forests, N addition often has no effect on plant [N] [9,10] and causes further N saturation, which is indicated by significant increases in soil inorganic N concentration, N_2O emission, and nitrate (NO_3^-) leaching [11,12]. N inputs may also lead to an imbalance of N and P, and affect plant P status due to the tight coupling of N and P.

However, the effects of N input on plant P are contradictory. N inputs are generally considered to aggravate ecosystem P limitations, particularly in tropical forests [13]. Two mechanisms have been proposed. First, N-induced soil acidification may enhance the immobilization of phosphate with iron or aluminum oxide in old soils, thus, decreasing P availability [13]. Second, N addition may stimulate plant growth and increase the P demand of microorganisms and plants. Increasing evidence shows that N addition increases phosphatase activities in soils and rhizospheres in tropical forests, which is considered a sign of increased P limitation [12,14]. However, the increased phosphatase activity with N addition is also supposed to alleviate P limitation by investing additional N in the synthesis of more phosphatases [15,16], as N is the main component (8–32% N) of the phosphatases [17]. In addition, plants may also adapt to P deficiency caused by N deposition via multiple mechanisms, such as increasing foliar P uptake and resorption [18,19]. To date, whether N addition induces P limitation in tropical forests remains unresolved.

Phosphorus enrichment due to P fertilization alters the cycling of P in tropical forests. P addition commonly increases foliar P concentration ([P]) and decreases plant N:P ratio [20–22]. Furthermore, studies have shown large interspecific variation of plant [P] responses to P addition, but few explanations are presented. For example, Ostertag [21] reported 1.7–5.1 times higher foliar [P] after P addition in Hawaiian forests. Mayor et al. [9] found that P addition increased foliar [P] in three eudicot tree species but had no effect on the palm species in Panama. There are also debates on the difference in P accumulation after P fertilization between shade-tolerant and light-demanding species [20,23]. Therefore, the interspecific variation in responses may be attributed to different taxonomic and functional groups. In contrast, emerging evidence shows that stems or roots are more sensitive than leaves to P fertilization [24,25]. If more P is stored in non-photosynthetic organs, using leaf nutrients and stoichiometry to evaluate P limitation may not reveal the real P status. Therefore, additional tests are required to address these uncertainties, which are fundamental to predict the acclimation of tropical forest ecosystems under elevated P supply.

The effects of P addition may increase plant N levels and alleviate N saturation. For example, the addition of P was found to increase soil microbial biomass [26], decrease NO_3^- leaching and N_2O emission [12], and mitigate the inhibition of N deposition on CH_4 uptake in tropical forests [27,28]. Therefore, P addition has been proposed as a potential management practice to improve N-induced negative effects in tropical forests [12]. Commonly held views suggest that increased foliar [P] would be accompanied by elevated foliar [N] due to the tight stoichiometric coupling of the two elements. First, alleviation of P limitation can promote plant growth and productivity and stimulate N demand. Second, a large proportion of P compounds in plants also contain N [21], such as nucleic acids; thus, increased plant [P] is expected to increase [N] as well. Third, increased soil P availability may increase plant N acquisition by increasing the rate of nitrate reduction, because the activity of nitrate reductase is controlled by phosphorylation [29]. Nevertheless, how P addition affects plant [N] is rarely reported in N-rich tropical forests; it remains inconclusive whether P addition promotes plant N uptake.

In this study, we aimed to explore how long-term N and P additions affect plant [N], [P], and the N:P ratio by using an in-situ N and P fertilization experiment in an N-rich primary tropical forest in Southern China. The understory plants were selected because they are critical in tree recruitment and contribute significantly to plant diversity in forests [30,31]. We attempted to test the following hypotheses: (I) Long-term N addition would cause no change in plant [N], because the forest has been rich in N. (II) N addition would decrease plant [P] owing to N-induced acidification and decreased P supply. (III) Long-term P addition would greatly increase plant [P], and the responses of plant nutrients to P addition would vary in different species and organs (e.g., leaves, branches, and fine roots). (IV) Phosphorus addition would increase plant [N], considering that increased P availability may facilitate plant N demand and acquisition.

2. Methods

2.1. Study Site

This study was conducted in the Dinghushan Biological Reserve (DHSBR) in South China ($23°10'12''$ N, $112°32'42''$ E). The region is characterized by a monsoon climate with a mean annual temperature of 21 °C and a mean annual precipitation of 1927 mm [32]. The soil in the forest is lateritic red earth (Oxisol, according to the U.S. soil taxonomy system), which developed from a sandstone regolith and extends to a depth of >60 cm [33]. The reserve has experienced high N deposition (>30 kg N ha^{-1} yr^{-1}) since the 1990s [34]. In the studied forest, the N leaching rate was as high as its inputs via rainfall, indicating the saturation status in N [11]. In contrast, the soil's available [P] (extracted by Bray I solution) was low (~4 mg P kg^{-1}) [35].

The experiment was established in a monsoon evergreen broadleaf forest (250–300 m above sea level), which had been protected from human disturbance for more than 400 years by a local temple and monks [36]. The aboveground community comprises canopy, subcanopy, and understory plants. Dominant species in the canopy and sub-canopy layers include *Castanopsis chinensis* Hance, *Machilus chinensis* Hemsl, *Schima superba* Gardn. et Champ, *Cryptocarya chinensis* Hemsl, and *Syzygium rehderianum* Merr. et Perry (see details of the tree structure of the studied forest in Liu et al. [26]). Tree seedlings, shrubs, and herbaceous plants constitute the understory layer, accounting for a high species richness. The understory layer is in deep shade, as canopy closure is commonly >90% without obvious forest gaps [31].

2.2. Experimental Design

A full-factorial experimental design was used, completely randomized with two main factors: N effects (two levels: without N addition and with N addition of 15 g N m^{-2} yr^{-1}) and P effects (two levels: without P addition and with P addition of 15 g P m^{-2} yr^{-1}). Accordingly, there were four (2 × 2) treatments: control, N addition (15 g N m^{-2} yr^{-1}), P addition (15 g P m^{-2} yr^{-1}), and NP addition (15 g N m^{-2} yr^{-1} plus 15 g P m^{-2} yr^{-1}), each with five replicates, thus, constituting a total of 20 plots. Each plot was 5 × 5 m and was surrounded by a 5–10 m wide buffer strip to avoid interference. All plots were laid out randomly on a slope with minor adjustment of position to avoid top-down runoff (Figure S1). The N and P addition experiment was established in 2007. N (as NH$_4$NO$_3$) and P (as NaH$_2$PO$_4$) were added to the forest floor every other month from February 2007 to July 2015. The fertilizers were weighed and dissolved in 5 L of water for each plot and sprayed on the forest floor using a backpack sprayer (equivalent to 0.5 mm rainfall each time), and the control plots received an equivalent volume of pure water. The N addition level was selected because inorganic N has high mobility and the forest is N-rich. We used a relatively high load of N addition over the year to sustain a significantly increased soil N availability. Similarly, high loads of P additions were used to ensure a significant increase in soil P availability, given that the input phosphate is readily immobilized by iron or aluminum oxides in tropical old soils (pH < 4) and is inaccessible to plants [2,35,37]. Soil colloids need to be saturated to test plant responses to elevated P availability.

2.3. Sampling and Analysis of Plants

We identified seven species that were co-occurring in all plots (Table 1), including *Ardisia lindleyana* D. Dietrich. (*ALI*), *Carallia brachiata* Merr. (*CBR*), *Calamus rhabdocladus* Burret (*CRH*), *Cryptocarya chinensis* Hemsl (*CCH*), *Cryptocarya concinna* Hance (*CCO*), *Aidia canthioides* Masam (*ACA*), and *Gnetum montanum* Markgr. (*GMO*). They displayed a range of functional groups, including tree seedlings, shrubs, palm, and gymnosperm liana. Plant samples were collected in July 2015 (in the eighth year of experimental treatment). Fully matured leaves were sampled from two to four individuals (height < 2 m) for all seven species in each plot, while the connected branches were collected for tree seedling of four species (*ACA*, *CBR*, *CCH* and *CCO*) because of their availability. Five mineral soil cores (10 cm in depth) per plot were sampled using an auger (2.5 cm in diameter), combined into a composite sample, and sieved through a 2 mm mesh to remove stones and roots. Soil pH was measured with soil and deionized water (1:2.5 ratio) using a glass electrode. Soil available N was extracted by 2 mol L^{-1} KCl and measured with a UV-8000 spectrophotometer (Metash Instruments Corp., Shanghai, China). Soil available P was extracted by 0.03 mol L^{-1} ammonium fluoride and 0.025 mol L^{-1} hydrochloric acid and then analyzed colorimetrically [38]. To evaluate the response of fine roots N and P to long-term fertilization, we collected enough fine root samples within the plot scale (0–10 cm soil layer). The live fine roots (defined as <2 mm in diameter) were sorted and washed in pure water. Finally, soil and all plant tissues, including leaves, branches, and fine root samples, were oven-dried for 72 h at 65 °C and ground to a fine and homogeneous powder. The powder was prepared to analyze soil organic matter (SOM), soil total [N], and plant [N] by using an elemental analyzer (Vario ISOTOPE cube, Elementar, Hanau, Germany) and to analyze soil [P] and plant [P] using inductively coupled plasma (ICP-ES, Optima 2000, Perkin Elmer, Norwalk, CT, USA) after nitric acid and perchloric acid dissolution.

Table 1. Traits of seven species in a primary tropical forest.

Species	Abbreviation	Family	Growth Form
Ardisia lindleyana D. Dietrich	*ALI*	Myrsinaceae	shrub
Carallia brachiata Merr.	*CBR*	Rhizophoraceae	subcanopy tree seedling
Calamus rhabdocladus Burret	*CRH*	Arecaceae	palm liana
Cryptocarya chinensis Hemsl	*CCH*	Lauraceae	canopy tree seedling
Cryptocarya concinna Hance	*CCO*	Lauraceae	canopy tree seedling
Aidia canthioides Masam	*ACA*	Rubiaceae	subcanopy tree seedling
Gnetum montanum Markgr.	*GMO*	Gnetaceae	woody liana

2.4. Statistical Analyses

We presented all [N], [P], and N:P ratios on a mass basis. One-way analysis of variance (ANOVA) was employed to examine the effect of differences in treatments (control, +N, +P, and +NP) on soil properties. Tukey's honestly significant difference test was used to compare the differences between treatments. Two-way ANOVA was performed to test the independent effect of N or P addition (the fixed factors) treatments and their interaction effect on plant [N], [P], and N:P ratios (the dependent variable). The ANOVA model included the main effects of N and P and the N × P interaction. General linear mixed effect models were used to evaluate the N or P concentration of three organs (leaf, branch, and fine root), with species as the random factor. Data were tested for normality and homogeneity, and logarithmic transformation was applied if the data failed to meet assumptions of the selected model. All analyses were performed using SPSS 19.0 for Windows (SPSS, Chicago, IL, USA). Statistically significant differences were set at p-values $\leq$ 0.05. Mean values $\pm$ standard errors of the mean are reported in the text.

3. Results

3.1. Responses of Soil Properties to N or P Addition

After 8 years of continuous fertilization, there was a mild but not significant increase in soil extractable N and a decrease in soil pH in N addition plots compared with those in control plots, but there were no responses for soil total N and P, extractable P, and SOM (Table 2). Long-term P addition significantly increased soil total P and extractable P to an abundant level (0.6 g kg^{-1} and 111.7 mg kg^{-1}, respectively) but did not significantly change other soil properties (Table 2). Both N and P addition greatly increased soil total P, extractable P, and soil extractable N, with a decreasing trend in soil pH (Table 2).

Table 2. Responses of soil properties to long-term N and P addition.

	Control	+N	+P	+NP
SOM (g kg^{-1})	31.15 (0.74)	29.28 (0.39)	30.50 (2.75)	30.28 (1.66)
TN (g kg^{-1})	1.88 (0.09)	1.82 (0.06)	1.79 (0.10)	1.76 (0.09)
TP (g kg^{-1})	0.20 (0.02) b	0.22 (0.02) b	0.62 (0.04) a	0.47 (0.06) a
Extractable N (mg kg^{-1})	12.70 (0.91) ab	14.93 (0.88) ab	11.51 (0.82) b	16.40 (1.57) a
Extractable P (mg kg^{-1})	3.26 (0.30) b	5.28 (1.25) b	111.76 (20.64) a	73.09 (10.83) a
pH$_{(H2O)}$	3.88(0.04) ab	3.70 (0.02) c	3.95 (0.04) a	3.77 (0.04) bc

Note: SOM, soil organic matter; TN, total N; TP, total P. Values are means with standard error in parentheses (n = 5). Different lowercase letters indicate significant differences at $p < 0.05$ level among treatments.

3.2. Effect of N and P Addition on the Nutrient Status of Understory Vegetation

Foliar [N] and branch [N] averaged 13.3 mg g^{-1} and 16.9 mg g^{-1}, respectively. The species with the highest foliar [N] (*CCO*, 16.4 mg g^{-1}) possessed the highest branch [N] (*CCO*, 24.3 mg g^{-1}). Stand-level fine root [N] averaged 10.6 mg g^{-1}. Overall, plant [N] was not different among the organs (leaves, branches, and fine roots) (Figure 1). The mean foliar [P] value was 0.51 mg g^{-1}, with significant differences among species ($p < 0.001$) (Figure 2). The gymnospermous liana *GMO* had an exceptionally high foliar [P] (1.09 mg g^{-1}). The mean values of branch [P] and fine root [P] were 0.35 mg g^{-1} and 0.46 mg g^{-1}, respectively. No significant difference in [P] observed among the organs (Figure S2). Foliar N:P ratio ranged from 9.8 to 49.7, with a mean value of 32.1. The N:P ratios of branches and fine roots averaged 53.9 and 23.3, respectively (Figure S2).

Nitrogen addition did not increase the foliar [N] except for that of *ACA*, with a 7.7% increase ($p < 0.01$, Table S1 and Figure 1). Similarly, N had no effect on branches [N] or fine roots [N] (Table S1 and Figure 1). Moreover, N addition did not change the [P] and N:P ratio, with a few exceptions (Table S1). For example, N addition decreased the foliar N:P ratio of *CCH* ($p = 0.04$) and *GMO* ($p = 0.04$) and the fine roots N:P ratio ($p < 0.001$) (Table S1 and Figure 3).

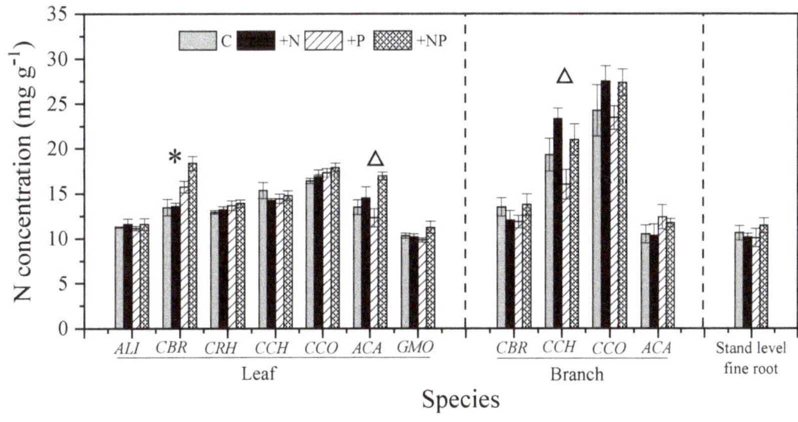

Figure 1. Effect of long-term N and/or P addition on N concentration of different species and organs. Notes: Error bars indicate standard error (SE; n = 5). Triangles (Δ) and asterisks (*) above the column indicate significant effect of N addition and P addition, respectively (two-way analysis of variance (ANOVA), $p < 0.05$); Species abbreviations are the same as in Table 1.

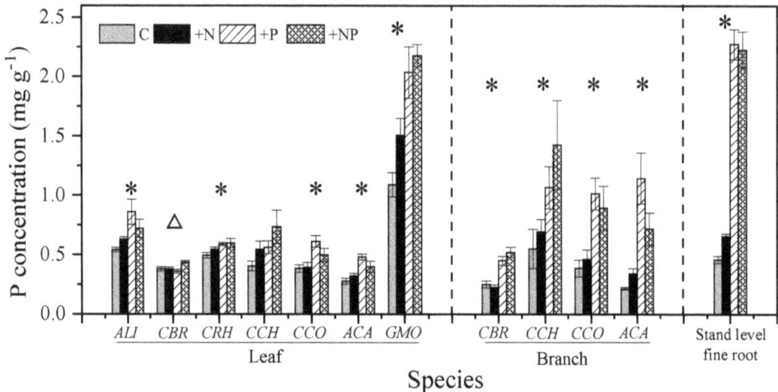

Figure 2. Effect of long-term N and/or P addition on P concentration of different species and organs. Notes: Error bars indicate standard error (SE; n = 5); Triangles (Δ) and asterisks (*) above the column indicate significant effect of N addition and P addition, respectively (two-way analysis of variance (ANOVA), $p < 0.05$); Species abbreviations are the same as in Table 1.

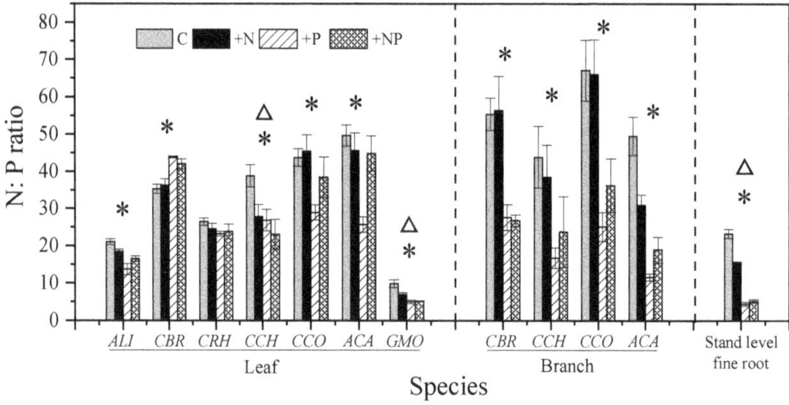

Figure 3. Effect of long-term N and/or P addition on N:P ratio of different species and organs. Notes: Error bars indicate standard error (SE; n = 5). Triangles (Δ) and asterisks (*) above the column indicate significant effect of N addition and P addition, respectively (two-way analysis of variance (ANOVA), $p < 0.05$); Species abbreviations are same as in Table 1.

Phosphorus addition significantly increased foliar [P], branches [P], and fine roots [P] by 47%, 192%, and 396%, respectively (Figure 2). In addition, the responses of foliar [P] to P addition varied significantly among co-occurring species and their organs (Figure 4). For example, the foliar [P] of *CBR* was not changed by P addition, while its branch [P] increased by 79%. In comparison, the increases in foliar [P] and branch [P] of *ACA* were as high as 74% and 432%, respectively. Therefore, branch [P] and leaf [P] increased remarkably with P addition for all species (Figure S3). P addition showed no effect on foliar [N], except in *CBR* ($p < 0.001$), which show a 13.7% increase (Figure 1). P addition also had no effect on branches [N] and fine roots [N] (Table S1). P addition significantly decreased the foliar N:P ratio in all species, except in *CBR* (Table S1). After P addition, the foliar N:P ratio was still >20 in five of the seven species (Figure 3). In addition, significant interaction of N and P addition was only observed in plant [P] of a few species ($p < 0.05$) and fine root N:P ratio ($p < 0.01$; Table S1).

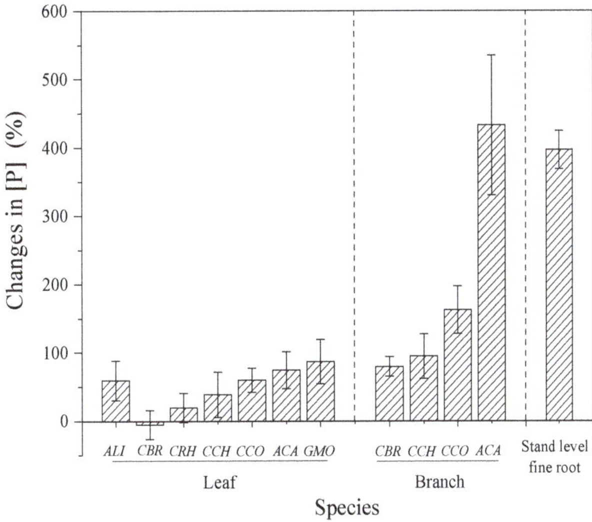

Figure 4. Effect of P addition on the changes of P concentration [P] in different species and organs compared with their control counterparts. Notes: Error bars indicate standard error (SE; *n* = 5). Species abbreviations are the same as in Table 1.

4. Discussion

4.1. Effects of N Addition

Consistent with the first hypothesis, we found that long-term N addition did not increase the tissue [N] and N:P ratio in most species. Similar results have been reported for many other lowland tropical forests [9,24]. Several reasons may account for the mute responses in the present study. First, plants adapt to N-rich environments and have no selection pressure for additional N storage. Generally, N is abundant and open cycled in humid tropical forests because of the vigorous biological N fixation [39]. The amount of input N was found to equally leak out this N-rich forest ecosystems [40]. Second, the studied forest may be more limited by other resources such as K, Ca, and Mg, which hinder additional N demand. A study in a Panama forest showed that tree seedling growth is stimulated by K addition rather than N addition [41]. Third, no responses of plant N may be a strategy for avoiding herbivory risk, given that herbivory is generally N-limited [42]. Increased herbivory risk after N input may constrain N accumulation.

Contrary to hypothesis II, we did not find a decrease in plant [P] under long-term N addition, and even *CBR* showed increased foliar [P] (Figure 2 and Table S1). Traditionally, it has been suggested that elevated N input aggravates P limitation by acidifying soils or stimulating plant growth [28,43]. However, the evidence supporting these points largely derives from the increase in phosphatase activity caused by N input. Indeed, increased phosphatase activity after N addition was found at our site and in many other tropical forests [14,44,45]. However, phosphatase activity is an indirect parameter that only reflects the enhanced capability of P acquisition; it cannot demonstrate that plants are P limited. Our results suggest that N addition does not decrease plant [P], or that N addition does not aggravate plant P limitation. Conversely, N inputs mitigate P limitation for a few species according to the slight decrease in N:P ratios (Figure 3). This may be because N addition does not stimulate plant growth [31], and dilute plant [P], while these may have compensation mechanisms, such as enhanced phosphatase activity and leaf resorption, to adjust to the decreased P supply [15,16].

4.2. Effects of P Addition

The increased soil P availability causes significant increases in plant [P], and the magnitude of plant [P] responses varies among different species and organs, supporting hypothesis III. Increased foliar [P] under P addition has been frequently reported in many studies [9,24], whereas the present study may provide an extreme example under high loads of P addition. This is because P addition did not stimulate photosynthesis and growth of the studied plants (unpublished data) and of plants in some other tropical forests [46,47]. The remarkable increases in plant [P] could be attributed to the storage function. Indeed, we found that the foliar N:P ratio was above 20 (P limitation) in half of the included species even after long-term P addition (Figure 2), indicating a strong impetus for P acquisition. This may be because the large amount of P storage is beneficial when considering the uncertainty in P supply or confers advantages to understory plant to help them grow rapidly when forest gaps occur.

Although studies have shown the species-specific responses of plant [P] and disproportionate P storage in different organs following P addition, explanations are rarely provided. The species-specific responses may be attributed to their phylogenetic constraints. It is common for a community to be constituted by species with different biogeochemical niches, which are closely related to their shared ancestry [48]. Two examples in the present study support this point. We found that *CCH* and *CCO* were quite similar in foliar nutrient concentrations and responses to P addition (Figure 4), as both belong to the same genus (*Cryptocarya*). In contrast, we found that *GMO* was distinct from other species in its elemental composition and showed notably high foliar [P] but the lowest foliar [N]; thus, the largest increase in foliar [P] was observed after P addition. This seems reasonable because among studied species, *GMO* is the only gymnosperm (Table 1). Fine roots and branches were much more sensitive to P addition than leaves (Figure 4), suggesting the main function of P storage in non-photosynthetic organs. These results are consistent with those of other studies. One explanation is that the relatively lower increase in foliar [P] may protect leaves from negative effects (such as toxicity level or increased herbivory) by isolating excess P away from the leaves to other organs [49]. Therefore, branches and roots may serve as nutrient reservoirs that support optimal N:P balance in the leaves, suggesting the capability to regulate P distribution within the whole plant.

Contrary to hypothesis IV, P addition had no effect on foliar [N] for most of the studied species (Figure 1 and Table S1). Commonly held views are that increased foliar [P] would be accompanied by elevated foliar [N] owing to the tight stoichiometric coupling of two elements. A large proportion of P compounds in plants also contain N, such as nucleic acids. However, our results suggest that the acquisition of plant N and P may be independent processes. The data of only one species, *CBR*, with the lowest foliar [P], seemed to support the view that increased foliar [P] accompanies elevated foliar [N]. One plausible explanation is that the forest has already been N-saturated, and plant [N] has been at the optimal level. Therefore, no factors, such as P addition, could further elevated plant [N], as a high load of N addition has no effect on plant [N], discussed above. Another reason may be that plants of tropical forests are P limited, based on their high foliar N:P ratio [50]. P addition alleviates the inherent demand of P by means of storage. In brief, our results suggest that P addition may have little effect on N retention by increasing plant uptake in N-rich tropical forests.

5. Conclusions

We conclude with the following main findings from the long-term N and P addition experiment in an N-rich tropical forest: (1) N addition has no effect on plant [N] and does not alter plant [P], suggesting that plants have already been N saturated and that further N deposition would not aggravate plant P limitation. (2) P addition increases plant [P], whereas the magnitude of responses varies in different species and organs. Increased P storage occurs in non-photosynthetic organs. P addition also has no effect on plant [N], suggesting that plant P storage would not keep pace with the increased uptake of N.

Our findings provide new insights into plant nutrient responses under disturbed nutrient cycling and indicate that P inputs, as practical management to mitigate N-induced negative effects, may have profound effects on plant communities in tropical forests.

Supplementary Materials: The following are available online at www.mdpi.com/xxx/s1, Figure S1: Layout of plots and treatments in our study forest, Figure S2: The N concentration, P concentration and N: P ratio in different organs in blank control, Figure S3: Effect of N and/or P addition on ratios of branch to leaf in N and P concentration of different species, Table S1: The *p*-value of two-way ANOVA in effect of N and P addition on N concentration, P concentration.

Author Contributions: Writing—original draft preparation, Q.M.; writing—review and editing, X.L., H.C., C.W., Z.P. and J.M. All authors have read and agreed to the published version of the manuscript.

Funding: Financial support was provided by the National Natural Science Foundation of China (No. 31700422, 41731176, 41922056, 31872691), Guangdong Basic and Applied Basic Research Foundation (2019A1515011642).

Data Availability Statement: The data presented in this study are available on request from the corresponding author.

Acknowledgments: We wish to thank Shaoming Cai and Lijie Deng for their skillful assistance in field work, Hui Mo, Xiaoying You and Xiaoping Pan for their assistance in laboratory work. We would like to thank the two anonymous reviewers and the editor for their insightful comments.

Conflicts of Interest: The authors declare no conflict of interest.

References

1. Raven, P.H.; Gereau, R.E.; Phillipson, P.B.; Chatelain, C.; Jenkins, C.N.; Ulloa, C.U. The distribution of biodiversity richness in the tropics. *Sci. Adv.* **2020**, *6*, eabc6228. [CrossRef]
2. Vitousek, P.; Porder, S.; Houlton, B.; Chadwick, O. Terrestrial phosphorus limitation: Mechanisms, implications, and nitrogen-phosphorous interactions. *Ecol. Appl.* **2010**, *20*, 10. [CrossRef]
3. Galloway, J.N.; Dentener, F.J.; Capone, D.G.; Boyer, E.W.; Howarth, R.W.; Seitzinger, S.P.; Asner, G.P.; Cleveland, C.C.; Green, P.A.; Holland, E.A.; et al. Nitrogen Cycles: Past, Present, and Future. *Biogeochemistry* **2004**, *70*, 153–226. [CrossRef]
4. Yuan, Z.; Jiang, S.; Sheng, H.; Liu, X.; Hua, H.; Liu, X.; Zhang, Y. Human perturbation of the global phosphorus cycle: Changes and consequences. *Environ. Sci. Technol.* **2018**, *52*, 2438–2450. [CrossRef]
5. Swaine, M.D. Rainfall and Soil Fertility as Factors Limiting Forest Species Distributions in Ghana. *J. Ecol.* **1996**, *84*, 419. [CrossRef]
6. Zalamea, P.; Turner, B.L.; Winter, K.; Jones, F.A.; Sarmiento, C.; Dalling, J.W. Seedling growth responses to phosphorus reflect adult distribution patterns of tropical trees. *New Phytol.* **2016**, *212*, 400–408. [CrossRef] [PubMed]
7. Aber, J.; McDowell, W.; Nadelhoffer, K.; Magill, A.; Berntson, G.; Kamakea, M.; McNulty, S.; Currie, W.; Rustad, L.; Fernandez, I. Nitrogen Saturation in Temperate Forest Ecosystems. *BioScience* **1998**, *48*, 921–934. [CrossRef]
8. Bleeker, A.; Hicks, W.; Dentener, F.; Galloway, J.; Erisman, J. N deposition as a threat to the World's protected areas under the Convention on Biological Diversity. *Environ. Pollut.* **2011**, *159*, 2280–2288. [CrossRef]
9. Mayor, J.R.; Wright, S.J.; Turner, B.L. Species-specific responses of foliar nutrients to long-term nitrogen and phosphorus additions in a lowland tropical forest. *J. Ecol.* **2013**, *102*, 36–44. [CrossRef]
10. Mao, Q.; Lu, X.; Mo, H.; Gundersen, P.; Mo, J. Effects of simulated N deposition on foliar nutrient status, N metabolism and photosynthetic capacity of three dominant understory plant species in a mature tropical forest. *Sci. Total Environ.* **2018**, *610-611*, 555–562. [CrossRef]
11. Fang, Y.; Gundersen, P.; Mo, J.M.; Zhu, W.X. Input and output of dissolved organic and inorganic nitrogen in subtropical forests of South China under high air pollution. *Biogeosciences* **2008**, *5*, 339–352. [CrossRef]
12. Chen, H.; Gurmesa, G.A.; Zhang, W.; Zhu, X.; Zheng, M.; Mao, Q.; Zhang, T.; Mo, J. Nitrogen saturation in humid tropical forests after 6 years of nitrogen and phosphorus addition: Hypothesis testing. *Funct. Ecol.* **2015**, *30*, 305–313. [CrossRef]
13. Deng, Q.; Hui, D.F.; Dennis, S.; Reddy, K.C. Responses of terrestrial ecosystem phosphorus cycling to nitrogen addition: A meta-analysis. *Glob. Ecol. Biogeogr.* **2017**, *26*, 713–728. [CrossRef]
14. Wang, C.; Mori, T.; Mao, Q.; Zhou, K.; Wang, Z.; Zhang, Y.; Mo, H.; Lu, X.; Mo, J. Responses of soil microbial community to continuous experimental nitrogen additions for 13 years in a nitrogen-rich tropical forest. *Soil Biol. Biochem.* **2018**, *121*, 103–112. [CrossRef]
15. Deng, M.; Liu, L.; Sun, Z.; Piao, S.; Ma, Y.; Chen, Y.; Wang, J.; Qiao, C.; Wang, X.; Li, P. Increased phosphate uptake but not resorption alleviates phosphorus deficiency induced by nitrogen deposition in temperate Larix principisrupprechtii plantations. *New Phytol.* **2016**, *212*, 1019–1029. [CrossRef] [PubMed]

16. Chen, J.; Van Groenigen, K.J.; Hungate, B.A.; Terrer, C.; Van Groenigen, J.; Maestre, F.T.; Ying, S.C.; Luo, Y.; Jørgensen, U.; Sinsabaugh, R.L.; et al. Long-term nitrogen loading alleviates phosphorus limitation in terrestrial ecosystems. *Glob. Chang. Biol.* **2020**, *26*, 5077–5086. [CrossRef]

17. Pant, H.K.; Warman, P.R. Enzymatic hydrolysis of soil organic phosphorus by immobilized phosphatases. *Biol. Fertil. Soils* **2000**, *30*, 306–311. [CrossRef]

18. Zhou, K.; Lu, X.; Mori, T.; Mao, Q.; Wang, C.; Zheng, M.; Mo, H.; Hou, E.; Mo, J. Effects of long-term nitrogen deposition on phosphorus leaching dynamics in a mature tropical forest. *Biogeochemistry* **2018**, *138*, 215–224. [CrossRef]

19. You, C.; Wu, F.; Yang, W.; Xu, Z.; Tan, B.; Zhang, L.; Yue, K.; Ni, X.; Li, H.; Chang, C.; et al. Does foliar nutrient resorption regulate the coupled relationship between nitrogen and phosphorus in plant leaves in response to nitrogen deposition? *Sci. Total Environ.* **2018**, *645*, 733–742. [CrossRef]

20. Lawrence, D. The response of tropical tree seedlings to nutrient supply: Meta-analysis for understanding a changing tropical landscape. *J. Trop. Ecol.* **2003**, *19*, 239–250. [CrossRef]

21. Ostertag, R. Foliar nitrogen and phosphorus accumulation responses after fertilization: An example from nutrient-limited Hawaiian forests. *Plant Soil* **2010**, *334*, 85–98. [CrossRef]

22. Cárate-Tandalla, D.; Camenzind, T.; Leuschner, C.; Homeier, J. Contrasting species responses to continued nitrogen and phosphorus addition in tropical montane forest tree seedlings. *Biotropica* **2018**, *50*, 234–245. [CrossRef]

23. Brearley, F.Q.; Scholes, J.; Press, M.C.; Palfner, G. How does light and phosphorus fertilisation affect the growth and ectomycorrhizal community of two contrasting dipterocarp species? *Plant Ecol.* **2007**, *192*, 237–249. [CrossRef]

24. Burslem, D.F.R.P.; Grubb, P.J.; Turner, I.M. Responses to Nutrient Addition among Shade-Tolerant Tree Seedlings of Lowland Tropical Rain Forest in Singapore. *J. Ecol.* **1995**, *83*, 113. [CrossRef]

25. Schreeg, L.A.; Santiago, L.S.; Wright, S.J.; Turner, B.L. Stem, root, and older leaf N: P ratios are more responsive indicators of soil nutrient availability than new foliage. *Ecology* **2014**, *95*, 2062–2068. [CrossRef] [PubMed]

26. Liu, L.; Gundersen, P.; Zhang, T.; Mo, J. Effects of phosphorus addition on soil microbial biomass and community composition in three forest types in tropical China. *Soil Biol. Biochem.* **2012**, *44*, 31–38. [CrossRef]

27. Zhang, T.; Zhu, W.; Mo, J.; Liu, L.; Dong, S. Increased phosphorus availability mitigates the inhibition of nitrogen deposition on CH_4 uptake in an old-growth tropical forest, southern China. *Biogeosciences* **2011**, *8*, 2805–2813. [CrossRef]

28. Yu, L.; Wang, Y.; Zhang, X.; Dörsch, P.; Mulder, J. Phosphorus addition mitigates N_2O and CH_4 emissions in N-saturated subtropical forest, SW China. *Biogeosciences* **2017**, *14*, 3097–3109. [CrossRef]

29. Neufeld, H.S.; Lambers, H.; Chapin, F.S.; Pons, T.L. Plant Physiological Ecology. *Ecology* **1999**, *80*, 1785–1787. [CrossRef]

30. Gilliam, F.S. The Ecological Significance of the Herbaceous Layer in Temperate Forest Ecosystems. *Bioscience* **2007**, *57*, 845–858. [CrossRef]

31. Lu, X.; Mo, J.; Gilliam, F.S.; Zhou, G.; Fang, Y. Effects of experimental nitrogen additions on plant diversity in an old-growth tropical forest. *Glob. Chang. Biol.* **2010**, *16*, 2688–2700. [CrossRef]

32. Huang, Z.F.; Fan, Z.G. *The Climate of Dinghushan, Tropical and Subtropical Forest Ecosystem*; Science Press: Beijing, China, 1983.

33. Mo, J.; Brown, S.; Peng, S.; Kong, G. Nitrogen availability in disturbed, rehabilitated and mature forests of tropical China. *For. Ecol. Manag.* **2003**, *175*, 573–583. [CrossRef]

34. Lu, X.K.; Vitousek, P.M.; Mao, Q.G.; Gilliam, F.S.; Luo, Y.Q.; Zhou, G.Y.; Zou, X.; Bai, E.; Scanlon, T.M.; Hou, E.; et al. Plant acclimation to long-term high nitrogen deposition in an N-rich tropical forest. *Proc. Natl. Acad. Sci. USA* **2018**, *115*, 5187–5192. [CrossRef] [PubMed]

35. Mao, Q.; Lu, X.; Zhou, K.; Chen, H.; Zhu, X.; Mori, T.; Mo, J. Effects of long-term nitrogen and phosphorus additions on soil acidification in an N-rich tropical forest. *Geoderma* **2017**, *285*, 57–63. [CrossRef]

36. Shen, C.; Liu, D.; Peng, S.; Sun, Y.; Jiang, M.; Yi, W.; Xing, C.; Gao, Q.; Li, Z.; Zhou, G. ^{14}C measurement of forest soils in Dinghushan Biosphere Reserve. *Chin. Sci. Bull.* **1999**, *44*, 251–256. [CrossRef]

37. Mirabello, M.J.; Yavitt, J.B.; García, M.; Harms, K.E.; Turner, B.L.; Wright, S.J. Soil phosphorus responses to chronic nutrient fertilisation and seasonal drought in a humid lowland forest, Panama. *Soil Res.* **2013**, *51*, 215–221. [CrossRef]

38. Anderson, J.M.; Ingram, J.S.I. Tropical Soil Biology and Fertility. *Soil Sci.* **1994**, *157*, 265. [CrossRef]

39. Hedin, L.O.; Brookshire, E.N.J.; Menge, D.N.L.; Barron, A.R. The Nitrogen Paradox in Tropical Forest Ecosystems. *Annu. Rev. Ecol. Evol. Syst.* **2009**, *40*, 613–635. [CrossRef]

40. Gurmesa, G.A.; Lu, X.; Gundersen, P.; Mao, Q.; Zhou, K.; Fang, Y.; Mo, J. High retention of ^{15}N-labeled nitrogen deposition in a nitrogen saturated old-growth tropical forest. *Glob. Chang. Biol.* **2016**, *22*, 3608–3620. [CrossRef]

41. Wright, S.J.; Turner, B.L.; Sheldrake, M.; Garcia, M.N.; Yavitt, J.B.; Harms, K.E.; Kaspari, M.; Tanner, E.V.J.; Bujan, J.; Griffin, E.A.; et al. Plant responses to fertilization experiments in lowland, species-rich, tropical forests. *Ecology* **2018**, *99*, 1129–1138. [CrossRef]

42. Throop, H.L.; Lerdau, M.T. Effects of nitrogen deposition on insect herbivory: Implications for community and ecosystem processes. *Ecosystems* **2004**, *7*, 109–133. [CrossRef]

43. Lu, X.; Mao, Q.; Gilliam, F.S.; Luo, Y.; Mo, J. Nitrogen deposition contributes to soil acidification in tropical ecosystems. *Glob. Chang. Biol.* **2014**, *20*, 3790–3801. [CrossRef]

44. Treseder, K.K.; Vitousek, P.M. Effects of soil nutrient availability on investment in acquisition of N and P in Hawaiian rain forests. *Ecology* **2001**, *82*, 946–954. [CrossRef]

45. Zheng, M.; Huang, J.; Chen, H.; Wang, H.; Mo, J. Responses of soil acid phosphatase and beta-glucosidase to nitrogen and phosphorus addition in two subtropical forests in southern China. *Eur. J. Soil Biol.* **2015**, *68*, 77–84. [CrossRef]
46. Mo, Q.; Li, Z.; Sayer, E.J.; Lambers, H.; Li, Y.; Zou, B.; Tang, J.; Heskel, M.; Ding, Y.; Wang, F. Foliar phosphorus fractions reveal how tropical plants maintain photosynthetic rates despite low soil phosphorus availability. *Funct. Ecol.* **2019**, *33*, 503–513. [CrossRef]
47. Wright, S.J. Plant responses to nutrient addition experiments conducted in tropical forests. *Ecol. Monogr.* **2019**, *89*. [CrossRef]
48. Sardans, J.; Vallicrosa, H.; Zuccarini, P.; Farré-Armengol, G.; Fernández-Martínez, M.; Peguero, G.; Gargallo-Garriga, A.; Ciais, P.; Janssens, I.A.; Obersteiner, M.; et al. Empirical support for the biogeochemical niche hypothesis in forest trees. *Nat. Ecol. Evol.* **2021**, *5*, 184–194. [CrossRef]
49. McGroddy, M.E.; Daufresne, T.; Hedin, L.O. Scaling of C:N:P stoichiometry in forests worldwide: Implications of terrestrial redfield-type ratios. *Ecology* **2004**, *85*, 2390–2401. [CrossRef]
50. Townsend, A.R.; Cleveland, C.C.; Asner, G.; Bustamante, M.M.C. Controls over foliar N: P ratios in tropical rain forests. *Ecology* **2007**, *88*, 107–118. [CrossRef]

 *forests*

Article

Long- and Short-Term Inorganic Nitrogen Runoff from a Karst Catchment in Austria

Thomas Dirnböck [1,*], Heike Brielmann [1], Ika Djukic [1], Sarah Geiger [1], Andreas Hartmann [2,3], Franko Humer [1], Johannes Kobler [1], Martin Kralik [1,4], Yan Liu [2], Michael Mirtl [1,5] and Gisela Pröll [1]

[1] Environment Agency Austria, Spittelauer Lände 5, A-1090 Vienna, Austria; heike.brielmann@umweltbundesamt.at (H.B.); ika.djukic@umweltbundesamt.at (I.D.); sarah.geiger@umweltbundesamt.at (S.G.); franko.humer@umweltbundesamt.at (F.H.); johannes.kobler@umweltbundesamt.at (J.K.); martin.kralik@univie.ac.at (M.K.); michael.mirtl@umweltbundesamt.at (M.M.); gisela.proell@umweltbundesamt.at (G.P.)
[2] Chair of Hydrological Modeling and Water Resources, University of Freiburg, 79098 Freiburg, Germany; andreas.hartmann@hydmod.uni-freiburg.de (A.H.); yan.liu@hydmod.uni-freiburg.de (Y.L.)
[3] Department of Civil Engineering, University of Bristol, Tyndall Ave, Bristol, England BS8 1UG, UK
[4] Department of Environmental Geosciences, University of Vienna, Althanstraße 14, A-1090 Vienna, Austria
[5] Helmholtz Centre for Environmental Research GmbH–UFZ, Permoserstraße 15, 04318 Leipzig, Germany
* Correspondence: thomas.dirnboeck@umweltbundesamt.at; Tel.: +43-1-313043442

Received: 8 September 2020; Accepted: 16 October 2020; Published: 20 October 2020

Abstract: Excess nitrogen (N) deposition and gaseous N emissions from industrial, domestic, and agricultural sources have led to increased nitrate leaching, the loss of biological diversity, and has affected carbon (C) sequestration in forest ecosystems. Nitrate leaching affects the purity of karst water resources, which contribute around 50% to Austria's drinking water supply. Here we present an evaluation of the drivers of dissolved inorganic N (DIN) concentrations and fluxes from a karst catchment in the Austrian Alps (LTER Zöbelboden) from 27 years of records. In addition, a hydrological model was used together with climatic scenario data to predict expected future runoff dynamics. The study area was exposed to increasing N deposition during the 20th century (up to 30 to 35 kg N ha^{-1} y^{-1}), which are still at levels of 25.5 ± 3.6 and 19.9 ± 4.2 kg N ha^{-1} y^{-1} in the spruce and the mixed deciduous forests, respectively. Albeit N deposition was close to or exceeded critical loads for several decades, 70%–83% of the inorganic N retained in the catchment from 2000 to 2018, and NO$_3^-$ concentrations in the runoff stayed <10 mg L^{-1} unless high-flow events occurred or forest stand-replacing disturbances. We identified tree growth as the main sink for inorganic N, which might together with lower runoff, increase retention of only weakly decreasing N deposition in the future. However, since recurring forest stand-replacement is predicted in the future as a result of a combination of climatically driven disturbance agents, pulses of elevated nitrate concentrations in the catchment runoff will likely add to groundwater pollution.

Keywords: LTER; karst water; nitrogen deposition; nitrogen saturation; nitrate; ammonium; runoff; water quality

---

1. Introduction

Excess nitrogen (N) deposition and gaseous N emissions from industrial, domestic, and agricultural sources have led to increased nitrate leaching, the loss of biological diversity, and has affected carbon (C) sequestration in forest ecosystems [1–4]. Global mean N deposition is still increasing [5] while in Europe, N deposition peaked during the mid-1980s and was slowly going down until present day as a result of emission reductions (EU28) of NO$_x$ by 50% and NH$_3$ by 30% between the years 1990 and 2015 [6]. However, the amount of N deposition did not decrease in all areas in Europe [7] and the currently legislated emission reduction targets are too low to save ecosystems and biodiversity from further effects [8,9]. In order to assess the ecosystem effects of chronically high N deposition (among other air pollutants) in the forests of the Northern Limestone Alps in Austria, the long-term monitoring station "Zöbelboden" was set up in the year 1992. After 27 years of continuous observation and a number of experimental studies, we provide here an integrated view of these effects. Since half of the Austrian drinking water resources stem from karst areas [10] such as our study catchment, detailed knowledge about the drivers of N loss in upstream karst areas in the form of the water pollutant nitrate is pivotal. The more so because it may deviate from other catchments owing to a strong heterogeneity of subsurface flow and storage characteristics [11], and shallow, stony soils with low filtering capacity [12,13]. Nitrate may therefore additionally pollute downstream water sources already affected by agricultural fertilization [14,15]. An oversupply of N is thought to gradually diminish N retention in the living biomass and the soil so that surplus N leaves the ecosystem via leaching or gaseous emissions [16,17]. Cross-catchment studies have however shown that N retention is high even in areas with considerable N deposition [18,19], and signs of increasing oligotrophication [20].

N retention in trees only decreases beyond a certain level of N deposition while it increases at lower levels [21–23]. Since the deposition of N is in the range of expected growth reduction in European beech and Norway spruce [22], the dominating tree species at Zöbelboden, we expected such an effect. However, we were not able to single out this effect from the manifold factors of tree growth. Therefore, we explored our data on nutrient concentrations in leaves and needles. The down-regulation of tree growth can be related to nutrient deficiencies as a result of previous N-driven growth enhancement, reduced fine root biomass, and a change in the composition and abundance of mycorrhizal fungi [24,25].

Apart from the retention of deposited N in plant biomass, substantial amounts end up in the soil [26]. Fertilization experiments in forests show N addition generally increases soil organic matter (SOM) storage [27,28] but the amount of retained N depends upon the soil C:N ratio because microbial immobilization of added N may be low in soils with a C:N ratio < 20 [17].

The controlling role of the soil C:N ratio on N immobilization is also reflected in its tight relationship with the leaching of dissolved inorganic N (DIN), predominately in the form of nitrate [29,30]. The retention of DIN in catchments distributed across unmanaged European forest areas with no local N emission sources is high but decreased with increasing inorganic N deposition [18]. Since our catchment has been exposed to chronic N deposition for at least three decades before the monitoring started, and because DIN leaching seems to react very fast to the addition of DIN [31], we did not expect a long-term trend in DIN export within the timeframe of monitoring. However, apart from the long-term effects of high N deposition on DIN leaching, short-term disturbances such as forest stand-replacement, either naturally or due to management, can significantly increase nitrate loss with the seepage of water [32–34] thereby elevating the catchment runoff of DIN [35–37]. Knowledge about the interactive effects of long-term chronic N deposition with stand-replacing disturbances is important [38], because in Europe's forests, disturbances from wind, bark beetle and wildfires have already and are expected to increase further in response to climate change [39,40]. Beginning during the year 2006, storms and spruce bark beetle outbreaks have caused stand replacement in 5%–10% of the study area with an immediate increase in the DIN runoff [37]. Therefore, we expected that forthcoming disturbances will likely lead to pulses of nitrate loss in our study catchment, too.

Based on an analysis of the input-output budget of N, we explored vegetation, soil, and other catchment sinks in order to explain the likely effects of N deposition on nitrate leaching and to provide an outlook to the future. In detail, we (1) hypothesized increasing deficiency of P and K as found in other sites with high N deposition [7,22,41], and (2) increased soil N storage [27,28] and decreasing C:N ratios [17] in the soils, which we analyzed with long-term soil inventory data supplemented with a plot-scale N fertilization experiment focusing on SOM decomposition. With climate scenario data, the application of a hydrological model, and the knowledge we have gained through analyzing the long-term data from Zöbelboden, we discuss the potential impact of the manifold drivers on nitrate discharge in the future.

2. Methods

2.1. Study Site

LTER Zöbelboden has a size of 90 ha and is situated in the northern part of the national park "Northern Limestone Alps", approximately 50 km south of Linz (N 47°50′30″, E 14°26′30″) (https://deims.org/8eda49e9-1f4e-4f3e-b58e-e0bb25dc32a6) (Figure 1). The altitude ranges from 550 m to 956 m a.s.l. The main rock type is Norian dolomite (Hauptdolomit), which is partly overlain by limestone (Plattenkalk). Due to the dominating dolomite, the watershed is not as heavily karstified as limestone karst systems but shows typical karst features, such as conduits and sinkholes. These conduits and sinkholes provide pathways for rapid water flow and quick response times to water input at the soil-bedrock interface. The long-term average annual temperature is 7.2 °C. The coldest monthly temperature at 900 m a.s.l. is −1 °C in January, the highest is 15.5 °C in August. Annual rainfall ranges from 1500 to 1800 mm. Monthly precipitation ranges from 75 mm (February) to 182 mm (July). Snowfall occurs between October and May with an average duration of snow cover of about 4 months. From the start of the project in 1992 onwards, forest management has been restricted to single tree harvesting just in case of bark beetle infestation. Wind throw is frequent in the area with single tree events and events affecting larger areas.

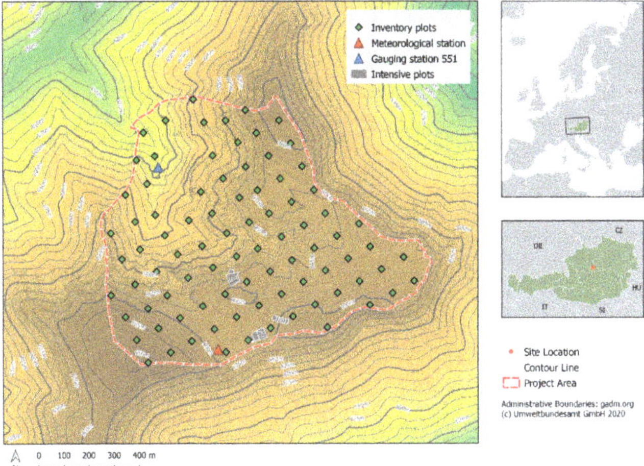

Figure 1. Location of LTER Zöbelboden with the main monitoring installations.

The catchment can be divided into two distinct sites: A very steep (30–70°) slope from 550–850 m a.s.l. and an almost flat plateau (850–956 m a.s.l.) on the top of the mountain. The areal coverage of each site is 50% of the watershed. At each site, one plot has been selected for intensive measurements of hydrochemical variables (Figure 1). Intensive plot I (IP I) is located on the plateau where Chromic Cambisols and Hydromorphic Stagnosols are found. This plot was moved to a nearby location (IP III) in 2008 with the same characteristics because of forest disturbance. Intensive plot II (IP II) is located on the slope and is dominated by Lithic and Rendzic Leptosols (FAO/ISRIC/ISSS, 2006). The mean slopes are 14° at IP I and 36° at IP II. IP I is dominated by Norway spruce (*Picea abies* (L.) H. Karst.) following plantation after a clear cut around the year 1910, whereas a mixed mountain forest with beech (*Fagus sylvatica* L.) as the dominant species, Norway spruce, sycamore (*Acer pseudoplatanus* L.), and ash (*Fraxinus excelsior* L.) covers IP II.

Wind throw and bark beetle disturbances started in 2006 with the damages caused by the storm Kyrill followed by two other storms in 2008. The subsequent spruce bark beetle outbreak peaked in the year 2011 (Figure 2). At Zöbelboden, these disturbances mostly damaged single trees and groups of trees while stand replacement of larger areas (>0.5 ha) only occurred in 5% to 10% of the catchment.

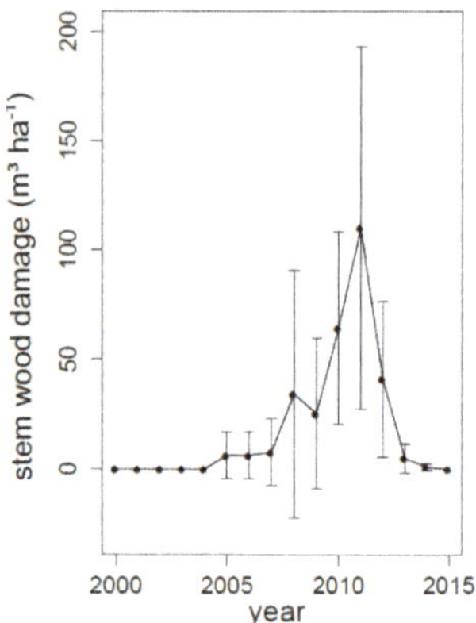

Figure 2. Wind throw and spruce bark beetle disturbance in the Kalkalpen National Park between 2000 and 2015.

2.2. Climate and Air Pollution

Meteorological data (air temperature, precipitation, vapor pressure, wind speed, and solar radiation) were recorded half-hourly at two climate stations, at a clearing area at the plateau (950 m a.s.l., 280 m distance to the monitoring plot) and at a nearby tower (40 m height, 60 m distance to the plot). The tower can be considered to represent the climatic situation at the monitoring plot, therefore data from the clearing area was regressed to the tower, resulting in a daily record from 1993 to 2019. Data gaps were filled with nearby climate stations based on linear regression (all within a radius of 10 km). Snow depth was measured weekly with a measuring stick during the sampling campaigns. In order to explore the climatic changes since 1950, long-term temperature and precipitation data from a nearby station (Reichraming, Hydrografischer Dienst) at 360 m a.s.l was bias-corrected with on-site meteorological data. We used the monthly mean difference between temperature and precipitation in the years 1993 to 2010, when data from both stations were available and adjusted for those of the entire data series of the Reichraming station.

Bulk precipitation was collected at the clearing (non-forested) area adjacent to the climate station with 5 to 10 (from 2005 onwards) bulk collectors, each with 20 cm in diameter. Water samples were pooled, filtered, and kept cool (4 °C) until sample preparation. Weekly samples were mixed (volume-weighted) biweekly or monthly (from 2008 onwards due to financial reasons). Nitrate was analyzed by ion chromatography with conductivity detection (Dionex IC System 4000 I until 2002, thereafter with Dionex IC System Serie DX 500). Total N was determined by means of spectrophotometric analysis (Abimed TN 05). NH_4^+ concentrations of the weekly samples were also measured by spectrophotometry (Milton Roy Spectronic).

We calculated the total deposition of N for the two intensively measured sites as the sum of throughfall and canopy exchange. The latter was based on a canopy exchange model according to Staelens et al. [42] with sodium as the tracer ion, bulk precipitation and throughfall, a yearly time step, and relative uptake efficiency of NH_4^+ of 6. In order to get a long-term deposition of N, we scaled reconstructed deposition from 1880 to 2000 [43] to the measurements using the mean bias during the years when both were available as a correction factor. Throughfall was collected at each intensively measured site with 15 regularly distributed bulk deposition samplers ($\emptyset$ = 20 cm). From 2006–2008, an additional two to five deposition samplers collected throughfall in the small bark beetle gap at IPI. While throughfall measurements ceased at IPI in September 2009 due to small-scale wind throw events and bark beetle infestations, measurements started at IPIII in August 2008. For chemical analyses, throughfall samples were pooled for each individual monitoring plot, except for bark beetle gap samples at IPI, which comprised individual samples. Subsequent throughfall sample preparation and analyses correspond to the described method for bulk precipitation.

2.3. Foliage Nutrient Concentrations

The dominant tree species, Norway spruce and European beech were sampled from a subsample of the forest and soil inventory plots distributed in a 100×100 m grid covering the entire study area (90 ha) and from the intensively monitored plots at the plateau (IP I and IP III) and the slope (IP II). Sampling began in the year 1992 for spruce needles and in 1993 for beech leaves. Between 36 and 52 dominant or predominant spruce trees and 16 to 17 beech trees were sampled annually until 2003. Thereafter sampling took place in the years 2004 (subsample of spruce trees), 2006, 2008, 2011, 2014, and 2017. In 2015, 2016, 2018, and 2019, only leaves and needles of the trees at IP I and IP III were collected and analyzed. All sample trees of the inventory plots were located as close as possible but outside the 10 (8) m circle of the forest inventory and the 100 m^2 vegetation plot. We chose new sample trees as close as possible to prior ones in case of damage or decreasing vitality. During each sampling campaign, we collected a 4 l plastic bag of mature leaves from the upper third of the north-west exposed crown of each beech tree. From spruce trees, the current and one-year-old needles were collected separately in a 0.5 L bag from as close as possible to the 7th whirl (counted from the top) of the north-west exposed crown. In some years until 2003, leaves and needles were collected separately from all four cardinal directions. Sampling took place in late August and September and in late September and October, respectively for beech and spruce. All specimen were labeled and then stored at $< 4\,^\circ$C until pretreatment and analysis in the lab.

The specimens were oven-dried (30 $^\circ$C) until constant weight, separated from the twigs, and grounded with an ultra-centrifugal grinding mill (ZM1, Retsch, D). Dry mass at 105 $^\circ$C was determined with a subsample of 100 undamaged spruce needles. Another subsample was digested ($HNO_3/HClO_4$) in glass vessels on a heating block (SMA 20 A, Gerhardt, D) to measure total residue with gravimetric vapor sorption at 105 $^\circ$C. Subsequent determination of Ca, K, Mg, Mn, and total P was carried out with inductively coupled plasma optical emission spectrometry (ICP-OES; Perkin-Elmer Optima 3000 XL, 3000 DV, and 7300 DV). Total N was measured by means of potentiometric titration (Kjeldatherm Vapotest 4S Gerhardt).

Foliage deficiency thresholds were taken from Mellert and Göttlein [44]. Temporal trends of single nutrients and ratios (N:Ca, N:Mg, N:K, and N:P) were tested using a Wilcoxon rank-sum test after outlier deletion comparing the first 4 years with the last 4 years (boxplot function).

2.4. Soil Chemistry

We collected mixed soil samples at 64 plots between July and August 2014 to compare with the soil survey data of the years 1992 and 2004. These plots were 100 m apart covering the entire catchment. For one mixed soil sample, three individual soil cores with 4 cm diameter were taken from the upper mineral soil layer (0–10 cm) after litter removal. The soil cores were taken with a 2.5 m distance to the 1992 and 2004 sampling sites. In 2004, the samples were taken 5 m apart from the samples in 1992. In all surveys, soil samples were dried at approximately 30 $^\circ$C, coarse aggregates crushed and dried again until constant weight. Then they were sieved through a 2 mm sieve. Soil suspensions were made by dissolving 5 g of each soil sample in 12.5 mL of 0.01 M $CaCl_2$ solution and pH was measured electrochemically with a pH electrode (Metrohm 654 pH meter in 1992 and 2004 and Argus X pH meter in 2014). For the determination of the C:N-ratio, the soil samples were ground and further decarbonized using 3M HCl solution. The organic carbon content was measured after dry combustion using isotope-ratio mass spectroscopy (IRMS). These methods deviate from the analysis in 1992 and 2004 where the total content of organic C (TOC) was calculated by subtracting the total content of $CaCO_3$ from the total content of C (TC-TIC). The total content of C (TC) was analyzed by dry combustion (1300 °C). Released CO_2 was detected coulometrically (Ströhlein Coulomat 702 and Si 111/6). The total content of $CaCO_3$ (TIC) was measured via the addition of HCl and volumetrically determination of the released CO_2 (Scheibler). The total content of N of the year's samples was determined by a modified Kjeldahl method. The organic N was converted to NH_4^+ by digestion with H_2SO_4 and a catalyst (Kjeldahlterm KT8 Gerhardt). The accumulated NH_4^+ was converted to NH_3 (distillation) and measured by potentiometric titration. To account for N oxygen compounds, salicylic acid was added prior to digestion. For the determination of total N stocks, the forest floor was sampled once with a 30 × 30 cm frame and the mineral soil with a metal pole with a 70 mm diameter at 3 points within a distance of two meters. Differences between years were tested using a paired Wilcoxon test.

2.5. Catchment Hydrology and N Measurements

Artificial tracer experiments, carried out in our study area, showed the large heterogeneity of the hydrologic system ranging from the retardation of 1 day to 10 years [45]. Water age dating with CFCs, ^3H, and $^3H/^3He$ even showed mean residence times of up to 20 years at one spring. Estimating the water balance of the larger system demonstrated that major parts of the rainfall input are transformed into the intermediate flow and deep percolation within the dolomite leaving the system as diffuse runoff instead of surface runoff (Humer and Kralik 2008). For this study, our main source of information was the gauging station (number 551) discharge and N concentration data between 2000 and 2018. Damped $\delta^{18}O$ variations at this spring indicated a delayed flow component, both fast and slow flow paths, and a considerable fraction of intermediate flow [46]. We disregarded the discharge data before and after this period because we deemed this data too uncertain. The discharge was calculated based on the gauging station's rating curve and 15 min water level sensor data. Missing values were gap-filled based on regression with nearby gauging station data. The recharge area of this spring was estimated based on a model calibration described in detail in [37,46,47].

Weekly observations of NO_3^--N, NH_4^+-N, and total N were available at the gauging station 551. From 2010 onwards, samples (ISO 5677–6) were filtered (0.45 µm) before the analysis. NH_4^+ concentrations were measured by spectrophotometry (Milton Roy Spectronic). Weekly NO_3^- and total N samples were pooled to provide volume-weighted biweekly (until March 2009) and monthly (thereafter) samples. NO_3^- concentrations were determined by ion chromatography with conductivity detection. DIN input was then calculated as the sum of NO_3^--N and NH_4^+-N. Dissolved organic nitrogen (DON) was calculated as the difference between total N and DIN. In order to calculate DIN discharge, we filled all missing 15 minutes values with the last weekly measurement value.

In 2018 and 2019, a spectral sensor probe (S:CAN spectro::lyser) with a UV-Vis 220–720 nm detector (15 mm path length) was installed in the measuring weir to obtain 15 minutes NO_3^- measurements.

2.6. Hydrochemical Modelling

We use the VarKarst model [47] for the discharge simulations, which was already performed satisfactorily in a previous study at this site [37]. The VarKarst model considers the variability of a karstic system reflected in the variability of (i) soil and epikarst depths, (ii) concentrated and diffuse recharge to the groundwater storage, and (iii) the epikarst and groundwater hydrodynamics. Spatial heterogeneity is accounted for by a set of N model compartments, which represent varying system properties over space. The influence of karst conduits is included by simulating concentrated recharge and fast conduit groundwater discharge. The VarKarst model is primarily running on a daily resolution. Details of the model can be found in [47,48].

For discharge projections, we perform the model simulations with the calibrated model parameters at this same site [37] using different climate projections.

We used eight RCP 8.5 scenarios of bias-adjusted regional climate model (RCM) data from the EURO-CORDEX initiative (Table 1), the European branch of the Coordinated Regional Downscaling Experiment (CORDEX) project [49,50], available through the data nodes of the Earth System Grid Federation (ESGF) model data dissemination system [51]. RCP 8.5 assumes emissions to rise throughout the 21st century. Because the actual altitude of a site did not match with the altitude of the closest RCM grid element, the 2 m air temperature was height corrected using a hypsometric lapse rate of 0.65 K per 100 m before temporal averaging was done.

Table 1. Description of the eight regional climate models that are used for the projections of the hydrological modeling.

RCM Model	Institute	Resolution
CLMcom-CCLM4-8-17	CNRM-CERFACS-CNRM-CM5	
SMHI-RCA4	CNRM-CERFACS-CNRM-CM5	
KNMI-RACMO22E	ICHEC-EC-EARTH	
IPSL-INERIS-WRF331F	IPSL-IPSL-CM5A-MR	12 km, daily
CLMcom-CCLM4-8-17	MPI-M-MPI-ESM-LR	
MPI-CSC-REMO2009	MPI-M-MPI-ESM-LR	
SMHI-RCA4	MPI-M-MPI-ESM-LR	
DMI-HIRHAM5	NCC-NorESM1-M	

Note: Details of the eight regional climate models (RCMs) under the RCP 8.5 can be found via https://b2share.eudat.eu/records/954510207378405aa897ec3b9f1f0c58.

3. Results

3.1. Climate and Deposition Trends

While precipitation remained stable since 1950, the temperature increased. The annual mean temperature was 6.56 °C between 1951 and 1970 and increased by 1.43 °C to 7.99 °C in the period 1991 to 2019. The warming started in the late 1980s (Figure 3).

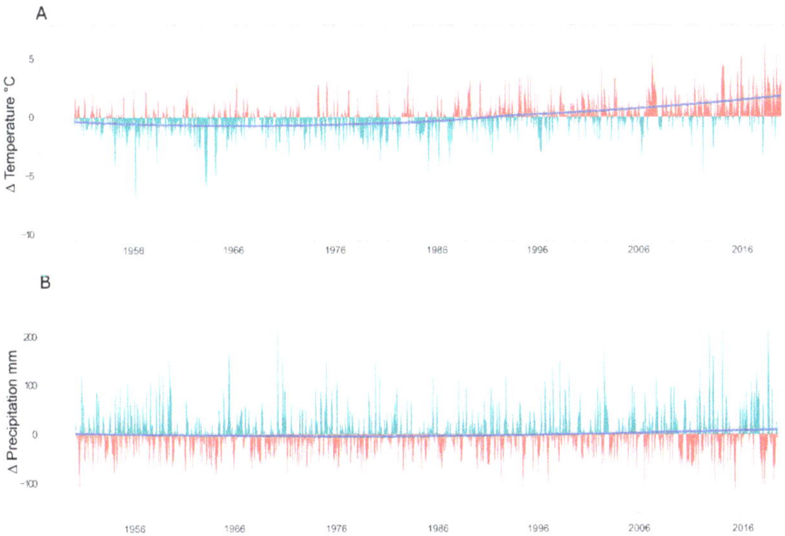

Figure 3. Reconstructed monthly anomalies of (**A**) temperature and (**B**) precipitation relative to the years 1961 to 1990 at LTER Zöbelboden. The blue solid lines represent the long-term trend and their confidence interval using a loess smoother.

The total annual inorganic N deposition between 2000 and 2018 was 25.5 ± 3.6 (s.d.) and 19.9 ± 4.2 kg N ha^{-1} y^{-1} in the spruce and the mixed deciduous forests, respectively. Two-thirds of the total inorganic N deposition was in the form of reduced N. The share of dry deposition was higher in the mixed deciduous forests (57% for reduced N and 15% for oxidized N) than the spruce forests (44% for reduced N and 10% for oxidized N) (Figure 4B). The annual deposition of DON was 4.8 ± 1.7 and 2.5 ± 1.1 kg N ha^{-1} y^{-1} in the spruce and the mixed deciduous forests, respectively.

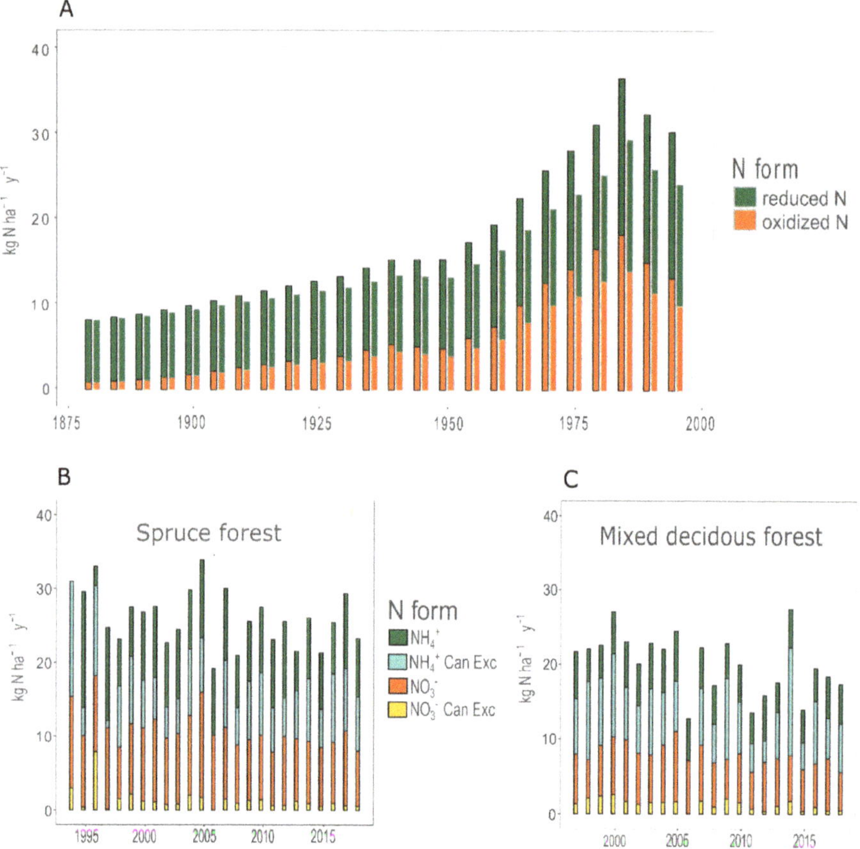

Figure 4. Inorganic N deposition in the study area. (**A**) reconstructed deposition between 1880 and 1990 and measured total NH$_4^+$-N and NO$_3^-$-N deposition in the two main forest ecosystems in the catchment: (**B**) spruce dominated forests, and (**C**) mixed deciduous forests. The pairs in A represent spruce dominated (left) and mixed deciduous forests (right). No shades in B and C indicate missing values.

Based on a reconstruction, total inorganic N deposition before 1900 was below 10 kg N ha^{-1} y^{-1} in the study area. Annual N deposition peaked in 1980 to approx. 30 to 35 kg N ha^{-1} y^{-1} and thereafter only slowly decreased to the current levels (Figure 4A).

3.2. Foliage Nutrient Concentrations

N concentration in beech foliage and spruce needles are given in Table 2. Between 1993 and 2018, the N concentrations decreased significantly ($p > 0.05$) in spruce (current and the one-year-old needles) but not in beech. Similarly, foliage concentration of P decreased in spruce (only one-year-old needles, $p = 0.024$), but not in beech. The beech foliar N:K ratio decreased significantly ($p = 0.07$) and the N:Ca ratio increased marginally ($p = 0.051$). In one-year-old spruce needles, the N:Mg ratio decreased ($p = 0.091$), and the N:Ca increased ($p = 0.028$). For all other nutrients and nutrient ratios, trends were not detected.

Table 2. Mean ± SE of tree foliage concentrations in Norway spruce and European beech at Zöbelboden (years 1992 to 2019). Nutrient deficiency or surplus according to [44] is given as red = deficient, green = normal, blue = surplus, and critical nutrient ratio levels as red = below limit, green = within limits, blue = above limit.

g kg^{-1}	Spruce		Beech
	Current Year Needles	**One-Year Needles**	
N	12.0 ± 0.08	11.4 ± 0.08	20.5 ± 0.13
P	1.1 ± 0.01	0.8 ± 0.01	0.7 ± 0.01
K	4.3 ± 0.08	3.4 ± 0.06	6.1 ± 0.1
Ca	5.1 ± 0.1	7.6 ± 0.14	12.8 ± 0.22
Mg	2.0 ± 0.03	2.2 ± 0.03	2.7 ± 0.5
N:P	11.3 ± 0.11	13.8 ± 0.14	29.0 ± 0.41
N:K	3.1 ± 0.07	3.5 ± 0.07	3.5 ± 0.06
N:Ca	2.5 ± 0.05	1.6 ± 0.03	1.7 ± 0.03
N:Mg	4.6 ± 0.09	5.6 ± 0.1	7.8 ± 0.14

3.3. Soil N, C, and C:N Ratio

Between 1992 to 2014, the mineral soil (0–10 cm) showed a long-term decrease in the total N content by 9% and in the organic C content by 2.5%. Accordingly, the C:N ratio decreased, but only after 2004 from 16 to 13.7 (Table 3).

Table 3. Soil total N and organic C content, the C:N ratio, and soil N stocks from resurveys of soil inventories in the years 1992, 2004, and 2014. Statistical significance is indicated with *** $p < 0.001$; ** $p < 0.005$; * $p < 0.05$.

	1992	2004	2014	Δ 1992 to 2004	Δ 1992 to 2014
	Mean ± SE concentrations [%] and C:N ratio in the mineral soil (0–10 cm)				
Total N	0.71 ± 0.04	0.62 ± 0.04	0.60 ± 0.03	↓ **0.11 **	↓ **0.9 ***
Organic C	10.44 ± 0.57	9.83 ± 0.55	7.98 ± 0.46	↓ 0.61	↓ **2.46 ***
C:N ratio	15.4 ± 0.5	16 ± 0.2	13.7 ± 0.5	↑ 0.6	↓ **1.6 ***
	Stocks [kg ha^{-1}] of total N (median and median absolute deviation)				
Organic layer	113 (79)	255 (151)	-	↑ **142.4 ***	-
Mineral layer 0–5	1644 (624)	1430 (433)	-	↓ **214.6 **	-
Mineral layer 5–10	1851 (747)	1599 (484)	-	↓ **251.4 **	-
Mineral layer 10–20	2414 (2197)	1918 (1074)	-	↓ 496.1	-

Total soil N stocks significantly increased from the year 1992 to 2004 in the organic layer and decreased in the mineral soil layers from 0–10 cm (Table 3).

3.4. Long Term Changes in N Discharge

The average runoff at the main gauging station was 7.24 ± 51.78 (SD) L s^{-1} totaling to a mean annual catchment runoff of 269 ± 96 mm with a maximum in the year 2009 (305 mm) and a minimum in the year 2015 (79 mm) (Figure 5A).

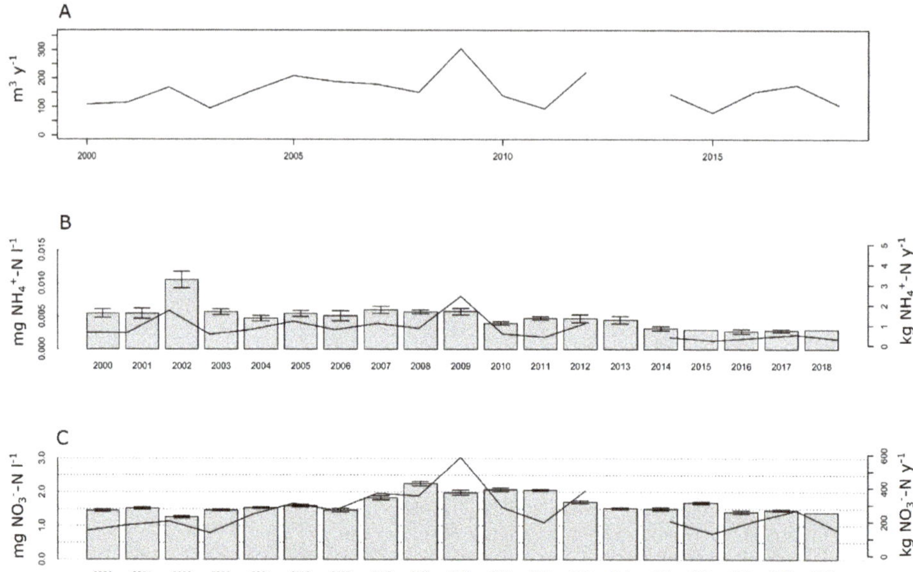

Figure 5. (**A**) Annual discharge, (**B**) NH_4^+-N and (**C**) NO_3^--N concentrations, and runoff at the gauging station 551 (bars: concentrations in mean ± SE, lines: fluxes).

The mean annual NH_4^+-N concentrations were 0.005 ± 0.002 mg L^{-1} with the highest values in 2002 (0.011 mg L^{-1}) and the lowest in the last years from 2015 to 2018 (0.003 mg L^{-1}). These concentrations total to an annual NH_4^+-N runoff of 0.8 ± 0.6 kg at the gauging station (Figure 5B). The annual discharge of NH_4^+-N from the catchment was 0.014 ± 0.91 kg ha^{-1} y^{-1}.

The mean annual NO_3^--N concentrations were 1.63 ± 0.27 mg L^{-1}. The highest annual mean occurred in the year 2008 (2.25 mg L^{-1}), the lowest in the year 2002 (1.26 mg L^{-1}). These concentrations total to an annual NO_3^--N runoff of 256 ± 115 kg at the gauging station (Figure 5C). The annual discharge of NO_3^--N from the catchment was 4.4 ± 2 kg ha^{-1} y^{-1}. As were the NO_3^--N concentrations, nitrate runoff peaked in the years of the forest disturbances (2007–2012).

Mean annual DON concentrations were 0.13 ± 0.07 mg L^{-1} and annual discharge was 0.33 ± 0.21 kg ha^{-1} y^{-1}.

During the upper quartile of discharge at the gauging station, 74% of the total annual DIN is leaving the catchment via runoff and 50% of the annual DIN during the highest 10% runoff events. For DON, these shares are 75% and 45%, respectively.

During a high-resolution automatic sampling approach in the year 2012, the following rain event-driven dynamics in NO_3^- at the gauging station could be seen (Figure 6A). Immediately after the start of the first rain event, with 9.6 mm precipitation, we recorded a decrease in NO_3^- concentration (Figure 6B). Once the water has passed the soil passage and reached the weir (after 12:00 o'clock on August 24), a small recovery has been noted. During the next stronger event (only two days later) with 19.4 mm precipitation, we measured an immediate dilution of NO_3^- caused by surface flow with subsequently a much stronger discharge of NO_3^-, which had passed soil and the karst aquifer. This process shows similar retention behavior as δ ^{18}O (Figure 6C). These characteristics differ in different springs of our karst catchment showing either no dilution effects or deviating storing capacities.

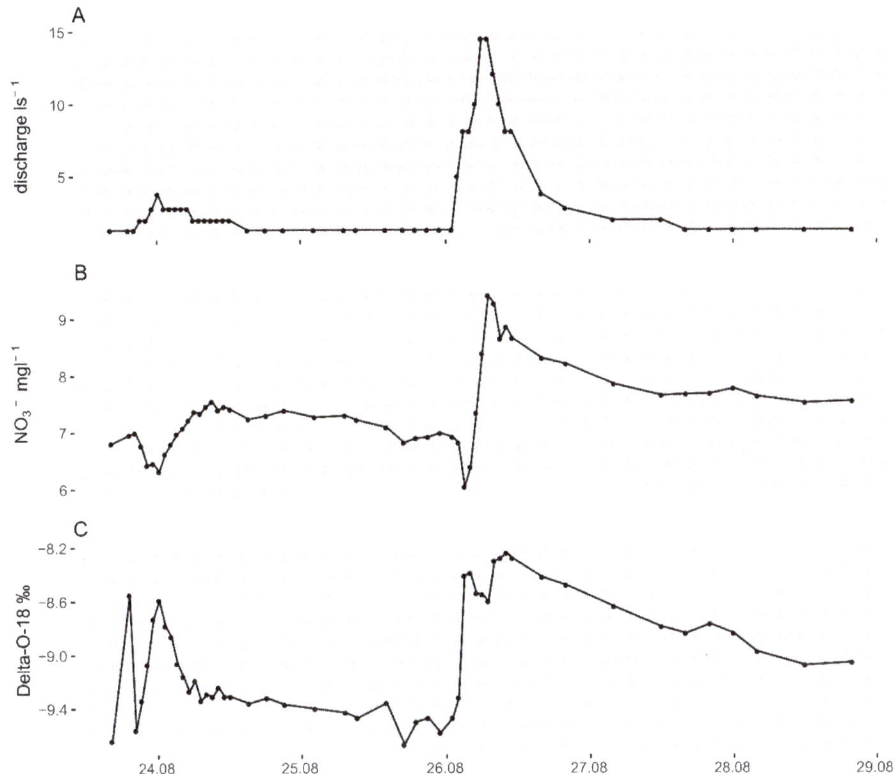

Figure 6. (**A**) Discharge, (**B**) NO$_3^-$ concentrations, and (**C**) δ ^{18}O at the gauging station 551 during a rainfall event in the year 2012.

When using high-resolution data from the optical probe over an entire year, the NO$_3^-$-N discharge of the year 2018 was in the range of the grab-sample data (high-resolution data: 2.61 kg ha^{-1} y^{-1}; weekly data: 2.55 kg ha^{-1} y^{-1}).

3.5. Future Projections of Climate and Discharge

From the climate projections under RCP 8.5, we see a general decrease in the monthly mean precipitation of the 8 climate models (Figure 7c). The projected monthly mean temperature rises particularly in 2060–2100 (Figure 7d3,d4). Compared to the historical mean monthly discharge, we see a general decreasing trend in the projections (Figure 7a). In the coming 40 years, the change of mean monthly discharge is not substantial (Figure 7a1,a2), whereas the model simulates a large decrease of discharge by the end of this century (Figure 7a4), especially for the high flow period in April and May. For actual evapotranspiration (Figure 7b), the model projects an increase in winter and spring seasons in the last 40 years of this century (Figure 7b3,b4), but a smaller AET in summer probably due to less available precipitation.

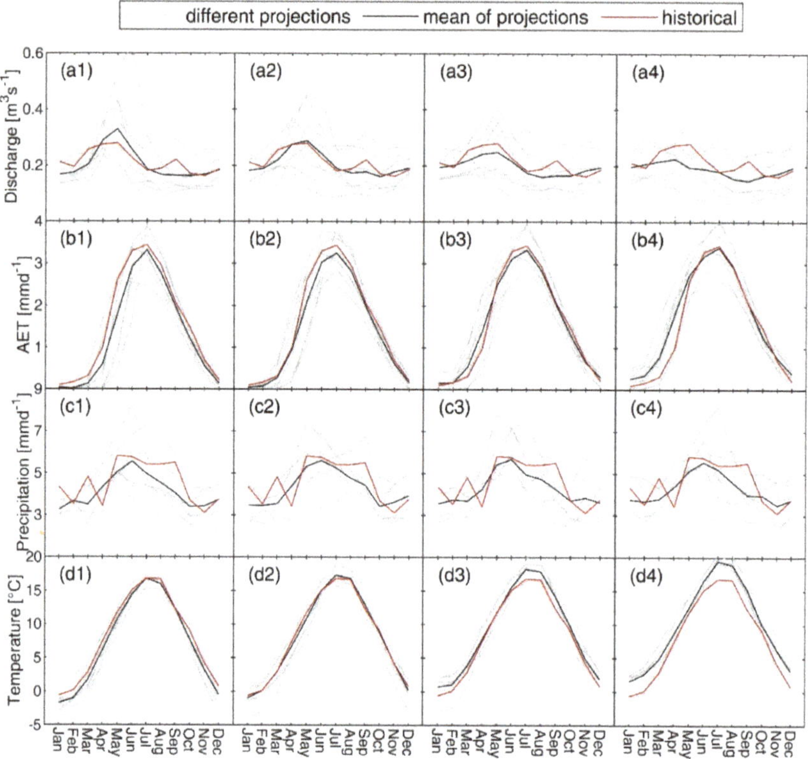

Figure 7. Comparisons between historical and projected mean monthly (**a**) discharge, (**b**) actual evapotranspiration, (**c**) precipitation, and (**d**) temperature, where the number 1–4 represent the four projected time periods: 2021–2040, 2041–2060, 2061–2080, and 2081–2100, respectively. The red line shows the historical simulations (discharge, AET) or observations (precipitation, temperature) for 1994–2019. The black line indicates the mean of the simulations obtained from the eight climate projections while the grey lines represent the results from the individual climate projections.

4. Discussion

Taking wet and dry deposition into account, the N deposition in the Zöbelboden catchment increased until the year 1980 to 30–35 kg N ha^{-1} y^{-1} and thereafter only slowly decreased to the current levels of 25.5 ± 3.6 and 19.9 ± 4.2 kg N ha^{-1} y^{-1} in the spruce and the mixed deciduous forests, respectively. Though there is quite some uncertainty regarding the calculation of dry deposition [52], this input of N was close to or exceeding the critical load (empirical CL at 10–20 kg N ha^{-1} y^{-1} according to [53]). In spite of these high inputs, during the last two decades, still 70–83% of the added DIN was retained in the catchment's vegetation, soils, the epikarst, the vadose or saturated zone. Retention of DIN only decreased during a period of forest stand-replacements in the catchment, which were caused by wind throw and bark beetle infestations at 5–10% of the area in the years 2007–2010. We infer from our results that DIN drainage must have increased with the increase in N deposition prior to the start of monitoring, staying at an elevated level until today. Future DIN discharge will depend upon a number of interacting factors, both potentially increasing or decreasing leaching loss of DIN.

4.1. Catchment N Retention

Several hydrological studies have been carried out to describe the dynamics of the Zöbelboden dolomite karst catchment. Water isotope studies found that the majority of the precipitation reaches the gauging station within a few months (Humer and Kralik, 2008; Kralik et al., 2009), which is in accordance with modeling results [37,47]. However, using age dating and artificial tracer experiments at individual springs within the study area, Kralik et al. (2009) also found ages from several days to several decades representing fast and slow flow paths also occurring in the catchment. By modeling DIN discharge, [37] inferred that slower flow paths do not have a significant impact on the retardation capacity of the hydrological system due to their small contribution (<5%). Fast flow paths become apparent during rain events and snow melting when NO_3^- concentrations increase within hours after short dissolution phases. Only during these events, NO_3^- concentration reaches >10 mg L^{-1}. High-flow events, therefore, dominate DIN discharge (e.g., 50% of the annual DIN discharge occurs during the highest 10% runoff events). We might have underestimated the contribution of these events to DIN discharge with the weekly grab sample data by systematically missing stronger flushing events (see the comparison between the grab sample and high-resolution data below).

Taking the difference between the DIN, which was deposited into the forests and soils, and the measured and modeled [37] DIN discharge, we inferred that approximately ~70–83% was retained in the catchment during the last two decades. This considerable retention capacity of high amounts of chronic N deposition is in line with observational and experimental studies. High N retention capacity (>80%) was found in many unmanaged (mostly forested) catchments in Europe during the period 1990–2012 [19]. Inorganic N retention also stayed relatively high in long-term whole catchment N addition experiments: 84–96% in Sweden [54], 1/3 additional leaching compared to a control catchment in Switzerland [38], and >81% in Maine, USA [55]. Obviously, only very high N deposition (100 kg N ha^{-1} y^{-1}) causes strong DIN leaching (>10 kg N ha^{-1} y^{-1}) in forests [17,56].

The N inputs to the catchment were strongly dominated by DIN and only 11% (deciduous forest) and 16% (coniferous forest) of the total N deposited as DON. Since DON discharge is most often related to DON deposition into forests [57], DON discharge also remained low. On average, 7.5% of the total N left the catchment in the form of DON.

Only a part of the deposited N reached the soil seepage water. When taking the difference between the sum of the annual mean aboveground litterfall of 15 kg N ha^{-1} y^{-1} [58] and the annual mean N deposition in the throughfall of the two dominant forest types, and the nitrate flux from the soils, which was on average 7.9 kg N ha^{-1} y^{-1} in a mixed deciduous forest [58] and 11.3 kg N ha^{-1} y^{-1} in a spruce forest [13], we found that 66–71% of the inorganic N was retained either in the soil or the vegetation. A modeling study [59] estimated that on average 3.4 kg N ha^{-1} y^{-1} was lost via gaseous emission in the forms of N_2O, N_2, and NO. Hence, N retention might have been slightly lower but note that we disregarded belowground litter input. In spite of these uncertainties, it seems safe to state that either the soils or the vegetation were the main sinks of deposited N and not sinks related to the vadose or saturated zone of the karst catchment. Since soil inventories clearly showed decreasing N stocks (approximately 19 kg N ha^{-1} y^{-1} when taking the organic layer and the first 10 cm of the mineral soil into account), N immobilization in plant biomass must have been the main N sink. Leitner et al. [58] calculated an annual tree N uptake of 61 kg N ha^{-1} y^{-1} for the mixed deciduous forests in the catchment. Apart from the general increase in CO_2 concentrations in the air, warming occurred particularly since the late 1980s and accumulated to an increase of the mean annual temperature of 1.4 °C between 1991 and 2019 as compared to the period between 1951 and 1970. All of these factors together with N fertilization led to an increase in tree growth all over Europe until about 2010 [60] and very likely have facilitated N uptake in trees in our study area.

In the catchment runoff, we found between 4.4 kg N ha^{-1} y^{-1} (measurements between 2000 and 2018) and 5.8 kg N ha^{-1} y^{-1} (model-based estimate according to Hartmann et al. [47]). Even though it is difficult to correctly measure both the soil seepage flux and the karst catchment runoff, mixing must have had a significant additional effect on DIN discharge besides the N immobilization in trees.

Apart from mixing, the attenuation of DIN in the karst aquifer might come from the use of deepwater of the vadose zone by trees [61] and microbial uptake on biofilms within the saturated zone [62].

4.2. Long-Term Trends Versus Short Pulses

Whole ecosystem fertilization experiments showed that an increase in nitrate discharge followed the onset of N addition from a few years and remained stable thereafter [31,38]. The monitoring at Zöbelboden started in the early 1990s when N deposition was already elevated for many years (in the 1950s N deposition exceeded 15 kg ha^{-1} y^{-1}). However, it is safe to assume that DIN leaching increased during these years of increasing N deposition. Many Austrian forests were and still are N deficient because of over-utilization in the past [63,64]. When focusing on nitrate, because it is by far the dominating N form in the catchment runoff (only 3‰ of DIN discharge was in the form NH_4^+-N), we did not find a long-term trend in its concentrations nor in its discharge (only NH_4^+-N decreased significantly). However, the wind throw and bark beetle disturbances to the forest caused a strong increase in nitrate concentrations and fluxes for some years. This is a known effect caused by less N uptake in trees and a surplus of mineralized soil N [32,33,65], which is subsequently leached to the aquifer. Hartmann et al. [37] estimated that an additional 2.7 kg N ha^{-1} y^{-1} leached annually during the disturbance period. Increased mineralization of soil N and subsequent leaching might be partly responsible for the strong decrease in soil N stocks.

4.3. Expected Future Pathways

The retention of deposited N in the catchment will depend upon a number of factors. First and foremost, DIN discharge will be regulated by climatic changes as was N leaching with soil seepage [13]. Soils on carbonate bedrock are often shallow and characterized by high stone content in the mineral soil, hence have a very high infiltration capacity [12]. The climate projections under RCP 8.5 indicates a decrease of 6–10% (100–166 mm y^{-1}) in the annual precipitation in the coming 80 years at this study site. There is no vast change in annual mean temperature in the near 40 years, but it will increase by 1.9 °C till 2100 according to the eight climate projections. The change in climate forcing will affect the hydrological response of the system leading to a decrease of 12% for the mean annual discharge until the end of this century. The projected mean annual AET shows a decrease in the coming 40 years since a precipitation decrease leads to less available water for evapotranspiration under similar temperature conditions (similar energy limit). However, the model projects an increase of 8% (48 mm y^{-1}) in AET in 2080–2100, which may be due to the obvious temperature increase throughout the whole year (Figure 7d4) that elevates the energy availability, particularly in winter and spring. We might expect lower N export in the catchment runoff but some uncertainty remains because high-flow conditions might increase even with lower average precipitation [66] and more frequent droughts additionally have the potential to increase NO_3^--N export in the area [58]. Changes in the amount and timing of winter snowfall and the subsequent changes in snowmelt patterns [67] will also affect future DIN leaching as shown in a number of studies [68]. The projected increase of temperatures will most likely lower the duration of snow coverage at the LTER Zöbelboden and may increase the rain and rain-on-snow events in winter with effects on discharge behavior [69]. Moreover, we expect higher fluctuation in soil climate during the winter, which will very likely affect N mineralization and N pools [70–72].

Climate might also indirectly elevate N retention, particularly through increased immobilization in trees due to warming [59]. However, this scenario could be hampered by limited tree growth due to nutrient deficiencies [22,41] and more frequent drought events. Deficiencies were found for some nutrients, particularly for N in Norway spruce but not so for European beech, and for P and K in both tree species. Ratios of N to other nutrients did not indicate increasing imbalances, which were partly made responsible for the tree growth reduction found in some regions in Europe and the US during recent years [21,22]. Although drought events in the past have had negative impacts on tree growth at Zöbelboden [73], precipitation is generally high and expected future decreases will unlikely be dramatic according to the simulations. However, water holding capacity is low in the shallow and stony soils, promoting water limitation for tree growth albeit relatively high precipitation [74]. Additionally, tree growth predominately occurs until the end of June so that lower soil moisture during spring due to changes in the snow cover and snow melting patterns may be particularly important [75]. Future effects of drought on tree growth and N cycling are therefore highly uncertain and a matter of current research in the study area.

We do not expect N oligotrophication in the near future, which occurred in the northern hardwood forests in the US due to a complex interplay of factors controlling forest production [20], simply because the currently legislated N emission reductions will only cause a slight decrease in the chronically high N deposition in the study area (in the range of ~2–3 kg N ha^{-1} y^{-1} in the year 2030 as compared to 2015 according to [8]).

The soils at Zöbelboden have top-soil C:N ratios <20 [76], where microbial immobilization of added N may be low [17]. The high amount of deposition in the form of NH_4^+ may also suppress microbial NO_3^- immobilization [77]. However, Brumme and Khanna [78] exemplified that soils on carbonate bedrock with high carbon contents and high pH values tend to retain large amounts of N. A large share of the high N leaching rates in the seepage water are likely directly from deposited NO_3^-, due to coarse-textured, shallow soils and preferential flow paths [12,13] rather than mineralized NO_3^- from the soil matrix. In a yet unpublished N addition experiment carried out in the study area, we found an increase in soil organic matter storage in the treatment plots as a result of reduced gross N mineralization rates and enzyme activity, which is in line with the findings from many other studies [27,28]. However, SOM increase only occurred in the organic layer and not in the mineral soil. Moreover, our soil inventory data showed that N stocks solely increased in the O horizon. Continued deposition of N in the current range will hence unlikely cause any major change in NO_3^- leaching.

Pulses of elevated NO_3^- runoff over some years have to be expected in the future as a consequence of climatically triggered forest disturbances. The abundance of Norway spruce has been artificially elevated in the study area as in all of Austria. With its vulnerability to a combination of disturbance agents, i.e., storms, drought, and spruce bark beetle [79], stand-replacing disturbances will occur more likely until deciduous trees have become more dominant [80].

5. Conclusions

By using the conceptual model of [17], the Zöbelboden catchment is experiencing kinetic N saturation, where soil and vegetation still immobilize N but the rate of N input exceeds their sink capacity. Thereby, N retention and leaching loss occur at the same time with elevated but stable DIN discharge and pulses of nitrate runoff during years with stand-replacing disturbances. Our data do not suggest any major change in the N retention in the near future. However, it also shows that discharge from unmanaged forests in upstream areas exposed to long-term, high N deposition can contribute to elevated nitrate concentrations of drinking water resources. Particular attention should be paid to forest management in order to reduce the risk of stand-replacing disturbances, because inorganic N deposition, which is usually retained, leaves the system during such periods via catchment runoff.

Author Contributions: Conceptualization, T.D., M.K., A.H., and M.M.; methodology, T.D. and Y.L.; software, A.H. and Y.L.; formal analysis, T.D., S.G., and Y.L.; data curation, I.D., H.B., G.P., S.G. and J.K.; writing—original draft preparation, T.D.; writing—review and editing, H.B., I.D., S.G., F.H., J.K., M.K., G.P., A.H., and Y.L.; visualization, T.D., S.G.; funding acquisition, T.D. and M.M. All authors have read and agreed to the published version of the manuscript.

Funding: H.B., I.D., T.D., J.K., S.G., G.P., and some measuring equipment were supported through LTER-CWN (FFG, F&E Infrastrukturförderung, project number 858024). T.D. and S.G. were also funded through the European Horizon 2020 Project eLTER-PLUS (INFRAIA-01-2018-2019). A.H. and Y.L. were supported by the Emmy-Noether-Programme of the German Research Foundation (DFG, grant number: HA 8113/1-1; project "Global Assessment of Water Stress in Karst Regions in a Changing World").

Acknowledgments: Long-term monitoring was over the years financed by several Austrian Federal Ministries through theircontribution to the UNECE-CLRTAP Integrated Monitoring Program. We acknowledge financial, technical, and laboratory support from the Kalkalpen National Park and the Austrian Federal Forests (ÖBf).

Conflicts of Interest: The authors declare no conflict of interest.

References

1. Bobbink, R.; Hicks, K.; Galloway, J.; Spranger, T.; Alkemade, R.; Ashmore, M.; Bustamante, M.; Cinderby, S.; Davidson, E.; Dentener, F.; et al. Global assessment of nitrogen deposition effects on terrestrial plant diversity: A synthesis. *Ecol. Appl.* **2010**, *20*, 30–59. [CrossRef] [PubMed]
2. de Vries, W.; Du, E.; Butterbach-Bahl, K. Short and long-term impacts of nitrogen deposition on carbon sequestration by forest ecosystems. *Curr. Opin. Environ. Sustain.* **2014**, *9*, 90–104. [CrossRef]
3. Gruber, N.; Galloway, J.N. An Earth-system perspective of the global nitrogen cycle. *Nature* **2008**, *451*, 293–296. [CrossRef] [PubMed]
4. Yue, K.; Peng, Y.; Peng, C.; Yang, W.; Peng, X.; Wu, F. Stimulation of terrestrial ecosystem carbon storage by nitrogen addition: A meta-analysis. *Sci. Rep.* **2016**, *6*, 19895. [CrossRef]
5. Vet, R.; Artz, R.S.; Carou, S.; Shaw, M.; Ro, C.-U.; Aas, W.; Baker, A.; van Bowersox, C.; Dentener, F.; Galy-Lacaux, C.; et al. A global assessment of precipitation chemistry and deposition of sulfur, nitrogen, sea salt, base cations, organic acids, acidity and pH, and phosphorus. *Atmos. Environ.* **2014**, *93*, 3–100. [CrossRef]
6. EMEP. *Transboundary Particulate Matter, Photo-Oxidants, Acidifying and Eutrophying Components*; EMEP Status Report; EMEP(MSC-W of EMEP): Oslo, Norway, 2017.
7. Schmitz, A.; Sanders, T.; Bolte, A.; Bussotti, F.; Dirnböck, T.; Johnson, J.; Penuelas, J.; Pollastrini, M.; Prescher, A.-K.; Sardans, J.; et al. Responses of forest ecosystems in Europe to decreasing nitrogen deposition. *Environ. Pollut.* **2019**, *244*, 980–994. [CrossRef]
8. Dirnböck, T.; Pröll, G.; Austnes, K.; Beloica, J.; Beudert, B.; Canullo, R.; de Marco, A.; Fornasier, M.F.; Futter, M.; Georgen, K.; et al. Currently legislated decreases in nitrogen deposition will yield only limited plant species recovery in European forests. *Environ. Res. Lett.* **2018**, *13*, 125010. [CrossRef]
9. Amann, M.; Anderl, M.; Borken-Kleefeld, J.; Cofala, J.; Heyes, C.; Höglund-Isaksson, L.; Kiesewetter, G.; Klimont, Z.; Moosmann, L.; Rafaj, P.; et al. *Progress towards the Achievement of the EU's Air Quality and Emissions Objectives*; IIASA: Laxenburg, Austria, 2018.
10. COST. *COST 65: Hydrogeological Aspects of Groundwater Protection in Karstic Areas—Final Report. EUROPEAN COMMISSION Directorate-General XII Science, Research and Development Environment Research Programme*; European Commission: Luxenbourg, 1995; p. 446.
11. Ford, D.; Wiliams, P. *Karst Hydrology and Geomorphology*; Wiley: New York, NY, USA, 2007.
12. Jandl, R.; Smidt, S.; Schindlbacher, A.; Englisch, M.; Zechmeister-Boltenstern, S.; Mikovits, C.; Schöftner, P.; Strebl, F.; Fuchs, G. The carbon and nitrogen biogeochemistry of a montane Norway spruce (Picea abies (L.) Karst.) forest: A synthesis of long-term research. *Plant Ecol. Divers.* **2012**, *5*, 105–114. [CrossRef]
13. Jost, G.; Dirnböck, T.; Grabner, M.-T.; Mirtl, M. Nitrogen Leaching of Two Forest Ecosystems in a Karst Watershed. *Water Air AMP Soil Pollut.* **2011**, *218*, 633–649. [CrossRef]
14. Grimvall, A.; Stålnacke, P.; Tonderski, A. Time scales of nutrient losses from land to sea—A European perspective. *Ecol. Eng.* **2000**, *14*, 363–371. [CrossRef]
15. Sutton, M.A.; Howard, C.M.; Erisman, J.W.; Billen, G.; Bleeker, A.; Grennfelt, P.; van Grinsven, H.; Grizzetti, B. (Eds.) *The European Nitrogen Assessment*; Cambridge University Press: Cambridge, UK, 2011.
16. Aber, J.; McDowell, W.; Nadelhoffer, K.; Magill, A.; Berntson, G.; Kamakea, M.; McNulty, S.; Currie, W.; Rustad, L.; Fernandez, I. Nitrogen saturation in temperate forest ecosystems: Hypotheses revisited. *BioScience* **1998**, *48*, 921–934. [CrossRef]
17. Lovett, G.M.; Goodale, C.L. A new conceptual model of nitrogen saturation based on experimental Nitrogen addition to an oak forest. *Ecosystems* **2011**, *14*, 615–631. [CrossRef]
18. Vuorenmaa, J.; Augustaitis, A.; Beudert, B.; Bochenek, W.; Clarke, N.; de Wit, H.A.; Dirnböck, T.; Frey, J.; Hakola, H.; Kleemola, S.; et al. Long-term changes (1990–2015) in the atmospheric deposition and runoff water chemistry of sulphate, inorganic nitrogen and acidity for forested catchments in Europe in relation to changes in emissions and hydrometeorological conditions. *Sci. Total Environ.* **2018**, *625*, 1129–1145. [CrossRef] [PubMed]

19. Vuorenmaa, J.; Augustaitis, A.; Beudert, B.; Clarke, N.; de Wit, H.A.; Dirnböck, T.; Frey, J.; Forsius, M.; Indriksone, I.; Kleemola, S.; et al. Long-term sulphate and inorganic nitrogen mass balance budgets in European ICP Integrated Monitoring catchments (1990–2012). *Ecol. Indic.* **2017**, *76*, 15–29. [CrossRef]

20. Groffman, P.M.; Driscoll, C.T.; Durán, J.; Campbell, J.L.; Christenson, L.M.; Fahey, T.J.; Fisk, M.C.; Fuss, C.; Likens, G.E.; Lovett, G.; et al. Nitrogen oligotrophication in northern hardwood forests. *Biogeochemistry* **2018**, *141*, 523–539. [CrossRef]

21. Horn, K.J.; Thomas, R.Q.; Clark, C.M.; Pardo, L.H.; Fenn, M.E.; Lawrence, G.B.; Perakis, S.S.; Smithwick, E.A.H.; Baldwin, D.; Braun, S.; et al. Growth and survival relationships of 71 tree species with nitrogen and sulfur deposition across the conterminous U.S. *PLoS ONE* **2018**, *13*, e0205296. [CrossRef] [PubMed]

22. Braun, S.; Schindler, C.; Rihm, B. Growth trends of beech and Norway spruce in Switzerland: The role of nitrogen deposition, ozone, mineral nutrition and climate. *Sci. Total Environ.* **2017**, *599–600*, 637–646. [CrossRef]

23. Solberg, S.; Dobbertin, M.; Reinds, G.J.; Lange, H.; Andreassen, K.; Fernandez, P.G.; Hildingsson, A.; de Vries, W. Analyses of the impact of changes in atmospheric deposition and climate on forest growth in European monitoring plots: A stand growth approach. *For. Ecol. Manag.* **2009**, *258*, 1735–1750. [CrossRef]

24. Lilleskov, E.A.; Kuyper, T.W.; Bidartondo, M.I.; Hobbie, E.A. Atmospheric nitrogen deposition impacts on the structure and function of forest mycorrhizal communities: A review. *Environ. Pollut.* **2019**, *246*, 148–162. [CrossRef]

25. van der Linde, S.; Suz, L.M.; Orme, C.D.L.; Cox, F.; Andreae, H.; Asi, E.; Atkinson, B.; Benham, S.; Carroll, C.; Cools, N.; et al. Environment and host as large-scale controls of ectomycorrhizal fungi. *Nature* **2018**. [CrossRef]

26. Templer, P.H.; Mack, M.C.; Iii, F.S.C.; Christenson, L.M.; Compton, J.E.; Crook, H.D.; Currie, W.S.; Curtis, C.J.; Dail, D.B.; D'Antonio, C.M.; et al. Sinks for nitrogen inputs in terrestrial ecosystems: A meta-analysis of 15N tracer field studies. *Ecology* **2012**, *93*, 1816–1829. [CrossRef] [PubMed]

27. Janssens, I.A.; Dieleman, W.; Luyssaert, S.; Subke, J.A.; Reichstein, M.; Ceulemans, R.; Ciais, P.; Dolman, A.J.; Grace, J.; Matteucci, G.; et al. Reduction of forest soil respiration in response to nitrogen deposition. *Nat. Geosci.* **2010**, *3*, 315–322. [CrossRef]

28. Forstner, S.J.; Wechselberger, V.; Müller, S.; Keibinger, K.M.; Díaz-Pinés, E.; Wanek, W.; Scheppi, P.; Hagedorn, F.; Gundersen, P.; Tatzber, M.; et al. Vertical Redistribution of Soil Organic Carbon Pools After Twenty Years of Nitrogen Addition in Two Temperate Coniferous Forests. *Ecosystems* **2019**, *22*, 379–400. [CrossRef] [PubMed]

29. Dise, N.B.; Wright, R.F. Nitrogen leaching from European forests in relation to nitrogen deposition. *For. Ecol. Manag.* **1995**, *71*, 153–161. [CrossRef]

30. MacDonald, J.A.; Dise, N.B.; Matzner, E.; Armbruster, M.; Gundersen, P.; Forsius, M. Nitrogen input together with ecosystem nitrogen enrichment predict nitrate leaching from European forests. *Glob. Chang. Biol.* **2002**, *8*, 1028–1033. [CrossRef]

31. Moldan, F.; Hruška, J.; Evans, C.; Hauhs, M. Experimental simulation of the effects of extreme climatic events on major ions, acidity and dissolved organic carbon leaching from a forested catchment, Gårdsjön, Sweden. *Biogeochemistry* **2011**, *107*, 455–469. [CrossRef]

32. Huber, C. Long Lasting Nitrate Leaching after Bark Beetle Attack in the Highlands of the Bavarian Forest National Park. *J. Environ. Qual.* **2005**, *34*, 1772–1779. [CrossRef]

33. Kohlpaintner, M.; Huber, C.; Weis, W.; Göttlein, A. Spatial and temporal variability of nitrate concentration in seepage water under a mature Norway spruce [Picea abies (L.) Karst] stand before and after clear cut. *Plant Soil* **2010**, *314*, 285–301. [CrossRef]

34. Rothe, A.; Mellert, K. Effects of Forest Management on Nitrate Concentrations in Seepage Water of Forests in Southern Bavaria, Germany. *Water Air Soil Pollut* **2004**, *156*, 337–355. [CrossRef]

35. Bernal, S.; Hedin, L.O.; Likens, G.E.; Gerber, S.; Buso, D.C. Complex response of the forest nitrogen cycle to climate change. *Proc. Natl. Acad. Sci. USA* **2012**, *109*, 3406–3411. [CrossRef]

36. Aber, J.D.; Ollinger, S.V.; Driscoll, C.T.; Likens, G.E.; Holmes, R.T.; Freuder, R.J.; Goodale, C.L. Inorganic nitrogen losses from a forested ecosystem in response to physical, chemical, biotic, and climatic perturbations. *Ecosystems* **2002**, *5*, 648–658. [CrossRef]

37. Hartmann, A.; Kobler, J.; Kralik, M.; Dirnböck, T.; Humer, F.; Weiler, M. Model-aided quantification of dissolved carbon and nitrogen release after windthrow disturbance in an Austrian karst system. *Biogeosciences* **2016**, *13*, 159–174. [CrossRef]

38. Schleppi, P.; Curtaz, F.; Krause, K. Nitrate leaching from a sub-alpine coniferous forest subjected to experimentally increased N deposition for 20 years, and effects of tree girdling and felling. *Biogeochemistry* **2017**, *134*, 319–335. [CrossRef]

39. Senf, C.; Pflugmacher, D.; Zhiqiang, Y.; Sebald, J.; Knorn, J.; Neumann, M.; Hostert, P.; Seidl, R. Canopy mortality has doubled in Europe's temperate forests over the last three decades. *Nat. Commun.* **2018**, *9*, 4978. [CrossRef] [PubMed]

40. Schelhaas, M.J.; Nabuurs, G.J.; Schuck, A. Natural disturbances in the European forests in the 19th and 20th centuries. *Glob. Chang. Biol.* **2003**, *9*, 1620–1633. [CrossRef]

41. Jonard, M.; Fürst, A.; Verstraeten, A.; Thimonier, A.; Timmermann, V.; Potočić, N.; Waldner, P.; Benham, S.; Hansen, K.; Merilä, P.; et al. Tree mineral nutrition is deteriorating in Europe. *Glob. Chang. Biol.* **2015**, *21*, 418–430. [CrossRef]

42. Staelens, J.; Houle, D.; de Schrijver, A.; Neirynck, J.; Verheyen, K. Calculating Dry Deposition and Canopy Exchange with the Canopy Budget Model: Review of Assumptions and Application to Two Deciduous Forests. *Water Air Soil Pollut.* **2008**, *191*, 149–169. [CrossRef]

43. Schöpp, W.; Posch, M.; Mylona, S.; Johansson, M. Long-term development of acid deposition (1880–2030) in sensitive freshwater regions in Europe. *Hydrol. Earth Syst. Sci.* **2003**, *7*, 436–446. [CrossRef]

44. Mellert, K.H.; Göttlein, A. Comparison of new foliar nutrient thresholds derived from van den Burg's literature compilation with established central European references. *Eur. J. For. Res.* **2012**, *131*, 1461–1472. [CrossRef]

45. Kralik, M.; Humer, F.; Papesch, W.; Tesch, R.; Suckow, A.; Han, L.F. (Eds.) *Karstwater-Ages in an Alpine Dolomite Catchment, Austria: Delta-18O, 3H, 3H/3He, CFC and Dye Tracer Investigations*; EGU General Assembly: Vienna, Austria, 2009.

46. Hartmann, A.; Kralik, M.; Humer, F.; Lange, J.; Weiler, M. Identification of a karst system's intrinsic hydrodynamic parameters: Upscaling from single springs to the whole aquifer. *Environ. Earth Sci.* **2011**, *65*, 2377–2389. [CrossRef]

47. Hartmann, A.; Weiler, M.; Wagener, T.; Lange, J.; Kralik, M.; Humer, F.; Mizyed, N.; Rimmer, A.; Barberá, J.A.; Andreo, B.; et al. Process-based karst modelling to relate hydrodynamic and hydrochemical characteristics to system properties. *Hydrol. Earth Syst. Sci.* **2013**, *17*, 3305–3321. [CrossRef]

48. Hartmann, A.; Mudarra, M.; Andreo, B.; Marín, A.; Wagener, T.; Lange, J. Modeling spatiotemporal impacts of hydroclimatic extremes on groundwater recharge at a Mediterranean karst aquifer. *Water Resour. Res.* **2014**, *50*, 6507–6521. [CrossRef]

49. Giorgi, F.; Jones, C.; Asrar, G.R. Addressing climate information needs at the regional level: The CORDEX framework. *Bull. World Meteorol. Organ.* **2009**, *58*, 175–183.

50. Gutowski, W.J., Jr.; Giorgi, F.; Timbal, B.; Frigon, A.; Jacob, D.; Kang, H.-S.; Raghavan, K.; Lee, B.; Lennard, C.; Nikulin, G.; et al. WCRP COordinated Regional Downscaling EXperiment (CORDEX): A diagnostic MIP for CMIP6. *Geosci. Model Dev.* **2016**, *9*, 4087–4095. [CrossRef]

51. Cinquini, L.; Crichton, D.; Mattmann, C.; Harney, J.; Shipman, G.; Wang, F.; Ananthakrishnan, R.; Miller, N.; Denvil, S.; Morgan, M.; et al. The Earth System Grid Federation: An open infrastructure for access to distributed geospatial data. *Future Gener. Comput. Syst.* **2014**, *36*, 400–417. [CrossRef]

52. Adriaenssens, S.; Staelens, J.; Baeten, L.; Verstraeten, A.; Boeckx, P.; Samson, R.; Verheyen, K. Influence of canopy budget model approaches on atmospheric deposition estimates to forests. *Biogeochemistry* **2013**, *116*, 215–229. [CrossRef]

53. Bobbink, R.; Hettelingh, J.P. *Review and Revision of Empirical Critical Loads and Dose-Response Relationships: Proceedings of the Expert Workshop, Noordwijkerhout 23–24 June 2010*; RIVM: Bilthoven, The Netherlands, 2011.

54. Moldan, F.; Jutterström, S.E.A.K.; Hruška, J.; Wright, R.F. Experimental addition of nitrogen to a whole forest ecosystem at Gårdsjön, Sweden (NITREX): Nitrate leaching during 26 years of treatment. *Environ. Pollut.* **2018**, *242*, 367–374. [CrossRef]

55. Patel, K.F.; Fernandez, I.J.; Nelson, S.J.; Gruselle, M.-C.; Norton, S.A.; Weiskittel, A.R. Forest N Dynamics after 25 years of Whole Watershed N Enrichment: The Bear Brook Watershed in Maine. *Soil Sci. Soc. Am. J.* **2019**, *83*, S161–S174. [CrossRef]

56. Magill, A.H.; Aber, J.D.; Currie, W.S.; Nadelhoffer, K.J.; Martin, M.E.; McDowell, W.H.; Melillo, J.M.; Steudler, P. Ecosystem response to 15 years of chronic nitrogen additions at the Harvard Forest LTER, Massachusetts, USA. *For. Ecol. Manag.* **2004**, *196*, 7–28. [CrossRef]

57. Michalzik, B.; Kalbitz, K.; Park, J.H.; Solinger, S.; Matzner, E. Fluxes and concentrations of dissolved organic carbon and nitrogen—A synthesis for temperate forests. *Biogeochemistry* **2001**, *52*, 173–205. [CrossRef]

58. Leitner, S.; Dirnböck, T.; Kobler, J.; Zechmeister-Boltenstern, S. Legacy effects of drought on nitrate leaching in a temperate mixed forest on karst. *J. Environ. Manag.* **2020**, *262*. [CrossRef] [PubMed]

59. Dirnböck, T.; Foldal, C.; Djukic, I.; Kobler, J.; Haas, E.; Kiese, R.; Kitzler, B. Historic nitrogen deposition determines future climate change effects on nitrogen retention in temperate forests. *Clim. Chang.* **2017**, *1*, 15. [CrossRef]

60. Pretzsch, H.; Biber, P.; Schütze, G.; Uhl, E.; Rötzer, T. Forest stand growth dynamics in Central Europe have accelerated since 1870. *Nat. Commun.* **2014**, *5*. [CrossRef]

61. Carrière, S.D.; Martin-StPaul, N.K.; Cakpo, C.B.; Patris, N.; Gillon, M.; Chalikakis, K.; Doussan, C.; Olioso, A.; Babic, M.; Jouineau, A.; et al. The role of deep vadose zone water in tree transpiration during drought periods in karst settings—Insights from isotopic tracing and leaf water potential. *Sci. Total Environ.* **2020**, *699*, 134332. [CrossRef]

62. Wilhartitz, I.C.; Kirschner, A.K.T.; Stadler, H.; Herndl, G.J.; Dietzel, M.; Latal, C.; Mach, R.L.; Farnleitner, A.H. Heterotrophic prokaryotic production in ultraoligotrophic alpine karst aquifers and ecological implications. *FEMS Microbiol. Ecol.* **2009**, *68*, 287–299. [CrossRef]

63. Jandl, R.; Smidt, S.; Mutsch, F.; Fürst, A.; Zechmeister, H.; Bauer, H.; Dirnböck, T. Acidification and Nitrogen Eutrophication of Austrian Forest Soils. *Appl. Environ. Soil Sci.* **2012**, *2012*, 9. [CrossRef]

64. Glatzel, G. The nitrogen status of Austrian forest ecosystems as influenced by atmospheric deposition, biomass harvesting and lateral organomass exchange. *Plant Soil* **1990**, *128*, 67–74. [CrossRef]

65. Dirnböck, T.; Kobler, J.; Kraus, D.; Grote, R.; Kiese, R. Impacts of management and climate change on nitrate leaching in a forested karst area. *J. Environ. Manag.* **2016**, *165*, 243–252. [CrossRef]

66. Yin, J.; Gentine, P.; Zhou, S.; Sullivan, S.C.; Wang, R.; Zhang, Y.; Guo, S. Large increase in global storm runoff extremes driven by climate and anthropogenic changes. *Nat. Commun.* **2018**, *9*, 4389. [CrossRef]

67. Barnett, T.P.; Adam, J.C.; Lettenmaier, D.P. Potential impacts of a warming climate on water availability in snow-dominated regions. *Nature* **2005**, *438*, 303–309. [CrossRef]

68. Crossman, J.; Catherine Eimers, M.; Casson, N.J.; Burns, D.A.; Campbell, J.L.; Likens, G.E.; Mitchell, M.J.; Nelson, S.J.; Shanley, J.B.; Watmough, S.A.; et al. Regional meteorological drivers and long term trends of winter-spring nitrate dynamics across watersheds in northeastern North America. *Biogeochemistry* **2016**, *130*, 247–265. [CrossRef]

69. Casson, N.J.; Eimers, M.C.; Buttle, J.M. The contribution of rain-on-snow events to nitrate export in the forested landscape of south-central Ontario, Canada. *Hydrol. Process.* **2010**, *24*, 1985–1993. [CrossRef]

70. Durán, J.; Morse, J.L.; Groffman, P.M.; Campbell, J.L.; Christenson, L.M.; Driscoll, C.T.; Fahey, T.J.; Fisk, M.C.; Likens, G.E.; Melillo, J.M.; et al. Climate change decreases nitrogen pools and mineralization rates in northern hardwood forests. *Ecosphere* **2016**, *7*, 53. [CrossRef]

71. Li, W.; Wu, J.; Bai, E.; Guan, D.; Wang, A.; Yuan, F.; Wang, S.; Jin, C. Response of terrestrial nitrogen dynamics to snow cover change: A meta-analysis of experimental manipulation. *Soil Biol. Biochem.* **2016**, *100*, 51–58. [CrossRef]

72. Schütt, M.; Borken, W.; Stange, C.F.; Matzner, E. Substantial net N mineralization during the dormant season in temperate forest soils. *J. Plant Nutr. Soil Sci.* **2014**, *177*, 566–572. [CrossRef]

73. Hartl-Meier, C.; Zang, C.; Büntgen, U.; Esper, J.; Rothe, A.; Göttlein, A.; Dirnböck, T.; Treydte, K. Uniform climate sensitivity in tree-ring stable isotopes across species and sites in a mid-latitude temperate forest. *Tree Physiol.* **2014**, *35*, 4–15. [CrossRef]

74. Kobler, J.; Jandl, R.; Dirnböck, T.; Mirtl, M.; Schindlbacher, A. Effects of stand patchiness due to windthrow and bark beetle abatement measures on soil CO_2 efflux and net ecosystem productivity of a managed temperate mountain forest. *Eur. J. For. Res.* **2015**, *134*, 683–692. [CrossRef]

75. Blankinship, J.C.; Meadows, M.W.; Lucas, R.G.; Hart, S.C. Snowmelt timing alters shallow but not deep soil moisture in the Sierra Nevada. *Water Resour. Res.* **2014**, *50*, 1448–1456. [CrossRef]

76. Pröll, G.; Dullinger, S.; Dirnböck, T.; Kaiser, B.; Richter, A. Effects of nitrogen on tree recruitment in a temperate montane forest as analysed by measured variables and Ellenberg indicator values. *Preslia* **2011**, *83*, 111–127.
77. Emmett, B.A. Nitrogen Saturation of Terrestrial Ecosystems: Some Recent Findings and Their Implications for Our Conceptual Framework. *Water Air Soil Pollut. Focus* **2007**, *7*, 99–109. [CrossRef]
78. Brumme, R.; Khanna, P.K. Ecological and site historical aspects of N dynamics and current N status in temperate forests. *Glob. Chang. Biol.* **2008**, *14*, 125–141. [CrossRef]
79. Jandl, R. Climate-induced challenges of Norway spruce in Northern Austria. *Trees For. People* **2020**, *1*, 100008. [CrossRef]
80. Thom, D.; Rammer, W.; Seidl, R. The impact of future forest dynamics on climate: Interactive effects of changing vegetation and disturbance regimes. *Ecol. Monogr.* **2017**, *87*, 665–684. [CrossRef] [PubMed]

Publisher's Note: MDPI stays neutral with regard to jurisdictional claims in published maps and institutional affiliations.

forests

Article

Effects of Moderate Nitrate and Low Sulphate Depositions on the Status of Soil Base Cation Pools and Recent Mineral Soil Acidification at Forest Conversion Sites with European Beech ("Green Eyes") Embedded in Norway Spruce and Scots Pine Stands

Florian Achilles [1,*], Alexander Tischer [1], Markus Bernhardt-Römermann [2], Ines Chmara [3], Mareike Achilles [1] and Beate Michalzik [1,4]

[1] Department of Soil Science, Friedrich-Schiller University Jena, Löbdergraben 32, 07743 Jena, Germany; alexander.tischer@uni-jena.de (A.T.); mareike.achilles@uni-jena.de (M.A.); beate.michalzik@uni-jena.de (B.M.)
[2] Institute of Ecology and Evolution, Friedrich-Schiller University Jena, Dornburger Str. 159, 07743 Jena, Germany; markus.bernhardt@uni-jena.de
[3] Referat Waldschutz, Standortskunde und Umweltmonitoring, Forstliches Forschungs- und Kompetenzzentrum Gotha, Jägerstraße 1, 99867 Gotha, Germany; ines.chmara@forst.thueringen.de
[4] German Center for Integrative Biodiversity Research (iDiv) Halle-Jena-Leipzig, 04103 Leipzig, Germany
*   Correspondence: florian.achilles@uni-jena.de

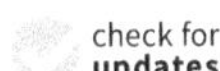

Citation: Achilles, F.; Tischer, A.; Bernhardt-Römermann, M.; Chmara, I.; Achilles, M.; Michalzik, B. Effects of Moderate Nitrate and Low Sulphate Depositions on the Status of Soil Base Cation Pools and Recent Mineral Soil Acidification at Forest Conversion Sites with European Beech ("Green Eyes") Embedded in Norway Spruce and Scots Pine Stands. *Forests* **2021**, *12*, 573. https://doi.org/10.3390/f12050573

Academic Editor: Tiina Maileena Nieminen

Received: 17 March 2021
Accepted: 28 April 2021
Published: 2 May 2021

Publisher's Note: MDPI stays neutral with regard to jurisdictional claims in published maps and institutional affiliations.

Abstract: High N depositions of past decades brought changes to European forests including impacts on forest soil nutrition status. However, the ecosystem responses to declining atmospheric N inputs or moderate N depositions attracted only less attention so far. Our study investigated macronutrient (N, S, Ca^{2+}, Mg^{2+}, K^+) pools and fluxes at forest conversion sites over 80 years old in Central Germany with European beech (so-called "Green Eyes" (GE)). The GE are embedded in large spruce and pine stands (coniferous stands: CS) and all investigated forest stands were exposed to moderate N deposition rates (6.8 ± 0.9 kg ha^{-1} yr^{-1}) and acidic soil conditions (pH$_{H2O}$ < 4.7). Since the understanding of forest soil chemical and macronutrient status is essential for the evaluation of forest conversion approaches, we linked patterns in water-bound nutrient fluxes (2001–2018) and in predicted macronutrient storage in the herbaceous and tree layer to patterns in litter fall (2016–2017) and in forest floor and mineral soil macronutrient stocks at GE and CS assessed in 2018. Our results exhibited 43% (N_t) and 21% (S) higher annual throughfall fluxes at CS than at GE. Seepage water at 100 cm mineral soil depth (2001–2018) of CS is characterized by up to fivefold higher NO_3^- (GE: 2 ± 0.7 µmol$_c$ L^{-1}; CS: 9 ± 1.4 µmol$_c$ L^{-1}) and sevenfold higher SO_4^{2-} (GE: 492 ± 220 µmol$_c$ L^{-1}; CS: 3672 ± 2613 µmol$_c$ L^{-1}) concentrations. High base cation ($\sum Ca^{2+}$, Mg^{2+}, K^+) concentrations in CS mineral soil seepage water (100 cm depth: 2224 ± 1297 µmol$_c$ L^{-1}) show significant positive correlations with SO_4^{2-}. Tree uptake of base cations at GE is associated especially with a Ca^{2+} depletion from deeper mineral soil. Foliar litter fall turns out to be the main pathway for litter base cation return to the topsoil at GE (>59%) and CS (>66%). The litter fall base cation return at GE (59 ± 6 kg ha^{-1} yr^{-1}) is almost twice as large as the base cation deposition (30 ± 5 kg ha^{-1} yr^{-1}) via throughfall and stemflow. At CS, base cation inputs to the topsoil via litter fall and depositions are at the same magnitude (24 ± 4 kg ha^{-1} yr^{-1}). Macronutrient turnover is higher at GE and decomposition processes are hampered at CS maybe through higher N inputs. Due to its little biomass and only small coverage, the herbaceous layer at GE and CS do not exert a strong influence on macronutrient storage. Changes in soil base cation pools are tree species-, depth- and might be time-dependent, with recently growing forest floor stocks. An ongoing mineral soil acidification seems to be related to decreasing mineral soil base cation stocks (through NO_3^- and especially SO_4^{2-} leaching as well as through tree uptake).

Keywords: nitrogen deposition; forest nutrient cycle; *Fagus sylvatica*; *Picea abies*; *Pinus sylvestris*; forest ground vegetation; litter fall; forest floor; fractional annual losses

1. Introduction

Large amounts of nitrogen (N) and sulfur (S) were emitted and deposited in Central Europe, mainly due to fossil fuel power stations, industrial productivity, traffic, agriculture, and livestock farming especially since the 1950s [1,2]. High N depositions caused various changes in European forests including impacts on the tree productivity, on the tree resilience against diseases [3,4], on the organic matter (OM) decomposition rate [5] and on the diversity of the herbaceous layer [6]. The latter is known to fulfil important ecosystem functions in macronutrient cycle regulation [7,8].

The deposition of acids, related to N and S emissions, and the deposition of protons (H^+) contributed to soil acidification [9] caused decreases in soil acid neutralizing capacity. Over 80% of German forest sites—investigated between 2006–2008—were characterized by an acidic pH < 6.2 in mineral soil (0–90 cm depth) [10]. Those forest soils are less resilient against additional acid inputs and are at a greater risk to macronutrient losses. Nitrate (NO_3^-) and sulphate (SO_4^{2-}) anions influenced the mineral soil cation exchange complex [11] by their neutralization through counter ions like calcium (Ca^{2+}), magnesium (Mg^{2+}) or potassium (K^+). As expressed by a pronounced NO_3^- leaching with soil solution, a high N deposition has also caused a N saturation of European forest sites [12]. In this context, the process of mineral soil base cation depletion was regional strengthened by losses as counter ions [13] due to N leaching via seepage water [14].

For Central Europe, joint monitoring programs (e.g., EMEP: Co-operative Programme for Monitoring and Evaluation of the Long-range Transmission of Air Pollutants in Europe or ICP Forests: International Co-operative Programme on Assessment and Monitoring of Air Pollution Effects on Forests) have impressively demonstrate the drastic decrease in SO_2 emissions (and to a lesser extend NO_x and NH_x) between 1990 and 2018 (Figure 1; SO_2 decreased by almost 95%, NO_2 by 58% and NH_3 by 16%) due to international conventions and stringent air pollution regulations [15]. Nevertheless, the trend of NH_3 deposition in bulk precipitation is stagnating over the last decade in Central Europe and the cumulative fluxes of N compounds in throughfall (TF) are often at a moderate (7–13 kg N ha^{-1} yr^{-1}) or even high (20–40 kg N ha^{-1} yr^{-1}) level [3,16]. Critical loads for forest ecosystem eutrophication and acidification [17] are reached or exceeded nearly at 65% of the area of European ecosystems in 2018 [15] and high exceedances at forest sites were found in Western Europe and Germany [18].

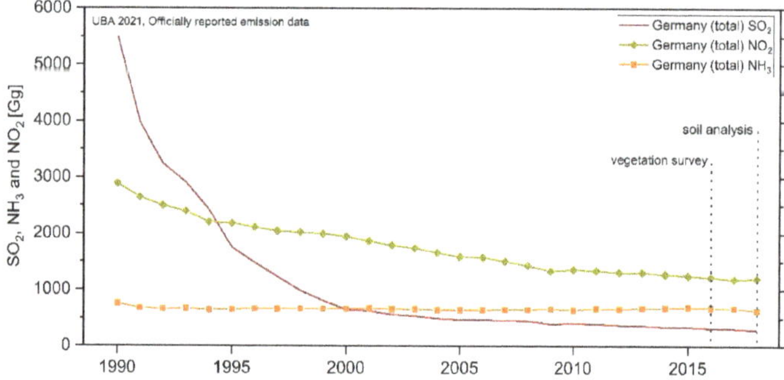

Figure 1. Trends in emissions of main eutrophying/acidifying air pollutants and years of vegetation and soil investigations, presented in this study. Total national emissions of NO_x (as NO_2), NH_x (as NH_3), and SO_x (as SO_2) from 1990 to 2018 at the area of Germany. Datasource: UBA 2021, officially reported emission data.

Although S depositions decreased strongly over the last decades, even today their legacies are found accumulated in the mineral soil as SO_4^{2-}, predominantly specific adsorbed to iron (Fe) and aluminum (Al) sesquioxides [19]. The desorption and the release of SO_4^{2-} into soil solution is held responsible for mineral soil base cation leaching losses with seepage water [10,20–22], which might decrease pH at deeper mineral soil layers and delay a recovery from formerly induced soil acidification through air pollutants, especially of acidic soils [23].

Tree species differ in their throughfall and consequently in their seepage flux at similar site conditions [24]. Coniferous forests are characterized by higher throughfall N and S fluxes, leading to on average 12 times higher NO_3^- and 1.6 times higher SO_4^{2-} leaching of macronutrients from mineral soil [25,26]. Trees link the below- and aboveground areas of an ecosystem by their water, macronutrient, and biomass cycling [27]. A characteristic usually of deciduous tree species like European beech (*Fagus sylvatica* L.) is the so-called "base pump effect" [28], expressed in the root uptake of Ca^{2+}, Mg^{2+} or K^+ from mineral soil, their storage in tree biomass and their partially return to the forest floor by a base cation-rich litter fall. The base cation sequestration into the tree biomass is associated with a mineral soil acidification ("bio-acidification") [29,30] as long as the assimilated cations are not released to the upper soil horizons by litter degradation [11,31]. Contrarily, the annual litter fall at deciduous stands contributes to higher forest floor pH values [32] and an enhanced organic matter turnover compared to coniferous reference stands [32–35]. As observed for German and Austrian forest stands, a recovery from acidification appears depth-depended, with increasing pH in the forest floor and upper soil horizons [10,20], whereas the deeper mineral soil remains acidified [20,36]. For Switzerland [37] no signs of a recovery resulting from decreasing acid depositions in mineral forest soils were found, whereas the trend on the European scale is non-uniform and appears to depend on the base saturation stage of investigated soil profiles [38]. In Central Europe, the conversion of Norway spruce (*Picea abies* L.) and Scots pine (*Pinus sylvestris* L.) forests into mixed forests is one primary aim of a sustainable forestry [39] and is assumed to mitigate the atmospheric input of eutrophying and potentially acidifying compounds to forest soils [26]. A major goal is the avoidance of unstable and disturbance-prone forest ecosystems with a decreasing forest vitality [4,12] under a changing climate with higher risks to drought and storm events in Central Europe [40,41]. Related aims for a forest conversion, for example with European beech, are numerous and include the improvement of soil functionality as expressed by the soil chemical and macronutrient status [42].

In the early 20th century, small groups (<1000 m^2) of European beech were planted, within Norway spruce- and Scots pine-stands (CS), across Central Germany to mitigate soil acidity stress (e.g., H^+ and Al toxicity or Mg^{2+} deficiency) under pure coniferous stands [43] and to improve decomposition processes in the respective forest floors [36,44]. The beech plantings were called "Grüne Augen" in German, hereinafter "Green Eyes" (GE). GE, established on former forested coniferous areas, offer the rare possibility of examining long-term forest conversion effects on soil functionality and associated ground vegetation communities without the influence of other soil-forming factors [36]. Since the understanding of forest soil chemical and macronutrient status is essential for the evaluation of forest conversion approaches, our study investigated important macronutrient pools and fluxes after about > 80 years of stand development at GE and CS.

The aim of our study was to compare input and output fluxes of N, S, Ca^{2+}, Mg^{2+}, and K^+ as well as the macronutrient stocks in forest floor and mineral soil at GE and CS, both subjected to moderate N depositions and decreasing mineral soil pH under acidic soil conditions. A further aim of this study was to identify the driving factors, internal (e.g., tree uptake) and external (e.g., air pollutant depositions), that are currently causing soil acidification. We investigated whether coniferous and deciduous forest stands differ in the explanatory processes and which factors favor or delay a soil recovery from acidification under declining atmospheric deposition of acidifying air pollutants.

2. Materials and Methods

2.1. Study Area

The study area (Figure 2, Table 1) is located in Eastern Thuringia (Thuringian Holzland, Central Germany) and the geological bedrock consists of quartz-rich lithologies [45] of the Triassic and Quaternary, which mainly comprises sandstones and silt-mud-stones (East Thuringian Buntsandstein Territory) [46]. Detailed site descriptions and pictures of the study sites are given in a previous study [36].

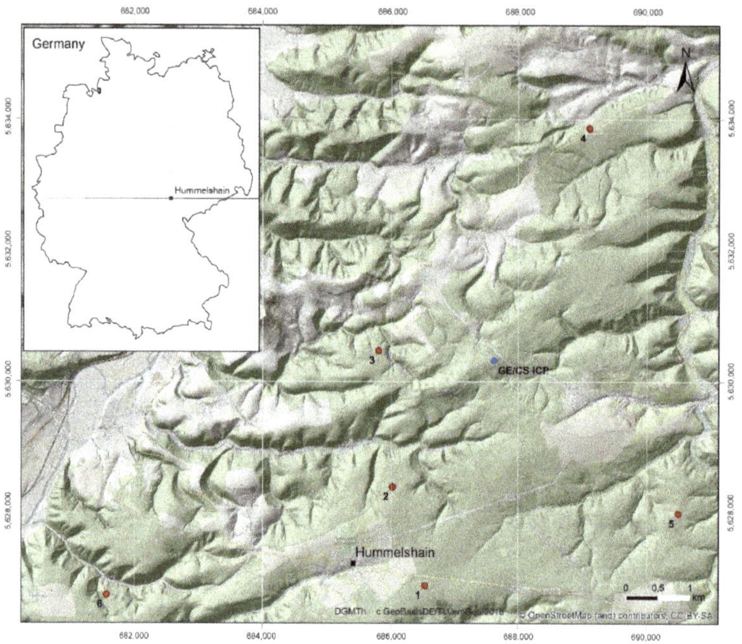

Figure 2. Location of the study area "Hummelshain" in Eastern Thuringia in Central Germany with the locations of the different exploration sites (red points, GE-CS pairs, 1–6) and the monitoring site (blue point, GE/CS ICP). GE: Groups of European beech within stands of Norway spruce and Scots pine (CS).

Table 1. Basic characteristics of investigated sites in Thuringia, Germany. GE: Groups of European beech within stands of Norway spruce and Scots pine (CS).

Plot	Coordinates [1]		Elevation	Soil Type	Sand	Silt	Clay
	E	N	(m a.s.l.)	(WRB, 2015)	(%)	(%)	(%)
GE 1	686,527	5,626,891	366	Dystric Planosol	4.0	83.1	12.9
CS 1	686,527	5,626,928	376	Dystric Planosol	9.1	79.6	11.3
GE 2	686,024	5,628,393	354	Dystric Planosol	18.1	69.3	12.3
CS 2	686,065	5,628,396	373	Dystric Planosol	15.4	72.3	12.4
GE 3	685,802	5,630,469	314	Stagnic Luvisol	7.9	80.3	11.8
CS 3	685,786	5,630,500	317	Stagnic Luvisol	8.1	79.3	12.6
GE 4	689,106	5,633,858	292	Stagnic Luvisol	15.3	74.0	10.7
CS 4	689,148	5,633,868	290	Stagnic Luvisol	8.8	79.0	12.2
GE 5	690,496	5,627,993	352	Haplic Podzol	48.3	42.2	9.5
CS 5	690,478	5,627,970	360	Haplic Podzol	50.7	39.7	9.6
GE 6	681,550	5,626,745	300	Podzol	55.1	37.2	7.6
CS 6	681,543	5,626,727	291	Haplic Podzol	48.4	42.4	9.2
GE ICP [2]	687,590	5,630,348	376	Dystric Planosol	64.9	23.4	11.7
CS ICP [2]	687,622	5,630,199	371	Dystric Planosol	77.3	16.7	6.0

[1] WGS 1984 UTM zone 32 north. [2] Data from Veckenstedt (1999) [47].

2.2. Forest History and Stand Description

In 2018, the mean stand age of the investigated beech groups was 82 years, whereas the surrounding coniferous stands were approximately 77 years old (Table 2). The younger age of CS arises from timber harvest of former coniferous stands during and after the Second World War. The clear-cut areas around the emergent GE stands, which were planted in the 1930s, were immediately replanted with spruce and pine.

For this study, plot-pairs each consisting of a European beech plantation of 800–1000 m^2 in size and the surrounding forest stands of Norway spruce and Scots pine (mainly mixtures of both species, Table 2), were investigated. The CS plots were selected with regard to comparable site conditions (topography, geology, soil development), macronutrient status, water availability, and forest management (Table 1). One plot-pair (GE/CS ICP) forms part of the Level II Intensive Monitoring network of ICP Forests Programme (International Co-operative Programme on Assessment and Monitoring of Air Pollution Effects on Forests) and is operated by the Thuringian State Forestry Institution (ThüringenForst) since the early 2000s. The GE/CS ICP plot-pair serves as a monitoring site, which provides exemplary information on the hydrology and water-bound element fluxes and complements the investigation of further input–output fluxes and pools of the forest macronutrient cycle, which were investigated at the exploration sites GE/CS 1–6. We assume that patterns in the different parts of the element cycle were transferable between the six exploration sites and one monitoring site due to strong similarities concerning the site and forest stand conditions of the respective GE or CS. In the cases of three sites, the soil of both the GE and CS was limed ($CaCO_3$ with 30% $MgCO_3$). This was done in 1988/1989 at site 2 and 4, in 1993 at GE/CS ICP and in case of site 4 again in 2007 [36,48].

Forest yield classes of the investigated tree stands (Table 2) were determined in 2018 [36,49]. Leaf area index (LAI), solar direct (DirRad [MJ m^{-2} yr^{-1}]), and solar diffuse (DifRad [MJ m^{-2} yr^{-1}]) radiation below the tree canopy were quantified at the exploration sites using the equipment of Hemiview (Delta-T devices Ltd., Cambridge, UK). The measurements were carried out at five representative points on each plot (GE/CS 1–6) under conditions of diffuse skylight at twilight in summer 2019, when deciduous trees were full in leaf. Digital hemispherical photos were taken with a fisheye lens attached to a digital camera (Canon EOS 70 D, Canon Inc., Tokyo, Japan) clamped in a self-levelling camera mount (Delta T Devices Ltd., Cambridge, UK). The camera mount was oriented to magnetic north at 1.3 m above ground level on a tripod. The camera settings followed the manufacturer's recommendations (Delta T Devices Ltd., Cambridge, UK). Canopy cover pictures were processed (minimum algorithm, blue color plane) in ImageJ [50] using the plugin Hemispherical 2.0 [51].

Table 2. Stand description. GE: Groups of European beech within stands of Norway spruce and Scots pine (CS).

Plot	Age of Stand in 2018 (yr)	Basal Area of Stand (m^2 ha^{-1})	LAI	DirRad [1] (MJ m^{-2} yr^{-1})	DifRad [2] (MJ m^{-2} yr^{-1})	Herb Layer Biomass [3] (Mg ha^{-1})	Tree Species	Proportion of Basal Area (%)	Tree Biomass [4] (Mg ha^{-1})	Mean Diameter DBH (cm)	Yield Class
GE 1	85	38.4	2.3	704	142	0.09	*Fagus sylvatica*	80	233	24.1	II
							Betula pendula	17	NA	36.7	II
							Picea abies	3	NA	20.1	III
CS 1	73	31.2	2.2	832	340	0.74	*Picea abies*	26	44	25.0	II
							Pinus sylvestris	74	106	35.5	I
GE 2	80	44.1	2.6	538	126	herb layer absent	*Fagus sylvatica*	100	286	17.0	III
CS 2	79	53.7	2.1	1196	286	0.04	*Picea abies*	44	127	26.9	II
							Pinus sylvestris	56	138	34.0	I
GE 3	80	73.6	2.7	343	94	0.001	*Fagus sylvatica*	100	558	21.5	II
CS 3	95	49.9	2.6	520	272	0.15	*Picea abies*	100	334	33.8	I
GE 4	82	78.4	1.7	406	241	0.004	*Fagus sylvatica*	100	674	25.2	I
CS 4	71	52.9	2.8	437	241	0.67	*Picea abies*	100	322	26.1	I
GE 5	80	54.2	2.5	438	93	0.001	*Fagus sylvatica*	82	288	17.8	III
							Picea abies	9	NA	26.9	II
							Pinus sylvestris	8	NA	40.5	I
							Larix decidua	1	NA	17.6	worse than III
CS 5	78	38.6	1.9	968	304	0.05	*Picea abies*	51	87	21.4	III
							Pinus sylvestris	49	77	27.0	II
GE 6	86	46.4	2.9	426	100	herb layer absent	*Fagus sylvatica*	100	247	14.5	IV
CS 6	63	35.9	1.7	1598	255	0.05	*Pinus sylvestris*	100	146	21.5	II
GE ICP	86	37.6	NA	NA	NA	NA	*Fagus sylvatica*	100	244	20.8	III
CS ICP	67	40.7	NA	NA	NA	NA	*Picea abies*	41	112	23.6	I
							Pinus sylvestris	59	110	28.3	I

[1] Direct solar radiation below the canopy. [2] Diffuse solar radiation below the canopy. [3] Predicted biomass of herbaceous ground vegetation (calculated with PhytoCalc 1.4 after Bolte (2006) [52]). [4] Predicted above-ground-tree-biomass (whole-tree, basal-area-weighted and calculated after Block et al. (2016) [53]).

2.3. ICP Forests Monitoring between 2001–2018

The bulk deposition data analyzed in this study, originated from measurements (pooled samples of 5 precipitation collectors) under the framework of the ICP Forests Intensive Monitoring Programme (Level-II) at a cleared area (non-forested, UTM 32 U: 687560/5630027) at 250 m distance to the monitoring plot-pair GE/CS ICP. The deposition of NH_4-N, NO_3-N, N_{org}, SO_4-S, and base cations was determined according to the ICP Forests [54] protocol. Throughfall deposition of N and S compounds and base cations at GE/CS ICP was measured using pooled samples from 20 throughfall collectors in summer time and 10 collectors in winter (100 cm^2 sampling area), respectively. Stemflow was collected for European beech at GE ICP in compliance with the guidelines of the Level-II-program [54]. Soil solution was collected at GE/CS ICP at the interface between forest floor and mineral soil (zero-tension plate lysimeters to collect forest floor solution), within the rooting zone in mineral soil (at 20 cm and 50 cm depth; 5 suction cups each) and below the rooting zone (100 cm depth; 5 suction cups) with tension lysimeters. Soil solution chemistry (pH, electrical conductivity (EC), ion concentrations) was determined following the standard protocols given in ICP Forests [54]. The biweekly deposition and solution data (2001–2018) were provided by the Thuringian State Forestry Institution (ThüringenForst FFK, Gotha, Germany). Critical loads for N (CL_{eutN}) and S (CL_{maxS}) at GE/CS ICP were calculated as mass depositions by [55] using the simple mass balance equation for managed woodland habitats [56] to allow for comparison with measured fluxes and investigated macronutrient pools.

2.4. Vegetation Survey

Forest vegetation (tree, shrub, herb, and bryophyte layer) was assessed according to Braun-Blanquet [57] in June 2016 at the exploration sites [36]. If possible, the elongated length of flowering and non-flowering shoots of 15 individuals of each herbaceous plant species were measured to calculate their mean shoot length (cm). The relationship between plant species biomass, cover, and mean shoot length was used to estimate macronutrient stocks for 13 morphological growths groups (see Section 2.7.1) [52,58].

2.5. Litter Fall, Forest Floor, and Soil Sampling

To quantify annual litter fall, litter collectors with an area of 616 cm^2 were installed in February 2016 in five replicates at each exploration plot. Litter was collected quarterly over one year until March 2017. The base of the litter traps was perforated allowing the litter to dry rapidly after rainfall. We therefore assumed that litter decomposition within the collectors was reduced to a negligible level.

In April 2018, forest floor was sampled in seven spatial replicates for each of the exploration plots. Humus classification was done according to Zanella, et al. [59] in accordance to the German standard protocol [60] and each forest floor replicate was morphologically divided into characteristic diagnostic horizons (Oi, Oe, Oa) [59]. Soil types were classified according to the Ad-Hoc-AG-Boden [61] and converted to WRB [62] nomenclature at representative soil pits from an earlier study [63]. Soil samples from three depth increments (0–10 cm, 10–20 cm and 20–40 cm) were taken by a soil corer (diameter 5.3 cm) at the same locations of forest floor sampling.

2.6. Laboratory Methods

The collected litter samples were oven-dried at 50 °C to constant weight and bulked to annual samples across the sampling dates per litter trap. The annual litter samples were sorted for leaves of beech, needles, bark, small twigs, reproductive parts (beechnuts, beech husks, cones), fine material (bud scales, flowers), and miscellaneous components. Non-plant components were discarded. The litter fractions considered in our study correspond to 79.8% (GE) and 76.1% (CS) of the total annual litter fall in the respective forest stands (we disregarded the fine material and the litter related to the admixed tree species). The dry-weight of each litter fraction was determined and the sorted samples were shredded separately to a final particle size < 5 mm. The oven-dried material (50 °C) of the Oi, Oe, and Oa layers was shredded to a final particle size < 5 mm. The mineral soil was oven dried (50 °C) and passed through a 2-mm sieve. An aliquot of the obtained fine forest floor material and fine-earth samples, respectively, was used to measure pH (WTW pH/Cond 340i with SenTix 41-3, Weilheim, Germany) and electrical conductivity (WTW pH/Cond 340i with TetraCon 325, Weilheim, Germany) in deionized water according to DIN ISO 10390:2005–12. Soil texture was determined on fine-earth samples by a Laser Particle Sizer (Particle Sizer LS 13,320–Laser Diffraction Particle Size Analyzer, Beckman Coulter, Brea, CA, USA).

For elemental analysis, an aliquot of each litter- (leaves, needles, beechnuts), forest floor- (Oi, Oe, Oa), and mineral soil- (0–10 cm, 10–20 cm, and 20–40 cm depth increments) sample was pulverized to 0.04 mm (Mixer Mill MM200 Retsch, Hahn, Germany) and analyzed for total N and S contents (CNS-analyser, vario EL cube, elementar, Langenselbod, Germany). The concentration of base cations in pulverized samples was determined after microwave digestion (ETHOS 1; Milestone S.r.l., Milano, Italy) in aqua regia-extracts by inductively coupled plasma optical emission spectrometry (725 ICP-OES system; Agilent Technologies, Santa Clara, CA, USA).

Macronutrient stocks of the different litter fractions were calculated by multiplying element contents and the dry mass weight of each litter fraction. Macronutrient contents of beech leaves and needles (pine and spruce) were measured plot-specific per litter collector, whereas contents of reproductive parts were based on measured data for beechnuts as well as literature values for beech husks and cones [64,65]. Macronutrient stocks of twigs and bark were calculated by multiplying the measured dry mass of the litter components and element contents based on literature values [66]. The assumptions made, are accessible in Table S1. Macronutrient stocks of single forest floor layers were calculated by multiplying element contents and the dry mass of the respective layer. The dry mass was determined by weighting the oven-dried forest floor material (sampled in a defined area of 255 cm^2) after sorting out living roots. Macronutrient stocks per mineral soil depth increment were calculated based on the fine-earth fraction (<2 mm) by multiplication of element contents and fine-earth stocks (determined by weighting the oven-dried mineral soil of each subsample, sampled in a defined area of 22.1 cm^2).

2.7. Data Analysis

2.7.1. Vegetation Data

Mean weighted (by coverage) ecological indicator values (EIV) for nutrient availability/eutrophication (N), soil reaction (R), and light (L) were calculated according to Ellenberg, et al. [67]. To predict the herbaceous aboveground biomass and macronutrient stocks of the herbaceous layer of GE and CS plots, we used the estimation model PhytoCalc 1.4 [52,58]. The estimations by PhytoCalc have been validated in independent studies for several German forest types [68–70]. The model is based on data from biomass harvests of 46 plant species of deciduous and coniferous forests and uses the relationship between plant species biomass, cover, and mean shoot length to estimate macronutrient stocks for 13 morphological growth groups [52,58]. We assigned the herbaceous plant species (vegetation survey in 2016) to that groups (Table S2) and estimated herbaceous layer biomass (dry matter; kg ha^{-1}) as well as macronutrient stocks for each exploration plot (GE/CS 1–6) using the growth group-specific functions of PhytoCalc 1.4.

The aboveground-tree-biomass and macronutrient stocks of the whole-tree ($\sum$ small wood, stemwood, and bark) of European beech (GE 1–6) and Scots pine or Norway spruce (CS 1–6) were estimated using the relationship of the respective forest stand age according to yield tables [53,60]. The predicted annual macronutrient uptake into stemwood and bark was calculated in accordance with Schlutow and Ritter [55] only for the intensively monitored plot-pair GE/CS ICP. The macronutrient uptake resulted from the derived annual tree increment [55], multiplied by the element contents of stemwood and bark based on De Vries et al. [71] and Jacobsen et al. [72].

2.7.2. Litter Fall, Soil Solution, Forest Floor, and Mineral Soil Data

Total forest floor stocks (dry matter and macronutrients) were analyzed as the sum of the Oi, Oe, and Oa layer stocks per exploration site and total mineral soil stocks were calculated as the sum of the single depth increment stocks per site. Correlations between macronutrient concentrations (NO_3^-, SO_4^{2-}, K^+, Ca^{2+}, Mg^{2+}), pH, and electrical conductivity in seepage water of forest floor and mineral soil at the plot-pair GE/CS ICP were analyzed using generalized pairs plots (Pearson correlation) in the R statistical environment, R 3.6.0 [73] with contributed package GGally (ggpairs function) [74]. Group wise summary statistics were used in R with the contributed package doBy (summaryBy function) to check litter fall and associated macronutrient return as well as forest floor and mineral soil data for differences between forest types (GE, CS) and sites. Differences in DirRad and DifRad between GE and CS were analyzed using a *t-test* (package car).

Interactions of forest type were analyzed as fixed effects with linear mixed effect models (LMM) [75] in R [73] with the package nlme [76] with site as random effect. All variables were described as the mean value with standard deviation (SD) or standard error (SE) as derived from the estimates of statistical models. Estimates were generated using emmeans package [77] and the model summary was extracted by stargazer package [78]. Data were checked for normality and homogeneity of model residues and, if necessary, were transformed using log10. The random effect term considers nested sampling within forest floor/soil cores and sites. The final model includes variance heterogeneity among sites and between forest type. Model parameters were estimated by Maximum-likelihood estimation [75,79]. The effect of improvement of random effect structure was tested based on residual plots and on Akaike Information Criteria [80]. The final model estimates (Table S3) were based on Restricted Maximum Likelihood estimates. Marginal and conditional coefficients of determination (r^2 m and r^2 c) were calculated with the package MuMIn [81].

To predict the annual base cation release from weathering processes in the main rooting zone of the soil matrix, we followed the soil type-texture approximation for non-calcareous soil profiles given in CLRTAP [56]. This method is based on the plot-specific soil texture and parent material (combination of both into a weathering rate class), the depth of the main rooting zone and the mean annual temperature [82]. The assignments made, are accessible in Table S4. As conversion factors between equivalent and molar mass we used: 0.02004 (Ca^{2+}), 0.01215 (Mg^{2+}), and 0.0391 (K^+). Forest stand input–output balances for Ca^{2+}, Mg^{2+}, and K^+ were calculated for GE and CS regarding important pathways of forest macronutrient cycling (e.g., extrapolated deposition, litter fall, estimated weathering release, and estimated tree uptake).

2.7.3. Fractional Annual Losses and Macronutrient Residence Time in Forest Floor

The fractional annual loss (FAL) or turnover rate of forest floor material was estimated after Olson [83] and Hobbie, et al. [84] per exploration plot based on the ratio between litter fall and forest floor mass under the assumption that the forest stands are in steady state (annual decomposition equals annual litter input; Vesterdal, et al. [85]). Although steady state assumptions in ecosystems can be prone to errors, the calculation of fractional annual losses still provides a useful indicator for comparative purposes of the fraction of material that leaves the forest floor pool. The investigated stands of GE and CS have already reached canopy closure in the late 1990s [86] and nearly steady state conditions were assumed due to relative constant litter fall data of the monitoring plot CS ICP ranging between 2750–3050 kg ha^{-1} for the years 2006–2010, except 2007 (caused by the European windstorm Kyrill). Fractional annual losses at CS were estimated for continuous litter fall by Equation (1) and in the case of GE for discrete litter fall in autumn by Equation (2) [85]. L is the dry mass of annual litter fall (kg ha^{-1} yr^{-1}) and F is the dry mass of total forest floor material ($\sum$ Oi, Oe, Oa) (kg ha^{-1} yr^{-1}). The residence time is the inverse of fractional annual losses.

$$k_1 = L/F \tag{1}$$

$$k_2 = L/(L + F) \tag{2}$$

Subsequently, based on the macronutrient stocks of the analyzed litter fractions, we estimated fractional annual losses and residence time of N, S, Ca^{2+}, Mg^{2+}, and K^+ stored in total forest floor per plot (GE/CS 1–6) referred to Equations (1) and (2) in case of continuous litter fall and discrete litter fall according to Cole and Rapp [87], Likens, et al. [88], and Hansen, et al. [89] by Equations (3) and (4). Lx is the annual amount of litter fall of element x (kg ha^{-1} yr^{-1}) and Fx is the amount of element x in total forest floor (kg ha^{-1} yr^{-1}). Because the analyzed litter fractions cover 79.8% of annual total litter fall at GE and 76.1% at CS, it is therefore possible that we slightly underestimated the fractional annual losses of macronutrients from forest floor in this study.

$$k_1\,x = Lx/Fx \tag{3}$$

$$k_2\,x = Lx/(Lx + Fx) \tag{4}$$

3. Results

3.1. Macronutrient Fluxes at the Monitoring Sites in Bulk Precipitation, Throughfall, and Stemflow between 2001–2018

Hydrological data recorded between 2001 and 2018 exhibited that throughfall at GE ICP amounts to 76% and at CS ICP to 65% of mean annual bulk precipitation (680 $\pm$ 141 mm), causing an annual interception loss of 24% (GE ICP) and 35% (CS ICP). Total nitrogen ($\sum$ NH$_4$-N, NO$_3$-N and N$_{org}$) depositions in bulk precipitation remained relatively constant at a moderate level of 6.8 $\pm$ 0.9 kg ha^{-1} yr^{-1} since 2011 (Figure 3A). The throughfall N$_t$ flux between 2001 and 2018 decreased (-4.2 kg ha^{-1}) for GE ICP, whereas it slightly increased ($+1.4$ kg ha^{-1}) for CS ICP. The forest-type-specific critical loads for N (CL$_{eutN}$; kg ha^{-1} yr^{-1}) were exceeded for throughfall fluxes over the whole investigation period at CS ICP but not after 2012 at GE ICP. The mean flux of N$_t$ in stemflow (GE ICP) decreased over time (Figure 3C). Mean bulk precipitation N$_t$ fluxes were mostly composed of NH$_4$-N and NO$_3$-N in equal parts until 2008, turning into a slight dominance of NH$_4$-N afterwards. As shown in Figure 3, throughfall N$_t$ fluxes at GE ICP as well as at CS ICP were dominated by NO$_3$-N (GE ICP: 51.5%; CS ICP: 49.2% of N$_t$), followed by NH$_4$-N (GE ICP: 32.7%; CS ICP: 38.9%). The mean annual throughfall N$_t$ flux was 43% higher at CS ICP (18.4 $\pm$ 1.7 kg ha^{-1} yr^{-1}) compared to GE ICP (12.9 $\pm$ 2.6 kg ha^{-1} yr^{-1}) (Figure 3E,G). The SO$_4$-S fluxes (bulk precipitation and throughfall at GE/CS ICP) decreased clearly over time by a third to a half and even by 92% in stemflow of GE ICP (Figure 3A,E,G). Although the throughfall SO$_4$-S flux in 2001 at CS ICP was still 32% higher than at GE ICP, it decreased sharper over time and reached similar levels of about 3.6 kg ha^{-1} yr^{-1} at GE and CS since 2015 (Figure 3E,G). The forest-type-specific critical loads for S (CL$_{maxS}$; kg ha^{-1} yr^{-1}) were clearly below the thresholds considering throughfall fluxes (Figure 3). In summary, there was a higher annual input of N and S compounds at CS ICP than at GE ICP between 2001–2018 with 67.4% more NH$_4$-N (4.3 to 7.2 kg ha^{-1}), 36.4% more NO$_3$-N (6.6 to 9.0 kg ha^{-1}), 10.0% more N$_{org}$ (2.0 to 2.2 kg ha^{-1}), and 20.9% more SO$_4$-S (4.3 to 5.2 kg ha^{-1}). At GE ICP an additional annual input of in average 0.4 kg ha^{-1} NH$_4$-N, 0.6 kg ha^{-1} NO$_3$-N, 0.3 kg ha^{-1} N$_{org}$, and 0.7 kg ha^{-1} SO$_4$-S must be taken into account due to stemflow fluxes (Figure 3C). Nevertheless, total input of N and S compounds was even higher at CS ICP.

Total base cation ($\sum$ K$^+$, Ca^{2+}, Mg^{2+}) deposition via bulk precipitation decreased since 2001 to a level of 4.9 kg ha^{-1} in 2018 (Figure 3B). The yearly throughfall fluxes of base cations decreased for GE ICP (-4.1 kg ha^{-1}) as well as for CS ICP (-6.7 kg ha^{-1}) (Figure 3). Stemflow base cation fluxes of GE ICP also showed a decreasing trend (-5.1 kg ha^{-1}). Annual fluxes of base cations in bulk precipitation were mostly composed of Ca^{2+} and to a lesser extent of K$^+$ and Mg^{2+} (Figure 3B). The throughfall fluxes of GE ICP as well as of CS ICP were dominated clearly by K$^+$ (64.9% and 65.0%), followed by Ca^{2+} (27.9% and 26.7%) and Mg^{2+} (7.2% and 8.3%). The throughfall and stemflow base cation flux (2001–2018) was slightly higher at GE ICP (30.1 $\pm$ 5.3 kg ha^{-1}) than at CS ICP (24.0 $\pm$ 4.2 kg ha^{-1}).

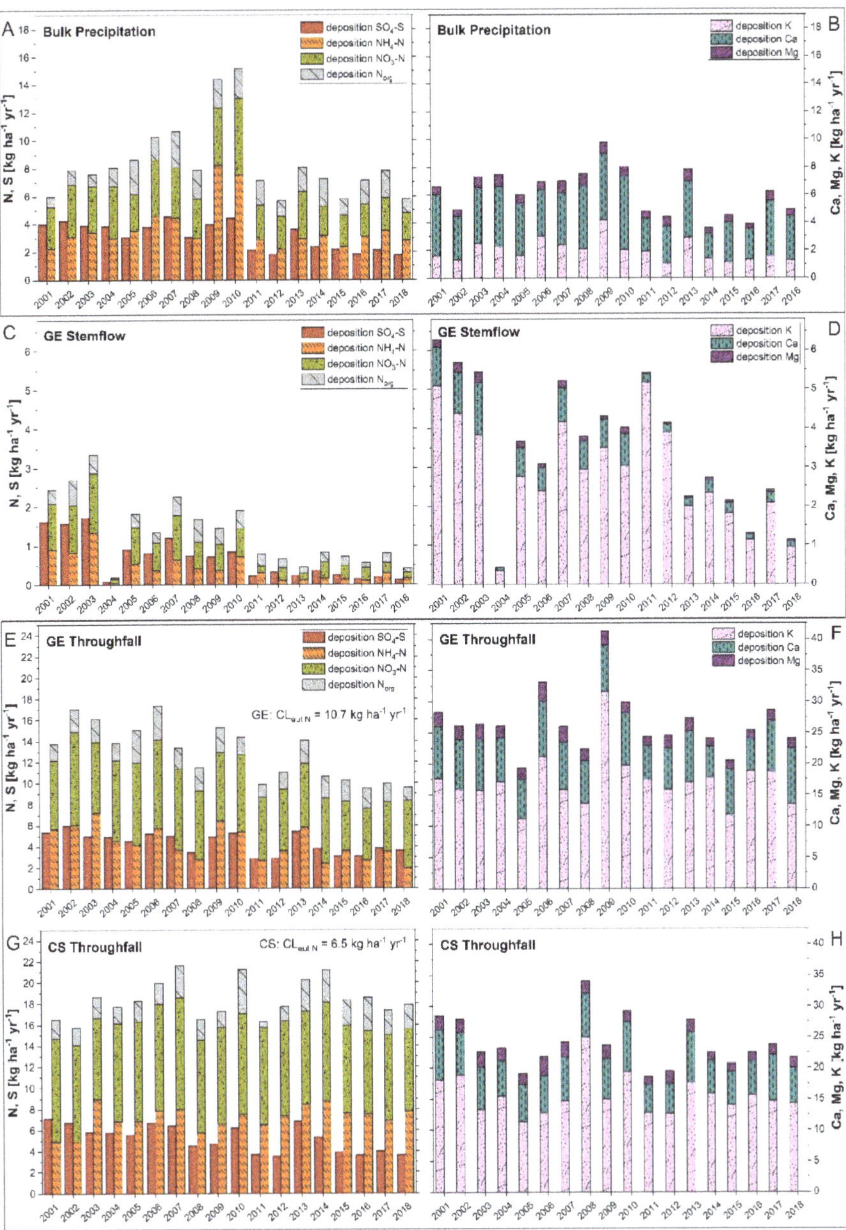

Figure 3. Deposition patterns of NH_4-N, NO_3-N, N_{org}, SO_4-S and base cations (K^+, Ca^{2+}, Mg^{2+}) of bulk precipitation (**A,B**), stemflow (**C,D**), and throughfall (**E–H**) within the forest hydrological cycle at the plot-pair GE/CS ICP between 2001 and 2018. The site was observed in the framework of the ICP Forests Intensive Monitoring Programme (Level-II; ThüringenForst). Forest-type-specific critical loads [17] for eutrophying nitrogen (CL_{eutN}; kg ha^{-1} yr^{-1}) and acidifying sulfur (CL_{maxS}; kg ha^{-1} yr^{-1}) were calculated after the simple mass balance model [55,56]. Total N-depositions in throughfall reach ecological relevant values (CL_{eutN} is exceeded mostly before 2012 for GE and is exceeded in the whole investigation period for CS), whereas SO_4-depositions are permanently clearly below the CL_{maxS} (GE CL_{maxS}: 17.5 kg ha^{-1} yr^{-1}; CS CL_{maxS}: 26.0 kg ha^{-1} yr^{-1}). GE: Groups of European beech within stands of Norway spruce and Scots pine (CS) near Hummelshain (Thuringia, Central Germany).

3.2. Solution Chemistry at the Monitoring Sites

3.2.1. pH and Electrical Conductivity

The bulk precipitation (2001–2018: pH 5.8 ± 0.4) was classified as moderately acidic [90], as well as the stemflow of GE ICP (pH 5.6 ± 0.4; Table 3). The pH of throughfall solution was slightly acidic (GE ICP) and strongly acidic (CS ICP) (Table 3). While forest floor seepage water showed no differences in pH between GE ICP and CS ICP (pH < 4.4 ± 0.4), mineral soil solutions of GE ICP (slightly decreasing with depth) were less acidic with 34.2% (20 cm), 17.1% (50 cm), and 14.6% (100 cm) higher pH values compared to CS ICP (Table 3). The electrical conductivity of throughfall between 2001–2018 at GE ICP was 32% lower compared to CS ICP, while the electrical conductivity of forest floor seepage water was similar between GE ICP and CS ICP (electrical conductivity < 84.2 ± 21.6). Differences between the forest type became apparent with depth (up to 65–79% lower electrical conductivity values at GE ICP than at CS ICP), especially below the rooting zone (100 cm depth; Table 3).

Table 3. Solution data (amount, pH, electrical conductivity) of bulk precipitation, stemflow (GE) and throughfall (GE and CS) as well as seepage water of forest floor and mineral soil (20 cm, 50 cm and 100 cm depth) at the plot-pair GE/CS ICP between 2001 and 2018. The site was observed in the framework of the ICP Forests Intensive Monitoring Programme (Level-II; ThüringenForst). GE: Groups of European beech within stands of Norway spruce and Scots pine (CS) near Hummelshain (Thuringia, Central Germany).

Plot	Solution Amount (mm)	pH	Electrical Conductivity (μS cm^{-1})
	Bulk precipitation		
	680 (141)	5.8 (0.4)	19.1 (2.4)
	Stemflow (SF)		
GE [1]	61 (38)	5.6 (0.4)	54.4 (10.4)
	Throughfall (TF)		
GE [1]	514 (112)	6.1 (0.3)	45.8 (7.4)
CS [1]	440 (122)	5.2 (0.2)	67.0 (12.2)
	Forest floor solution		
GE [1]	NA	4.4 (0.4)	77.9 (24.8)
CS [1]	NA	4.3 (0.3)	84.2 (21.6)
	Soil solution 20 cm		
GE [1]	NA	5.1 (0.6)	63.7 (12.0)
CS [1]	NA	3.8 (0.1)	182.1 (81.1)
	Soil solution 50 cm		
GE [1]	NA	4.8 (0.2)	59.4 (13.2)
CS [1]	NA	4.1 (0.1)	177.1 (69.5)
	Soil solution 100 cm		
GE [1]	NA	4.7 (0.2)	88.9 (30.8)
CS [1]	NA	4.1 (0.4)	431.6 (245.8)

[1] Based on biweekly data (2001–2018) from GE ICP and CS ICP. Standard deviation (SD) in brackets.

3.2.2. Macronutrient Concentrations in Forest Floor and Soil Solution between 2001–2018

The NO_3^- concentration of GE ICP cascaded downwards with highest values in forest floor (2.8 μmol$_c$ L^{-1}) and lowest at 100 cm depth (1.8 μmol$_c$ L^{-1}). NO_3^- only reached a level of 0.4–2.7% of the SO_4^{2-} concentration at GE ICP, which showed an increasing concentration with depth with highest concentrations (100 cm: 491.9 μmol$_c$ L^{-1}) below the main rooting zone (Figure 4A). Compared to GE ICP, the seepage water of CS ICP was characterized by up to sixfold higher NO_3^- concentrations (except of 20 cm depth with slightly higher concentrations at GE ICP). NO_3^- concentrations at CS ICP were highest in forest floor and lowest at 20 cm depth with increasing concentrations from 50 cm to 100 cm depth. Nevertheless, the NO_3^- concentration at CS ICP was small compared to SO_4^{2-} (forest floor: 14.9%, mineral soil: 0.2–0.3% of the SO_4^{2-} concentration). SO_4^{2-} at CS ICP showed lowest concentrations in forest floor (115.4 μmol_c L^{-1}) and highest at 100 cm depth (3672.0 μmol_c L^{-1}) (Figure 4B). The mineral soil depth trend in SO_4^{2-} concentration of CS ICP was similar to GE ICP, but the SO_4^{2-} concentration was five to sevenfold (755–3672 μmol_c L^{-1}) higher at CS ICP (Figure 4A,B). High SO_4^{2-} concentrations in the mineral soil solution of CS ICP were mainly caused by the time period between 2001–2010 (mostly > 1000 μmol_c L^{-1}), whereas a decrease in SO_4^{2-} concentrations was observable between 2011–2018 (mostly < 1000 μmol_c L^{-1}) (Figure S1). For GE ICP, no clear trend was noticeable over time for NO_3^- or SO_4^{2-} mineral soil solution concentrations (Figure S1).

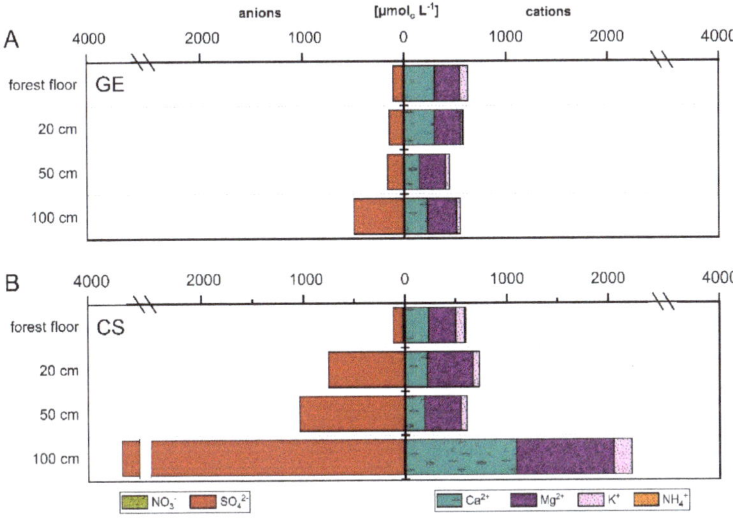

Figure 4. Depth patterns in macronutrient concentration (μmol_c L^{-1}) of forest floor and soil solution (20 cm, 50 cm, 100 cm depth) divided into anions (NO_3^-, SO_4^{2-}) and cations (NH_4^+, K^+, Ca^{2+}, Mg^{2+}) at the plot-pair GE/CS ICP. (**A**): GE; (**B**): CS. The rooting zone of main tree species covers a depth of 70 cm (GE) and 55 cm (CS) [47]. Nitrate (GE: < 3.5 μmol_c L^{-1}; CS: < 17.2 μmol_c L^{-1}) and ammonium (GE: < 7.4 μmol_c L^{-1}; CS: < 12.6 μmol_c L^{-1}) concentrations are low and not feasible. The site was observed in the framework of the ICP Forests Intensive Monitoring Programme (Level-II; ThüringenForst). GE: Groups of European beech within stands of Norway spruce and Scots pine (CS) near Hummelshain (Thuringia, Central Germany).

Mean total base cation concentrations in seepage water of forest floor and mineral soil at GE ICP stayed predominantly constant with depth (ranging between 446.0–632.0 μmol_c L^{-1}) and showed highest concentrations in forest floor solution (Figure 4A). Contrary to GE ICP, base cation concentrations in seepage water of CS ICP were lowest in the forest floor solution, followed by slightly increased concentrations in the upper mineral soil and threefold higher total base cation concentrations (2224.0 μmol_c L^{-1}) below the rooting zone at 100 cm depth (Figure 4B). This increase in mineral soil solution at CS ICP was highest for the Ca^{2+} concentration, which was nearly sixfold higher at 100 cm than at 50 cm depth (Figure 4B). Compared to GE ICP, the concentration of base cations was similar at CS ICP in forest floor and upper mineral soil solution (20–50 cm depth). At 100 cm depth, concentrations were threefold higher for Mg^{2+} and up to nearly five times higher for K^+ and Ca^{2+} at CS ICP (Figure 4). The base cation concentrations in mineral soil solution at GE/CS ICP showed lower concentrations in the last decade (Figure S1). High total base cation concentrations in seepage water were mainly caused by the time period between 2001–2009/10 (mostly > 600 μmol_c L^{-1}), whereas afterwards a decrease was observable (mostly < 600 μmol_c L^{-1}) (Figure S1).

Total base cation concentrations in seepage water between 2001–2018, showed a negative—but insignificant—correlation with NO_3^- concentrations (Figures S2–S5, except for GE at 100 cm depth). They were significantly positive correlated at most depth increments with SO_4^{2-} concentrations, especially in mineral soil of CS ICP (Figures S2–S5). At CS ICP and GE ICP, electrical conductivity (GE ICP: r > 0.696 and CS ICP: r > 0.766) was significantly (p < 0.01) positively correlated with base cation concentrations (primarily Ca^{2+} and Mg^{2+}) as well as SO_4^{2-} concentrations (Figures S2–S5). SO_4^{2-} concentrations correlated negatively with pH of forest floor and soil solution at GE ICP and CS ICP (Figures S2–S5, except for GE at 20 cm depth). Positive correlations between SO_4^{2-} and the sum of K^+, Ca^{2+}, and Mg^{2+} concentrations in mineral soil were stronger for CS ICP (r = 0.948–0.979) than for GE ICP (r = 0.288–0.937). They were mainly caused by Ca^{2+} and Mg^{2+}, whereas K^+ concentrations showed no clear trend and were even negatively correlated with SO_4^{2-} concentrations (e.g., GE ICP: 20 cm and 50 cm depth).

3.3. Canopy Characteristics and Litter Fall at the Exploration Sites

The investigated forest stands exhibited closed canopies conditions in the late 1990s, resulting in a dense foliage cover in 2018, especially for GE. Mean weighted (by coverage) ecological indicator values for light classified the CS as partly shaded and the GE as shady. The leaf area index (LAI) and the related solar direct (DirRad [MJ m^{-2} yr^{-1}]) and solar diffuse (DifRad [MJ m^{-2} yr^{-1}]) radiation below the tree canopy showed, that brighter conditions occurred at CS (Table 2). During the growing season DirRad (t F: 37 = 4.9, p < 0.001) was significantly lower at GE (474 ± 171 MJ m^{-2} yr^{-1}) than at CS (925 ± 470 MJ m^{-2} yr^{-1}), as well as DifRad (t F: 57 = 8.9, p < 0.001), which was significantly stronger at CS (283 ± 51 MJ m^{-2} yr^{-1}) than at GE (133 ± 74 MJ m^{-2} yr^{-1}).

The total annual litter fall at GE was dominated by beech leaves (43.7%), beech husks (20.7%) and beechnuts (10.8%), whereas twigs and bark had only a minor proportion (dry matter; Figure S6). At CS, total annual litter fall was mainly composed of needles of Scots pine and Norway spruce (33.9%), followed by cones (27.3%), twigs (9.8%), and bark (5.1%). Based on the analyzed litter, the annual input of organic matter via litter fall at GE (5.47 Mg ha^{-1}) was twice as high as at CS (2.67 Mg ha^{-1}) (p < 0.0001; Table S3). The annual macronutrient return at GE exceeded double or threefold the input at CS for Ca^{2+} and N and was even five to six times higher for K^+ and S (p < 0.0001; Figure S7). These patterns occurred at all investigated sites. The limed plots (GE/CS 2, GE/CS 4) showed no anomalies, except a slightly higher base cation litter flux at CS 4. The macronutrient input fluxes to forest floor via litter fall at GE were highest for N (61.7 ± 5.0 kg ha^{-1} yr^{-1}) and descended in the order N > Ca^{2+} > K^+ > S > Mg^{2+} (CS: N with 16.9 ± 1.4 kg ha^{-1} yr^{-1} > Ca^{2+} > K^+ > Mg^{2+} > S). Most of the nitrogen returned through leave litter (53%) and reproductive parts (46%) at GE as well as through needle litter fall (79%) at CS (Table S5). Foliar litter fall (leaves or needles) turned out to be the main pathway for litter base cation return at GE (>59%) and at CS (>66%) (Table S5). Reproductive parts were important for Ca^{2+} (15.8% of considered litter fall Ca fluxes), Mg^{2+} (31.1%) and K^+ (39.3%) fluxes at GE (Table S5).

3.4. Macronutrient Stocks in Tree and Herb Layer Biomass, Forest Floor and Mineral Soil at the Exploration Sites

The predicted above-ground biomass of the whole-tree was higher at GE than at CS (Table 2). European beech stored more N in above-ground biomass than the coniferous tree species (median GE: 559 ± 286 kg ha^{-1}; median CS: 368 ± 137 kg ha^{-1}), while predicted whole-tree S stocks were on a similar level for GE and CS (Table S6). On average, the predicted total base cation stocks in above-ground biomass were higher for GE, but they varied with site (Table S6). The predicted whole-tree stocks of Ca^{2+} were similar in both forest types, but predicted K^+ and Mg^{2+} storage was higher for GE (Table S6). The share of

K$^+$ (GE: 43.9%; CS: 35.0%) in total base cation stocks was almost as big as the share of Ca^{2+} at GE (GE: 45.6%; CS: 55.5%). GE stored proportionally more K$^+$ in above-ground biomass than CS, whereas Ca^{2+} storage was higher at CS.

In most GE, ground vegetation was missing or consisted of only a few species in the herbaceous layer with a low biomass (GE: 3.7 kg ha^{-1}; CS: 245.1 kg ha^{-1}). At CS, the predicted above-ground biomass of the herbaceous layer was 0.11% of the predicted tree biomass, whereas at GE, the proportion was substantially lower (0.01%). Thus, the predicted macronutrient storage in the herbaceous layer was considerably higher at CS (Table S6). The predicted herbaceous layer base cation stocks of GE as well as of CS were dominated by K$^+$ (GE: 70.0%; CS: 75.5%).

The cumulative forest floor organic matter stock ($\sum$ Oi, Oe, Oa) at GE (51 $\pm$ 7 Mg ha^{-1}) was half of the stock of CS (104 $\pm$ 14 Mg ha^{-1}) ($p < 0.0001$; Figure S8). The same was hold true for forest floor N stocks (GE: 641 $\pm$ 89 kg ha^{-1}; CS: 1358 $\pm$ 188 kg ha^{-1}; $p < 0.0001$), whereas S stocks were nearly threefold higher at CS (GE: 57 $\pm$ 8 kg ha^{-1}; CS: 157 $\pm$ 23 kg ha^{-1}; $p < 0.0001$) (Table S6). Forest floor stocks of Ca^{2+}, Mg^{2+}, and K$^+$ were higher at CS, following the trend of dry matter stocks (Table S6). Forest floor base cation stocks were highly affected by the site (Table S6). Omitting the limed exploration plots (GE/CS 2, GE/CS 4), forest floor Ca^{2+} and Mg^{2+} stocks of the un-limed plots were slightly higher at GE than at CS (Table S6). There were no differences between the forest types for mineral soil macronutrient stocks ($\sum$ 0–10 cm, 10–20 cm, 20–40 cm) (Tables S6 and S7; Figure S8). Furthermore, they showed no clear trend between the different sites, except at the limed plots (GE/CS 2, GE/CS 4), which were characterized—contrary to the trend in mineral soil nitrogen stocks—by higher N stocks at GE (Table S6).

3.5. Forest Floor Organic Matter Turnover and Related Macronutrient Residence Time at the Exploration Sites

The calculated turnover rates (k) of forest floor organic matter (Table S8) were higher for GE with a mean k value of 0.09 (0.09 $\pm$ 0.02 yrs^{-1}) and smaller for CS with k < 0.03 (0.03 $\pm$ 0.01 yrs^{-1}) ($p < 0.01$). The related fractional annual losses of N ($p < 0.0001$) and base cations ($p < 0.01$) stored in the forest floor showed a similar trend with nearly three to six times higher loss rates at GE (Figure S9). The turnover of S was even 14 times higher at GE (but differs stronger with site than the other nutrients; Table S8; $p < 0.0001$). The mean residence time for forest floor material and macronutrients was < 13 years at GE and lasted longer (>21 years) at CS (Table S8; Figure S9). The high residence time for S at CS was mainly affected by the site and caused by forest floor organic matter stocks (Table S9). Organic matter and macronutrients were retained nearly three times longer at CS 1, CS 2, and CS 6 than at the other coniferous sites. The respective sites were classified as the wettest (CS 1/2; seasonal wet) or the driest (CS 6; moderately dry) sites [36]. The accumulation of organic matter and related S stocks (and other macronutrients) at these sites was highest in the Oe and Oa horizons (Table S9). In general, dry matter stocks of Oi and Oa at CS were at a similar magnitude than at GE, but mean Oe DM stocks at GE were quite different with half of the stock of the CS (GE Oe: 23.0 $\pm$ 5.8 kg ha^{-1}; CS Oe: 54.6 $\pm$ 13.6 kg ha^{-1}).

3.6. Predicted Weathering Rate, Predicted Tree Uptake, and Input–Output Balances

The predicted annual base cation release from weathering processes in the main rooting zone of the soil matrix was similar between the forest types (Table S10). Nevertheless, slightly higher amounts of base cations—dominated by K$^+$—were released at GE (median: 21 $\pm$ 5 kg ha^{-1} yr^{-1}) compared to CS (median: 17 $\pm$ 6 kg ha^{-1} yr^{-1}). The estimated annual macronutrient uptake into tree stemwood and bark at the monitoring plot-pair GE/CS ICP, showed a twofold higher N (GE: 6.9 kg ha^{-1} yr^{-1}; CS: 3.4 kg ha^{-1} yr^{-1}) and an up to threefold higher base cation uptake at GE than at CS (Table S10). The uptake was higher for Ca^{2+} and K$^+$ than for Mg^{2+} (Table S10).

Forest stand input–output balances for N, Ca^{2+}, Mg^{2+}, and K$^+$ indicated, that tree macronutrient uptake at GE and CS was balanced by mean annual deposition (throughfall and stemflow), mean annual weathering of soil matrix and especially yearly macronutrient return via litter fall (Table S10). In case of Ca^{2+}, balances considering the input due to deposition or weathering and the output via tree uptake were negative for GE (Table S10). Nevertheless, in summary the approximately calculated budget was positive for N, Ca^{2+}, Mg^{2+}, and K^+ at GE and CS, indicating growing macronutrient pools, especially in forest floor and topsoil. Due to missing seepage fluxes, the possible macronutrient storage can be overestimated, although forest floor and mineral soil solution data (e.g., depth patterns in macronutrient concentration) showed no pronounced leaching losses.

4. Discussion

4.1. Moderate N Deposition at GE and CS

Results from joint European monitoring programs [15,91] and other studies [12] express concern, that moderate to high N bulk depositions still threaten soil chemistry and quality, alter herbaceous flora and may lead to declines in forest productivity. Since 2011, N_t bulk deposition in the studied area was relatively constant at a moderate level of 6.8 ± 0.9 kg ha^{-1} yr^{-1}. The higher mean annual throughfall N_t deposition between 2001–2018 at CS ICP can be explained by the stronger filtering effect of coniferous canopies [32] and throughfall at CS is more enriched with N compounds due to canopy exchange processes like the leaching of accumulated deposits or the leaching from internal plant tissues [92,93] than at GE.

Our results offer only little evidence for herbaceous layer diversity losses or distinct mineral soil base cation losses concerning the moderate N bulk deposition at the investigated forest stands. The present study may have a certain validity for Central European forests under comparable climatic, topographic and geological settings as well as under a similar atmospheric deposition and forest history. The GE and CS form part of a contiguous forest area (Thuringian Holzland) and our results are in line with recent studies, which reviewed the influence of N deposition on European forests [3,9,12,94]. There it was reported that low to moderate N bulk depositions (<7–13 kg N ha^{-1} yr^{-1}) currently do not affect most European forests on a large-scale in terms of tree growth, tree vitality, soil chemistry, or soil water quality [9,94–96].

A strong effect of N deposition on the herbaceous layer of GE and CS cannot be exhibited by plant species occurrence or richness as well as ecological indicator values (EIV N). An increase in eutrophication indicators such as *Urtica dioica, Galium aparine,* and *Epilobium angustifolium* was limited to CS 4 and not widespread, indicating a response to site-specific regeneration-oriented liming rather than to extensive eutrophication through N depositions. The herbaceous layer plays an important role for the forest ecosystem stability and within the macronutrient cycle, besides its small biomass [7]. The importance for regulating forest ecosystem processes is in detail unknown [97], but the herbaceous layer seems to enhance forest floor organic matter decomposition rates, due to its production of short-lived (often with annual turnover) and easily degradable aboveground biomass [7,98]. Depending on the species composition and the degree of cover, the herbaceous layer can represent a significant source or even a sink of macronutrients (e.g., for moderate or high N deposition) [99]. Most of the GE are characterized by a low to very low predicted herb layer biomass and at some GE the herbaceous layer is even missing (Table 2). Thus, the macronutrient regulatory function is rather low at GE and the predicted nutrient storage in the herbaceous layer is greater at CS (Table S6), expressed, for example, by a 23-fold higher predicted N storage in the herbaceous biomass at CS. Nevertheless, compared to the predicted macronutrient storage in the tree biomass (Table S6), the predicted storage in the herbaceous layer is low and the predicted herbaceous biomass at GE is only 0.01% of the predicted tree biomass and 0.11% at CS, respectively. This proportion is typically higher (~0.1–1%) for northern hemisphere deciduous forests [7], indicating that the herbaceous layer at GE and CS do not exert a strong influence on macronutrient storage (or N retention) there.

4.2. Forest Floor Organic Matter and Macronutrient Turnover

GE are characterized by more bioactive humus forms than CS (mull, moder at GE and moder, mor at CS) [36], underlined by half of the size of forest floor organic matter stocks (Table S6) and despite a litter fall that is twice as high than at CS (Table S10). A low residence time of organic matter and macronutrients, also expressed in higher turnover rates (fractional annual losses; Table S8) at GE, indicate a higher biomass or activity of decomposers and detritivores as well as better decomposition conditions at GE. A less acidic pH in forest floor [36], topsoil, and soil seepage water (Table 3), probably favored the presence of a decomposing soil fauna at GE (e.g., epigeic/endogeic earthworms) [100,101]. The forest type effect can be masked by site as well as forest management, visible in case of the fractional annual loss of organic matter at GE 1, GE 6, and CS 4 (Table S8). GE 1 (Dystric Planosol) is one of the wettest, whereas GE 6 (Podzol) is one of the driest sites and both site conditions may delay decomposition [102], even under deciduous tree species [28]. Conversely, lime applications can favor the organic matter turnover at CS, leading to a higher fractional annual loss similar to that of the GE (Table S8). The degradation of organic matter is diminished at CS, although the amount of light during the growing season is higher at CS (Table 2). The soil-biological activity, as well as the rate of litter decomposition during the vegetation period, is a function of temperature at the local scale [103] and thereby of solar direct radiation below the tree canopy [104]. A lower water supply within the CS (higher mean annual interception loss at CS, Table 3) and a high polyphenol content of the needle litter [105] may hamper turnover at CS. The needle litter is not per se decay resistant, but the biochemical transformations during the decomposition process form novel stable compounds [106], that contribute to recalcitrant forest floor organic matter stocks. In CS, the litter decomposition is expected to be facilitated primarily through microorganisms like fungi or bacteria [102], since the abundance of earthworms at CS might be very low due to soil acidity [33,101]. Coinciding, the annual throughfall N_t flux is increasing at CS and higher than under the canopy of European beech (Figure 3). Knorr, et al. [107] showed, that litter decomposition is stimulated by minor N fluxes (<5 kg ha^{-1} yr^{-1}), but hampered by higher N depositions (>5 kg ha^{-1} yr^{-1}). It is at least conceivable that the decomposition of the coniferous litter is further inhibited at the CS due to the higher throughfall N fluxes, leading to a suppressed activity of fungal and microbial extracellular oxidative enzymes (lignolytic enzymes) [5,108,109].

4.3. Soil Acidification and Its Modulation at GE and CS

4.3.1. The Role of Litter Fall, Forest Hydrology and Weathering Processes

In a former study [36] we observed no significant change in forest floor pH values between 1999 and 2018, in both the GE and CS. We suggested, that in the case of GE, the supply of base cation-rich litter buffer forest floor acidification, which may occur during decomposition and organic matter turnover processes [32]. The present study supports this hypothesis and shows that GE are characterized by twofold (Ca^{2+}) to even fivefold (K^+) higher annual macronutrient return via litter fall than CS (Table S10), especially due to leaves and reproductive parts (Table S5). This annual return of base cations likely feedbacks to the forest floor and topsoil acid neutralizing capacity. Additionally, throughfall fluxes directly provide base cations to the available macronutrient pool of the forest floor [93,110] and hence buffering decreasing pH values. Between 2001 and 2018, the mean annual base cation input via throughfall and stemflow was slightly higher at GE (30.1 ± 5.3 kg ha^{-1}) than at CS (24.0 ± 4.2 kg ha^{-1}) and the mean throughfall fluxes are dominated by K^+ and Ca^{2+} (Figure 3).

In contrast to the forest floor, the mineral soil of GE and CS acidified during the last 19 years by 0.2 to 0.5 pH units [36]. Weathering processes of the soil matrix in the main rooting zone of GE and CS contribute to a higher acid neutralizing capacity in the mineral soil and the predicted annual base cation flux (Table S10) is almost similar between the forest types with slightly higher amounts at GE. The base cation release is dominated by K^+ (Table S10). Nevertheless, we found slightly lower total base cation stocks in mineral soil of GE (Table S6; Figure S8), matching the result of a stronger mineral soil acidification. Independently, Großherr [48] found that the exchangeable base cation mineral soil pool in a depth of 15–28 cm, is also smaller at the plot GE ICP (234.0 kg ha^{-1}) than at CS ICP (334.4 kg ha^{-1}). This is especially true for Ca^{2+} and K^+.

4.3.2. The Role of Seepage Water and Nitrate

Nitrate is considered as a critical ion for the process of base cation depletion (counter ion for Ca^{2+}, Mg^{2+}, and K^+ in soil solution) through leaching from the mineral soil via seepage water [13]. Nevertheless, our results suggest an existing, but minor risk for base cation leaching losses associated with NO_3^- at GE and CS. Unfortunately, macronutrient seepage fluxes were not determined at the investigated monitoring sites in the past and we estimate the role of NO_3^- for mineral soil acidification with the use of different predictors for N seepage fluxes at GE and CS in our study.

The depth curve of the NO_3^- seepage water concentration cascades downwards for GE with highest concentrations in the forest floor (Figure 4), indicating that most of the NO_3^- from mineral soil solution is consumed by European beech. N is required in large amounts by higher plants as a constituent of proteins, nucleic acids, coenzymes or secondary metabolites [111] and it is taken up by tree roots as NO_3^- as one major inorganic N source [11,111,112]. The NO_3^- seepage water concentration is up to sixfold higher at CS, increases below 50 cm depth and it seems, that at CS—where throughfall N_t fluxes are higher—not all of the deposited NO_3^- is consumed by Norway spruce and Scots pine. The N uptake rate into tree coarse wood and bark is only half the size of GE (Table S10). The remaining NO_3^- may reduce soil base cation stocks, but nevertheless the base cation concentration in seepage water, show a negative correlation with NO_3^- concentration (Figures S2–S5). Other indicators for NO_3^- leaching in European forests confirm a lower risk for nitrate seepage water losses at CS. The forest floor C:N ratio can function as such an indicator for coniferous stands [113,114]. The mean forest floor C:N ratio at the exploration sites is between 26–36 (Oa-Oi) at CS and between 22–35 (Oa-Oi) at GE [36], pointing on an intermediate N status with a low or at most moderate risk for NO_3^- leaching [113]. Throughfall N_t fluxes can serve additionally predictive, using a relationship relating N leaching fluxes to throughfall N fluxes, which provide good overall predictions of seepage N fluxes for European forest sites [115]. However, throughfall fluxes are a weak predictor at sites with low leaching fluxes [115] but throughfall N_t fluxes at GE and at CS indicate possible seepage N_t fluxes of 2 kg ha^{-1} yr^{-1} (GE) respectively 4 kg ha^{-1} yr^{-1} (CS). As annual leaching rates of 2–4 kg N ha^{-1} yr^{-1} (temperate deciduous forests) and 1–3 kg N ha^{-1} yr^{-1} (intensive coniferous plantations) are set as a threshold for acceptable leaching of inorganic N [56], the estimated seepage N_t fluxes at CS might be sometimes above the critical limits and may be an indicator for nutrient imbalances caused by nitrate leaching. Nevertheless, an assignment of the investigated GE and CS to the N accumulation type with growing N stocks or at least unchanged N stocks (quasi-steady-state Type) in forest floor and mineral soil [116] seems plausible based on the predicted N input–output balance (Table S10) and the estimated seepage N fluxes [114,115,117]. A synthesis of the predicted N status classification [116] and the N saturation stage [94] supply evidence that GE and CS are not N saturated hence being of minor risk for N-induced distinct herbaceous layer biodiversity losses, strong soil acidification, soil base cation losses, and soil water quality degradation [94]. Plausible N seepage water fluxes of < 4 kg ha^{-1} yr^{-1}, group in the lower range of N fluxes usual for Europe (1–30 kg ha^{-1} yr^{-1}) [114].

4.3.3. The Role of Seepage Water and Sulphate

Besides NO_3^-, SO_4^{2-} function as a counter ion for base cations in soil solution and can contribute to strong mineral soil base cation losses [10,20,21]. In some European and North American forested catchments, SO_4^{2-} accounts even for 68–100% of the acid leaching from forest soils [118]. Because atmospheric S deposition decreased strongly over time (Figure 3), the desorption of water-soluble SO_4^{2-} from Fe and Al sesquioxides seems to be responsible for the high SO_4^{2-} concentrations in mineral soil seepage water at GE and CS, with higher concentrations at CS (Figure 4). The desorption process in acidic mineral soil is probably stimulated through dissolved organic matter seepage fluxes [22,119]. Organic anions and dissolved organic carbon (DOC) compete with SO_4^{2-} for adsorption sites [120] at acidic soil pH with DOC adsorption being preferred over that of SO_4^{2-} [120]. This assumption is confirmed by lower DOC concentrations in mineral soil solutions at GE and CS at 50 cm and 100 cm depth, than at 20 cm depth, which may indicate DOC adsorption (GE/CS ICP, data not shown). In contrast to NO_3^-, base cation concentrations are significantly positive correlated with SO_4^{2-} concentrations in seepage water of most mineral soil depth increments, especially in those of the CS (Figures S2–S5), suggesting a remobilization of previously retained historic SO_4^{2-} which in turn causes base cation leaching losses. This process might decrease pH at the deeper mineral soil and delay a recovery from soil acidification. Nevertheless, a decrease in SO_4^{2-} concentrations is observable between 2011–2018 (Figure S1). Base cation concentrations in mineral soil solution at the GE and the CS also show a similar trend over time like SO_4^{2-}, with lower concentrations in the last decade (Figure S1).

4.3.4. The Role of the Tree Species: Base Pump Effect and Base Cation Depletion

Tree macronutrient uptake affect the chemical conditions and nutrient distribution in forest soils [11]. At sites under nutrient deficiency and acidic soil conditions, this process contributes to a further mineral soil acidification, as free acid is produced during the base cation uptake via the tree fine root system and the acid neutralizing capacity of the mineral soil is depleted [29,121]. This "bio-acidification" [30] is a main process for decreasing pH values in forest mineral soil [29] as long as the sequestrated cations are not returned to the topsoil by litter and wood decomposition [11] and mineral soil base cation stock losses are not balanced. At GE and CS, mineral soil acidification traces the main rooting zone typical for European beech, Norway spruce, and Scots pine and the strongest acidification occurred for GE, which were characterized by an up to threefold higher estimated annual uptake of base cations into tree stemwood and bark than CS (Table S10). Although our study did not directly determine the fate of base cations from mineral soil via root uptake and biomass formation to the forest floor, the base pump effect serves as one explanation for the observed soil acidification combined with a relatively low base cation pool restoration by rock weathering (Table S10). Recent input–output balances for Ca^{2+}, Mg^{2+}, and K^+ indicate, that the calculated budgets for GE and CS are positive (Table S10). Nevertheless, the base cation balance seems to be depth-dependent with growing stocks in forest floor and topsoil contrary to base cation losses from deeper mineral soil.

5. Conclusions

Our results offer no evidence for herb layer diversity losses and little evidence for distinct mineral soil base cation losses resulting from the moderate N bulk deposition at the investigated forest stands. The application of predictors for NO_3^- leaching showed, that neither GE nor CS are N saturated and that both only bear low risks for soil base cation (Ca^{2+}, Mg^{2+}, K^+) depletion or soil water quality degradation induced by N depositions. Nevertheless, processes like nitrification can contribute to a soil acidification and thus to a decrease in base cation saturation. An ongoing pressure from N deposition inputs on the forest soil at GE and CS can change the role of nitrogen compounds to become the main driver for forest soil acidification at the investigated stands.

In contrast to NO_3^-, SO_4^{2-} seems to be more important for the status of soil base cation pools at GE and CS, although S depositions decreased strongly over the last decades. The remobilization of previously specifically adsorbed SO_4^{2-} into soil solution causes base cation leaching losses at the investigated forest stands. This process might decrease the pH at deeper mineral soil and delay a recovery from soil acidification. Nevertheless, high SO_4^{2-} concentrations in mineral soil solution mainly occurred in the time period between 2001–2010, whereas a decrease in SO_4^{2-} and base cation concentrations is observable between 2011–2018. Based on this observation, base cation leaching losses from mineral soil as counter ions for SO_4^{2-} associated with an acid neutralizing capacity depletion and mineral soil acidification may be of lesser relevance for GE and CS, today. This points to tree growth and the uptake of Ca^{2+}, Mg^{2+}, and K^+ into tree biomass, likely before sulphate and nitrate leaching occur, as one main driver for recent mineral soil acidification under the current N deposition conditions at GE and CS. The "bio-acidification" traces the main rooting zone typical for European beech, Norway spruce and Scots pine at GE and CS. The base cation balance as well as soil acidification seems to be depth-dependent with growing base cation stocks and unchanged pH values over time in forest floor and topsoil contrary to base cation losses and decreasing pH values at deeper mineral soil. Due to the natural process of soil acidification and tree macronutrient uptake, it is questionable whether the investigated soils will recover in the long-term if the assimilated cations are not released to the topsoil by litter decomposition, even if the input of acidifying air pollutants is reduced.

The plantation of Green Eyes provides positive effects on the forest floor quality. The high rate of nutrient return via litter fall and the intensified turnover processes at GE may contribute to vital and stable forest ecosystems due to the activated nutrient cycling in forest floor and topsoil. Nevertheless, our results suggest that the bio-acidification of the mineral soil at 20–40 cm depth is higher at GE, which can affect future ecosystem dynamics and processes. In a wider sense and to predict nutrient imbalances (especially for Ca^{2+}, Mg^{2+}, and K^+), we recommend to forest managers that fine and coarse woody debris should remain in the forest and accumulate to function as a temporary reservoir or even a base cation source during its decomposition and as a hotspot for biodiversity at the forest sites. It remains a field of further research to determine the extent to which the effects of GE impinge on the surrounding CS and in which ways GE can be a part of future, vital forests in Germany and Central Europe.

Supplementary Materials: The following are available online at https://www.mdpi.com/article/10.3390/f12050573/s1, **Figure S1.** Macronutrient concentration (NO_3^-, SO_4^{2-}, Ca^{2+}, Mg^{2+}, K^+; $\mu mol_c\ L^{-1}$) in seepage water of forest floor (A, B), as well as of mineral soil at 20 cm (C, D), 50 cm (E, F) and 100 cm (G, H) depth at the plot-pair GE/CS ICP. **Figure S2.** Correlations between element concentrations (NO_3^-, SO_4^{2-}, K^+, Ca^{2+}, Mg^{2+}), pH and electrical conductivity in seepage water of forest floor at the plot-pair GE/CS ICP. **Figure S3.** Correlations between element concentrations (NO_3^-, SO_4^{2-}, K^+, Ca^{2+}, Mg^{2+}), pH and electrical conductivity in seepage water of mineral soil (20 cm depth) at the plot-pair GE/CS ICP. **Figure S4.** Correlations between element concentrations (NO_3^-, SO_4^{2-}, K^+, Ca^{2+}, Mg^{2+}), pH and electrical conductivity in seepage water of mineral soil (50 cm depth) at the plot-pair GE/CS ICP. **Figure S5.** Correlations between element concentrations (NO_3^-, SO_4^{2-}, K^+, Ca^{2+}, Mg^{2+}), pH and electrical conductivity in seepage water of mineral soil (100 cm depth) at the plot-pair GE/CS ICP. **Figure S6.** Composition of total annual litter fall (dry matter, %) at GE (A) and CS (B) as well as analyzed litter (79.8% at GE and 76.1% at CS of the total annual litter fall). **Figure S7.** Dry matter and macronutrient fluxes in annual litter fall for Green Eyes and coniferous sites. **Figure S8.** Dry matter stocks (in forest floor) and macronutrient stocks (in forest floor and mineral soil) for Green Eyes and coniferous sites. **Figure S9.** Fractional annual loss (turnover rate) and residence time in years of the forest floor organic material and of macronutrients stored in forest floor ($\sum$ Oi, Oe, Oa) for Green Eyes and coniferous sites. A steady state condition was assumed. **Table S1.** Macronutrient concentration of litter fall fractions. **Table S2.** Vegetation data used for the prediction of the herbaceous aboveground biomass and macronutrient stocks with the estimation model PhytoCalc 1.4 at GE and CS. **Table S3.** Summary statistics and ANOVA results of the linear mixed-effects models. **Table S4.** Soil data used for the prediction of the annual base cation release (Ca^{2+}, Mg^{2+}, K^+; kg ha^{-1}) from weathering processes in the main rooting zone of the soil matrix at GE and CS according to the soil type-texture approximation for non-calcareous soil profiles given in CLRTAP (2017). **Table S5.** Litter fractions and their share in the macronutrient return within the analyzed litter at GE and CS. **Table S6.** Macronutrient stocks (in above-ground-tree-biomass, herbaceous ground vegetation, forest floor, and mineral soil) for Green Eyes and coniferous sites. **Table S7.** Macronutrient stocks in mineral soil ($\sum$ 0–10 cm, 10–20 cm, 20–40 cm) and divided into single depth increments for Green Eyes and coniferous sites. **Table S8.** Fractional annual loss (turnover rate) and residence time in years of the forest floor organic matter (OM) and of macronutrients stored in forest floor ($\sum$ Oi, Oe, Oa) for Green Eyes and coniferous sites. A steady state condition was assumed. **Table S9.** Forest floor organic matter (OM) stocks (dry matter) at Green Eyes and coniferous sites. **Table S10.** Macronutrient fluxes (deposition, litter fall, predicted weathering in soil matrix) and predicted macronutrient budgets for Green Eyes and coniferous sites.

Author Contributions: Conceptualization, F.A., A.T., and B.M.; methodology, F.A., B.M., and I.C.; investigation, F.A. and M.A.; formal analysis, F.A. and A.T.; visualization, F.A. and M.A.; validation, F.A., A.T., and M.B.-R.; writing—original draft preparation, F.A.; writing—review and editing, B.M., A.T., M.B.-R., M.A., and I.C.; resources, B.M.; supervision, B.M.; funding acquisition, B.M. All authors have read and agreed to the published version of the manuscript.

Funding: The research of F.A. was supported by the scholarship "Landesgraduiertenstipendium 2018" (Friedrich-Schiller-University Jena, State of Thuringia, Germany) as well as budgetary resources of the Department of Soil Science of the FSU Jena. A.T. and M.A. were supported by budgetary resources of the Department of Soil Science of the FSU Jena.

Data Availability Statement: The data presented in this study are available in the Supplementary Materials as a R-project. Forests data of the Level II Intensive Monitoring Site (International Co-operative Programme on Assessment and Monitoring of Air Pollution Effects on Forests-ICP Forests) is available for use upon request: Bundesministerium für Ernährung, Landwirtschaft und Verbraucherschutz, National Focal Centre (NFC), E-Mail: 535@bmel.bund.de.

Acknowledgments: The authors are thankful to Bernhard Zeiss from the Thuringian forest office Jena-Holzland (Germany) for field work permits. We thank also Nico Frischbier and Sven Merten from the forest research institution "FFK Gotha" for providing location-, regional climate-, and air pollutant deposition data. We thank the two anonymous reviewers for their valuable comments and notes which clearly improved the manuscript quality.

Conflicts of Interest: The authors declare no conflict of interest.

References

1. Engardt, M.; Simpson, D.; Schwikowski, M.; Granat, L. Deposition of sulphur and nitrogen in Europe 1900–2050. Model calculations and comparison to historical observations. *Tellus B Chem. Phys. Meteorol.* **2017**, *69*. [CrossRef]
2. Kopáček, J.; Veselý, J. Sulfur and nitrogen emissions in the Czech Republic and Slovakia from 1850 till 2000. *Atmos. Environ.* **2005**, *39*, 2179–2188. [CrossRef]
3. Binkley, D.; Högberg, P. Tamm Review: Revisiting the influence of nitrogen deposition on Swedish forests. *For. Ecol. Manag.* **2016**, *368*, 222–239. [CrossRef]
4. Puhe, J.; Ulrich, B. *Global Climate Change and Human Impacts on Forest Ecosystems*; Springer: Berlin/Heidelberg, Germany, 2001; Volume 143.
5. Bonner, M.T.L.; Castro, D.; Schneider, A.N.; Sundström, G.; Hurry, V.; Street, N.R.; Näsholm, T. Why does nitrogen addition to forest soils inhibit decomposition? *Soil Biol. Biochem.* **2019**, *137*. [CrossRef]
6. Dirnböck, T.; Grandin, U.; Bernhardt-Romermann, M.; Beudert, B.; Canullo, R.; Forsius, M.; Grabner, M.T.; Holmberg, M.; Kleemola, S.; Lundin, L.; et al. Forest floor vegetation response to nitrogen deposition in Europe. *Glob Chang. Biol.* **2014**, *20*, 429–440. [CrossRef] [PubMed]
7. Gilliam, F.S. The Ecological Significance of the Herbaceous Layer in Temperate Forest Ecosystems. *BioScience* **2007**, *57*, 845–858. [CrossRef]
8. Knapp, S. The Link Between Diversity, Ecosystem Functions, and Ecosystem Services. In *Atlas of Ecosystem Services Drivers, Risks, and Societal Responses*; Schröter, M., Bonn, A., Klotz, S., Seppelt, R., Baessler, C., Eds.; Springer International Publishing: Cham, Switzerland, 2019; pp. 13–15. [CrossRef]
9. De Vries, W.; Dobbertin, M.H.; Solberg, S.; van Dobben, H.F.; Schaub, M. Impacts of acid deposition, ozone exposure and weather conditions on forest ecosystems in Europe: An overview. *Plant Soil* **2014**, *380*, 1–45. [CrossRef]
10. Wellbrock, N.; Bolte, A. *Status and Dynamics of Forests in Germany-Results of the National Forest Monitoring*; Springer Open: Berlin/Heidelberg, Germany, 2019; Volume 237, p. 388.
11. Binkley, D.; Fisher, R.F. *Ecology and Management of Forest Soils*, 5th ed.; Wiley-Blackwell: Hoboken, NJ, USA, 2019. [CrossRef]
12. De Vries, W.; Schulte-Uebbing, L. Impacts of Nitrogen Deposition on Forest Ecosystem Services and Biodiversity. In *Atlas of Ecosystem Services: Drivers, Risks, and Societal Responses*; Schröter, M., Bonn, A., Klotz, S., Seppelt, R., Baessler, C., Eds.; Springer International Publishing: Cham, Switzerland, 2019; pp. 183–189. [CrossRef]
13. Reuss, J.; Johnson, D. Effect of Soil Processes on the Acidification of Water by Acid Deposition. *J. Environ. Qual.* **1985**, *14*, 26–31. [CrossRef]
14. Rothe, A.; Kreutzer, K.; Küchenhoff, H. Influence of tree species composition on soil and soil solution properties in two mixed spruce-beech stands with contrasting history in Southern Germany. *Plant Soil* **2002**, *240*, 47–56. [CrossRef]

15. Fagerli, H.; Tsyro, S.; Jonson, J.E.; Nyíri, A.; Simpson, D.; Wind, P.; Benedictow, A.; Klein, H.; Mu, Q.; Rolstad Denby, B.; et al. *EMEP-Status Report 1/2020-Transboundary Particulate Matter, Photo-Oxidants, Acidifying and Eutrophying Components*; Norwegian Meteorological Institute: Oslo, Norway, 2020; p. 270.

16. Köchy, M.; Bråkenhielm, S. Separation of effects of moderate N deposition from natural change in ground vegetation of forests and bogs. *For. Ecol. Manag.* **2008**, *255*, 1654–1663. [CrossRef]

17. Nilsson, J.; Grennfelt, P. *Critical Loads for Sulphur and Nitrogen. Report from a Workshop Held at Skokloster, Sweden 19-24 March, 1988*; Nordic Council of Ministers; Nordisk Ministerraad: Koebenhavn, DK, USA, 1988; p. 418.

18. Lorenz, M.; Nagel, H.D.; Granke, O.; Kraft, P. Critical loads and their exceedances at intensive forest monitoring sites in Europe. *Environ. Pollut.* **2008**, *155*, 426–435. [CrossRef] [PubMed]

19. Walker, T.A.B.; McMahon, R.C.; Hepburn, A.; Ferrier, R. Sulphate dynamics of podzols from paired impacted and pristine catchments. *Sci. Total Environ.* **1990**, *92*, 235–247. [CrossRef]

20. Berger, T.W.; Turtscher, S.; Berger, P.; Lindebner, L. A slight recovery of soils from Acid Rain over the last three decades is not reflected in the macro nutrition of beech (Fagus sylvatica) at 97 forest stands of the Vienna Woods. *Environ. Pollut.* **2016**, *216*, 624–635. [CrossRef]

21. Růžek, M.; Myška, O.; Kučera, J.; Oulehle, F. Input-Output Budgets of Nutrients in Adjacent Norway Spruce and European Beech Monocultures Recovering from Acidification. *Forests* **2019**, *10*, 68. [CrossRef]

22. Näthe, K.; Levia, D.F.; Tischer, A.; Potthast, K.; Michalzik, B. Spatiotemporal variation of aluminium and micro- and macronutrients in the soil solution of a coniferous forest after low-intensity prescribed surface fires. *Int. J. Wildland Fire* **2018**, *27*, 471–489. [CrossRef]

23. Iwald, J. *Acidification of Swedish Forest Soils-Evaluation of Data from the Swedish Forest Soil Inventory*; Licentiate Swedish University of Agricultural Sciences: Uppsala, Sweden, 2016.

24. De Schrijver, A.; Staelens, J.; Wuyts, K.; Van Hoydonck, G.; Janssen, N.; Mertens, J.; Gielis, L.; Geudens, G.; Augusto, L.; Verheyen, K. Effect of vegetation type on throughfall deposition and seepage flux. *Environ. Pollut.* **2008**, *153*, 295–303. [CrossRef]

25. De Schrijver, A.; Geudens, G.; Augusto, L.; Staelens, J.; Mertens, J.; Wuyts, K.; Gielis, L.; Verheyen, K. The Effect of Forest Type on Throughfall Deposition and Seepage Flux: A Review. *Oecologia* **2007**, *153*, 663–674. [CrossRef]

26. Lorenz, M.; Clarke, N.; Paoletti, E.; Bytnerowicz, A.; Grulke, N.; Lukina, N.; Sase, H.; Staelens, J. Air Pollution Impacts on Forests in Changing Climate. In *Forest and Society–Responding to Global Drivers of Change*; IUFRO World Series; Mery, G.E.A., Ed.; International Union of Forest Research Organizations: Vienna, Austria, 2010; pp. 55–74.

27. Binkley, D.; Giardina, C. Why do tree species affect soils? *The Warp and Woof of tree soil interactions. Biogeochemistry* **1998**, *42*, 89–106. [CrossRef]

28. Wittich, W. *Die Heutigen Grundlagen der Holzartenwahl. Dargestellt am Beispiel des Nordwestdeutschen Waldgebietes*, 2nd ed.; M. & H. Schaper: Hannover, Germany, 1948.

29. Hallbäcken, L. Long Term Changes of Base Cation Pools in Soil and Biomass in a Beech and a Spruce Forest of Southern Sweden. *Z. Pflanz. Bodenk.* **1992**, *155*, 51–60. [CrossRef]

30. Westman, C.J.; Jauhiainen, S. Soil acidity in 1970 and 1989 in a coniferous forest in southwest Finland. *Can. J. Soil Sci.* **1998**, *78*, 477–479. [CrossRef]

31. Johnson, D.W.; Cresser, M.S.; Nilsson, S.I.; Turner, J.; Ulrich, B.; Binkley, D.; Cole, D.W. Soil changes in forest ecosystems: Evidence for and probable causes. *Proc. R. Soc. Edinb. Sect. B. Biol. Sci.* **1991**, *97*, 81–116. [CrossRef]

32. Binkley, D. The influence of Tree Species on Forest Soils: Processes and Patterns. *Agron. Soc. N. Z. Spec. Publ.* **1995**, *7*, 1–33.

33. Ammer, S.; Weber, K.; Abs, C.; Ammer, C.; Prietzel, J. Factors influencing the distribution and abundance of earthworm communities in pure and converted Scots pine stands. *Appl. Soil Ecol.* **2006**, *33*, 10–21. [CrossRef]

34. Bens, O.; Buczko, U.; Sieber, S.; Hüttl, R.F. Spatial variability of O layer thickness and humus forms under different pine beech–forest transformation stages in NE Germany. *J. Plant Nutr. Soil Sci.* **2006**, *169*, 5–15. [CrossRef]

35. Zederer, D.P.; Talkner, U.; Spohn, M.; Joergensen, R.G. Microbial biomass phosphorus and C/N/P stoichiometry in forest floor and A horizons as affected by tree species. *Soil Biol. Biochem.* **2017**, *111*, 166–175. [CrossRef]

36. Achilles, F.; Tischer, A.; Bernhardt-Römermann, M.; Heinze, M.; Reinhardt, F.; Makeschin, F.; Michalzik, B. European beech leads to more bioactive humus forms but stronger mineral soil acidification as Norway spruce and Scots pine–Results of a repeated site assessment after 63 and 82 years of forest conversion in Central Germany. *For. Ecol. Manag.* **2021**, *483*, 118769. [CrossRef]

37. Aherne, J.; Braun, S.; Tresch, S.; Augustin, S. Soil solution in Swiss forest stands: A 20 year's time series. *PLoS ONE* **2020**, *15*. [CrossRef]

38. Johnson, J.; Graf Pannatier, E.; Carnicelli, S.; Cecchini, G.; Clarke, N.; Cools, N.; Hansen, K.; Meesenburg, H.; Nieminen, T.M.; Pihl-Karlsson, G.; et al. The response of soil solution chemistry in European forests to decreasing acid deposition. *Glob Chang. Biol.* **2018**, *24*, 3603–3619. [CrossRef]

39. Ammer, C.; Bickel, E.; Kölling, C. Converting Norway spruce stands with beech-A review of arguments and techniques. *Austrian J. For. Sci.* **2008**, *125*, 3–26.

40. DWD. German Climate Atlas-Thuringia. Available online: https://www.dwd.de/EN/ourservices/germanclimateatlas/germanclimateatlas.html (accessed on 6 March 2020).

41. Kirilenko, A.; Sedjo, R.A. Climate change impacts on forestry. *Proc. Natl. Acad. Sci. USA* **2007**, *104*, 19697–19702. [CrossRef]

42. Augusto, L.; Ranger, J.; Binkley, D.; Rothe, A. Impact of several common tree species of European temperate forests on soil fertility. *Ann. Sci.* **2002**, *59*, 233–253. [CrossRef]
43. Graser, H. *Die Bewirtschaftung des Erzgebirgischen Fichtenwaldes. Erster Band*; Hofbuchhandlung H.Burdach: Dresden, Germany, 1928; Volume 1, p. 98.
44. Bärthel, E. *Holzart und Betriebsart im Gebiete des Ehemaligen Herzogtums Sachsen-Altenburg-Eine Darstellung und Untersuchung der Ursachen der Mannigfachen Verschiebungen der Holzartenverteilung und des Wechsels der Betriebsart*; Ludwigs-Universität Gießen: Gießen, Germany, 1926.
45. Hoppe, W.; Seidel, G. *Geologie von Thüringen: [Bezirke Erfurt, Gera, Suhl]*; Haack: Gotha, Germany, 1974.
46. Burse, K.; Neumann, T. Die Forstlichen Wuchsbezirke Thüringens. *Mitt. 37/2019 ThüringenForst* **2019**, *37*, 129–139.
47. Veckenstedt, T. *Vergleich von Buchenhorsten Mit Standortsgleichen Kiefern-Fichten-Mischbeständen Hinsichtlich der Wuchsleistungen, des Bodenzustandes und der Durchwurzelung im Forstamt Hummelshain*; Thüringer Fachhochschule für Forstwirtschaft: Schwarzburg, Germany, 1999.
48. Großherr, M. *Auswirkungen des Waldumbaus auf Bodenchemische Eigenschaften (KAK und Acidität) in Bodenprofilen des Thüringer Forstreviers Leuchtenburg*; Friedrich-Schiller-Universität Jena: Jena, Germany, 2011.
49. Schober, R. *Ertragstafeln Wichtiger Baumarten bei Verschiedener Durchforstung*; J.D. Sauerländer's Verlag: Frankfurt a. Main, Germany 1995.
50. Glatthorn, J.; Beckschäfer, P. Standardizing the protocol for hemispherical photographs: Accuracy assessment of binarization algorithms. *PLoS ONE* **2014**, *9*, e111924. [CrossRef] [PubMed]
51. Beckschäfer, P. *Hemispherical_2.0–Batch Processing Hemispherical and Canopy Photographs with Imagej–User Manual*; Georg-August-Universität Göttingen: Göttingen, Germany, 2015; pp. 1–6.
52. Bolte, A. *Biomasse-und Elementvorräte der Bodenvegetation auf Flächen des Forstlichen Umweltmonitorings in Rheinland-Pfalz (BZE, Level II)*; Universität Göttingen: Göttingen: Göttingen, Germany, 2006; p. 81.
53. Block, J.; Dieler, J.; Gauer, J.; Greve, M.; Moshammer, R.; Schuck, J.; Schwappacher, V.; Wunn, U. *Gewährleistung der Nachhaltigkeit der Nährstoffversorgung bei der Holz- und Biomassenutzung im Rheinland-Pfälzischen Wald*; Forschungsanstalt für Waldökologie und Forstwirtschaft Rheinland-Pflaz (FAWF): Trippstadt, Germany, 2016.
54. ICP Forests. *Manual on Methods and Criteria for Harmonized Sampling, Assessment, Monitoring and Analysis of the Effects of Air Pollution on Forests*; Thünen Institute of Forest Ecosystems: Eberswalde, Germany, 2016.
55. Schlutow, A.; Ritter, A. *Aktualisierung der Berechnung von Ökologischen Belastungsgrenzen (Critical Loads) und Ihren Überschreitungen für 14 Thüringer Wald-und Hauptmessstationen*; FFK Gotha-Forstliches Forschungs-und Kompetenzzentrum Gotha: Gotha, Germany, 2018; pp. 1–54.
56. CLRTAP. Latest Update of Guidance on Mapping Concentrations Levels and Deposition Levels-Manual on Methodologies and Criteria for Modelling and Mapping Critical Loads and Levels and Air Pollution Effects, Risks and Trends. 2017. Available online: https://www.umweltbundesamt.de/en/manual-for-modelling-mapping-critical-loads-levels (accessed on 10 March 2021).
57. Braun-Blanquet, J. *Pflanzensoziologie-Grundzüge der Vegetationskunde*; Springer: Berlin/Heidelberg, Germany, 1928; p. 330.
58. Bolte, A. *Abschätzung von Trockensubstanz-, Kohlenstoff-und Nährelementvorräten der Waldbodenflora-Verfahren, Anwendung und Schätztafeln (Assessment of Dry Weight and Storage of Carbon and Nutrients in Forest Ground Vegetation in the North-Eastern German Lowlands–Method, Application and Classification Tables)*; Tharandt: Saxony, Germany, 1999; Volume 7, p. 320.
59. Zanella, A.; Ponge, J.-F.; Jabiol, B.; Sartori, G.; Kolb, E.; Le Bayon, R.-C.; Gobat, J.-M.; Aubert, M.; De Waal, R.; Van Delft, B.; et al. Humusica 1, article 5: Terrestrial humus systems and forms-Keys of classification of humus systems and forms. *Appl. Soil Ecol.* **2018**, *122*, 75–86. [CrossRef]
60. AK-Standortskartierung. *Forstliche Standortsaufnahme: Begriffe, Definitionen, Einteilung, Kennzeichnungen, Erläuterungen*; IHW: Eiching, Germany, 2016; p. 400.
61. Ad-Hoc-AG-Boden. *Bodenkundliche Kartieranleitung (KA5)*; Schweizerbart: Hannover/Stuttgart, Germany, 2005; p. 438.
62. WRB, I.W.G. *World Reference Base for Soil Resources 2014, Update 2015 International Soil Classification System for Naming Soils and Creating Legends for Soil Maps*; FAO: Rome, Italy, 2015.
63. Kunze, A. *Bodenkundlicher Vergleich von Buchen-Laubholzinseln (sog. "Grünen Augen") mit gleichaltrigen Nadelholzreinbeständen im Wuchsbezirk „Ostthüringer Buntsandstein"*; Technische Universität Dresden: Dresden, Germany, 2000.
64. Hågvar, S. From Litter to Humus in a Norwegian Spruce Forest: Long-Term Studies on the Decomposition of Needles and Cones. *Forests* **2016**, *7*, 186. [CrossRef]
65. Lim, M.T.; Cousens, J.E. The Internal Transfer of Nutrients in a Scots Pine Stand 2. The Patterns of Transfer and the Effects of Nitrogen Availability. *Forestry* **1986**, *59*, 17–27.
66. Lyr, H.; Fiedler, H.J.; Tranquillini, W. *Physiologie und Ökologie der Gehölze*; Gustav Fischer Verlag Jena: Stuttgart, Germany, 1992; p. 620.
67. Ellenberg, H.; Weber, H.; Düll, R.; Wirth, V.; Werner, W.; Paulißen, D. *Zeigerwerte von Pflanzen in Mitteleuropa*; Goltze: Göttingen, Germany, 2001; Volume 18, p. 262.
68. Mölder, A.; Bernhardt-Römermann, M.; Schmidt, W. Herb-layer diversity in deciduous forests: Raised by tree richness or beaten by beech? *For. Ecol. Manag.* **2008**, *256*, 272–281. [CrossRef]

69. Schulze, I.M.; Bolte, A.; Schmidt, W.; Eichhorn, J. Phytomass, Litter and Net Primary Production of Herbaceous Layer. In *Functioning and Management of European Beech Ecosystems. vol. 208 ed.*; Brumme, R., Khanna, P.K., Eds.; Springer: Berlin/Heidelberg, Germany, 2009.

70. Heinrichs, S.; Bernhardt-Römermann, M.; Schmidt, W. The estimation of aboveground biomass and nutrient pools of understorey plants in closed Norway spruce forests and on clearcuts. *Eur. J. For. Res.* **2010**, *129*, 613–624. [CrossRef]

71. De Vries, W.; Hol, A.; Tjalma, S.; Voogd, J.C. *Literatuurstudie Naar Voorraden en Verblijftijden van Elementen in Bosecosystemen (Amounts and Turnover Rates of Elements in Forest Ecosystems: A Literature Study)*; Winand Staring Center: Wageningen, The Netherlands, 1990.

72. Jacobsen, C.; Rademacher, P.; Meesenburg, H.; Meiwes, K.-J. *Gehalte Chemischer Elemente in Baumkronenkompartimenten. Literaturstudie und Datensammlung*; Forschungszentrum Waldökosysteme der Universität Göttingen: Göttingen, Germany, 2003.

73. R Core Team. *R: A Language and Environment for Statistical Computing*; R Foundation for Statistical Computing: Vienna, Austria, 2020.

74. Emerson, J.W.; Green, W.A.; Schloerke, B.; Crowley, J.; Cook, D.; Hofmann, H.; Wickham, H. The Generalized Pairs Plot. *J. Comput. Graph. Stat.* **2012**, *22*, 79–91. [CrossRef]

75. Pinheiro, J.; Bates, D.M. *Mixed-Effects Models in Sand S-PLUS*; Springer: New York, NY, USA, 2000.

76. Pinheiro, J.; Bates, D.; DebRoy, S.; Sarkar, D. Nlme: Linear and Nonlinear Mixed Effects Models. R package Version 3.1-142; R Core Team. 2019. Available online: https://svn.r-project.org/R-packages/trunk/nlme/ (accessed on 10 March 2021).

77. Lenth, R.V. Least-Squares Means: The R Package lsmeans. *J. Stat. Softw.* **2016**, *69*, 1–33. [CrossRef]

78. Hlavac, M. Stargazer: Well-Formatted Regression and Summary Statistics Tables. R package version 5.2.2.; R Core Team. 2018. Available online: https://cran.r-project.org/web/packages/stargazer/stargazer.pdf (accessed on 10 March 2021).

79. Zuur, A.F.; Ieno, E.N.; Walker, N.; Saveliev, A.A.; Smith, G.M. *Mixed Effects Models and Extensions in Ecology with R.*; Springer: New York, NY, USA, 2009. [CrossRef]

80. Akaike, H. A new look at the statistical model identification. *IEEE Trans. Autom. Control.* **1974**, *19*, 716–723. [CrossRef]

81. Bartoń, K. MuMIn: Multi-Model Inference, R Package Version 0.12.0; R Core Team. 2009. Available online: https://cran.r-project.org/web/packages/MuMIn/index.html (accessed on 10 March 2021).

82. Sverdrup, H. *The Kinetics of base Cation Release due to Chemical Weathering*; Lund University Press: Lund, Sweden, 1990.

83. Olson, J.S. Energy storage and the balance of producers and decomposers in ecological systems. *Ecology* **1963**, *44*, 322–331. [CrossRef]

84. Hobbie, S.E.; Reich, P.B.; Oleksyn, J.; Ogdahl, M.; Zytkowiak, R.; Hale, C.; Karolewski, P. Tree species effects on decomposition and forest floor dynamics in a common garden. *Ecology* **2006**, *87*, 2288–2297. [CrossRef]

85. Vesterdal, L.; Schmidt, I.K.; Callesen, I.; Nilsson, L.O.; Gundersen, P. Carbon and nitrogen in forest floor and mineral soil under six common European tree species. *For. Ecol. Manag.* **2008**, *255*, 35–48. [CrossRef]

86. Huhn, J.; Reinhardt, F. Vegetationsaufnahmen im Forstamt Hummelshain. 2000; unpublished.

87. Cole, D.W.; Rapp, M. Elemental cycling in forest ecosystems. In *Dynamic Properties of Forest Ecosystems*; Reiche, D.E., Ed.; Cambridge University Press: Cambidge, UK, 1981; pp. 341–409.

88. Likens, G.E.; Drsicoll, C.T.; Buso, D.C.; Mitchell, M.J.; Lovett, G.M.; Bailey, S.W.; Siccama, T.G.; Reiners, W.A.; Alewell, C. The biogeochemistry of sulfur at Hubbard Brook. *Biogeochemistry* **2002**, *60*, 235–316. [CrossRef]

89. Hansen, K.; Vesterdal, L.; Schmidt, I.K.; Gundersen, P.; Sevel, L.; Bastrup-Birk, A.; Pedersen, L.B.; Bille-Hansen, J. Litterfall and nutrient return in five tree species in a common garden experiment. *For. Ecol. Manag.* **2009**, *257*, 2133–2144. [CrossRef]

90. Soil Science Division Staff. *Soil Survey Manual*; Government Printing Office: Washington, DC, USA, 2017; Volume 18.

91. Sutton, M.A.; Howard, C.M.; Erisman, J.W.; Billen, G.; Bleeker, A.; Grennfelt, P.; van Grinsven, H.; Grizzetti, B. *The European Nitrogen Assessment: Sources, Effects and Policy Perspectives*; Sutton, M.A., Howard, C.M., Erisman, J.W., Billen, G., Bleeker, A., Grennfelt, P., van Grinsven, H., Grizzetti, B., Eds.; Cambridge University Press: Cambridge, UK, 2011. [CrossRef]

92. Berger, T.W.; Untersteiner, H.; Schume, H.; Jost, G. Throughfall fluxes in a secondary spruce (Picea abies), a beech (Fagus sylvatica) and a mixed spruce–beech stand. *For. Ecol. Manag.* **2008**, *255*, 605–618. [CrossRef]

93. Levia, D.F.; Carlyle-Moses, D.; Iida, S.; Michalzik, B.; Nanko, K.; Tischer, A. *Forest-Water Interactions*; Springer: Berlin/Heidelberg, Germany, 2020; Volume 240, p. 624.

94. Schmitz, A.; Sanders, T.G.M.; Bolte, A.; Bussotti, F.; Dirnbock, T.; Johnson, J.; Penuelas, J.; Pollastrini, M.; Prescher, A.K.; Sardans, J.; et al. Responses of forest ecosystems in Europe to decreasing nitrogen deposition. *Env. Pollut.* **2019**, *244*, 980–994. [CrossRef]

95. Forsmark, B.; Nordin, A.; Maaroufi, N.I.; Lundmark, T.; Gundale, M.J. Low and High Nitrogen Deposition Rates in Northern Coniferous Forests Have Different Impacts on Aboveground Litter Production, Soil Respiration, and Soil Carbon Stocks. *Ecosystems* **2020**, *23*, 1423–1436. [CrossRef]

96. Bobbink, R.; Hicks, K.; Galloway, J.; Spranger, T.; Alkemade, R.; Ashmore, M.; Bustamante, M.; Cinderby, S.; Davidson, E.; Dentener, F.; et al. Global assessment of nitrogen deposition effects on terrestrial plant diversity: A synthesis. *Ecol. Appl.* **2010**, *20*, 30–59. [CrossRef]

97. Elliott, K.J.; Vose, J.M.; Knoepp, J.D.; Clinton, B.D.; Kloeppel, B.D. Functional Role of the Herbaceous Layer in Eastern Deciduous Forest Ecosystems. *Ecosystems* **2014**, *18*, 221–236. [CrossRef]

98. Welch, N.T.; Belmont, J.M.; Randolph, J.C. Summer Ground Layer Biomass and Nutrient Contribution to Above-ground Litter in an Indiana Temperate Deciduous Forest. *Am. Midl. Nat.* **2007**, *157*, 11–26. [CrossRef]

99. Bolte, A.; Lambertz, B.; Steinmeyer, A.; Kallweit, R.; Meesenburg, H. Zur Funktion der Bodenvegetation im Nährstoffhaushalt von Wäldern-Studien auf Dauerbeobachtungsflächen des EU Level II-Programms in Norddeutschland. *Forstarchiv* **2004**, *75*, 207–220.

100. Muys, B.; Granval, P. Earthworms as bio-indicators of forest site quality. *Soil Biol. Biochem.* **1997**, *29*, 323–328. [CrossRef]

101. Heinze, M.; Tomczyk, S.; Nicke, A. Vergleich von Rot-Buche (Fagus sylvatica L.) in sogennanten Grünen Augen mit benachbarten standortsgleichen Fichtenbeständen (Picea abies [L.] KARST.) des Thüringer Vogtlandes bezüglich Eigenschaften und Durchwurzelung des Bodens sowie Baumwachstum. *Forstw. Cbl.* **2001**, *120*, 139–153. [CrossRef]

102. Berg, B.; McClaugherty, C. *Plant Litter. Decomposition, Humus Formation, Carbon Sequestration*; Springer: Berlin/Heidelberg, Germany, 2014. [CrossRef]

103. Wang, L.; D'Odorico, P. Decomposition and Mineralization. In *Encyclopedia of Ecology*, 2nd ed.; Fath, B., Ed. Elsevier: Oxford, UK, 2013; pp. 280–285. [CrossRef]

104. Nihlgård, B. Pedological Influence of Spruce Planted on Former Beech Forest Soils in Scania, South Sweden. *Oikos* **1971**, *22*, 302–314. [CrossRef]

105. Albers, D.; Migge, S.; Schaefer, M.; Scheu, S. Decomposition of beech leaves (Fagus sylvatica) and spruce needles (Picea abies) in pure and mixed stands of beech and spruce. *Soil Biol. Biochem.* **2004**, *36*, 155–164. [CrossRef]

106. Prescott, C.E. Litter decomposition: What controls it and how can we alter it to sequester more carbon in forest soils? *Biogeochemistry* **2010**, *101*, 133–149. [CrossRef]

107. Knorr, M.; Frey, S.D.; Curtis, P.S. Nitrogen Additions and Litter Decomposition: A Meta-Analysis. *Ecology* **2005**, *86*, 3252–3257. [CrossRef]

108. Carreiro, M.M.; Sinsabaugh, R.L.; Repert, D.A.; Parkhurst, D.F. Microbial Enzyme Shifts Explain Litter Decay Responses to Simulated Nitrogen Deposition. *Ecology* **2000**, *81*, 2359–2365. [CrossRef]

109. Janssens, I.A.; Dieleman, W.; Luyssaert, S.; Subke, J.A.; Reichstein, M.; Ceulemans, R.; Ciais, P.; Dolman, A.J.; Grace, J.; Matteucci, G.; et al. Reduction of forest soil respiration in response to nitrogen deposition. *Nat. Geosci.* **2010**, *3*, 315–322. [CrossRef]

110. Salehi, M.; Zahedi Amiri, G.; Attarod, P.; Salehi, A.; Brunner, I.; Schleppi, P.; Thimonier, A. Seasonal variations of throughfall chemistry in pure and mixed stands of Oriental beech (Fagus orientalis Lipsky) in Hyrcanian forests (Iran). *Ann. For. Sci.* **2015**, *73*, 371–380. [CrossRef]

111. Marschner, H. *Mineral Nutrition of Higher Plants*, 3rd ed.; Academic Press: San Diego, CA, USA, 2012; p. 672.

112. Schulze, E.D. *Carbon and Nitrogen Cycling in European Forest Ecosystems*; Springer: Berlin/Heidelberg, Germany, 2000; Volume 142, p. 506.

113. Gundersen, P.; Callesen, I.; De Vries, W. Nitrate leaching in forest ecosystems is related to forest floor C/N ratios. *Environ. Pollut.* **1998**, *102*, 403–407. [CrossRef]

114. MacDonald, J.A.; Dise, N.B.; Matzner, E.; Armbruster, M.; Gundersen, P.; Forsius, M. Nitrogen input together with ecosystem nitrogen enrichment predict nitrate leaching from European forests. *Glob. Chang. Biol.* **2002**, *8*, 1028–1033. [CrossRef]

115. Van der Salm, C.; De Vries, W.; Reinds, G.J.; Dise, N.B. N leaching across European forests: Derivation and validation of empirical relationships using data from intensive monitoring plots. *For. Ecol. Manag.* **2007**, *238*, 81–91. [CrossRef]

116. Brumme, R.; Khanna, P.K. *Functioning and Management of European Beech Ecosystems*; Springer: Berlin/Heidelberg, Germany, 2009; Volume 208, p. 501.

117. Gundersen, P.; Emmet, B.A.; Kjonaas, O.J.; Koopmans, C.J.; Tietema, A. Impact of nitrogen deposition on nitrogen cycling in forests: A synthesis of NITREX data. *For. Ecol. Manag.* **1998**, *101*, 37–55. [CrossRef]

118. Watmough, S.A.; Aherne, J.; Alewell, C.; Arp, P.; Bailey, S.; Clair, T.; Dillon, P.; Duchesne, L.; Eimers, C.; Fernandez, I.; et al. Sulphate, nitrogen and base cation budgets at 21 forested catchments in Canada, the United States and Europe. *Env. Monit Assess* **2005**, *109*, 1–36. [CrossRef]

119. Gmach, M.R.; Cherubin, M.; Kaiser, K.; Cerri, C.E. Processes that influence dissolved organic matter in the soil: A review. *Sci. Agric.* **2020**, *77*, e20180164. [CrossRef]

120. Vance, G.F.; David, M.B. Dissolved organic carbon and sulfate sorption by spodosol mineral horizons. *Soil Sci.* **1992**, *154*, 136–144. [CrossRef]

121. Berger, T.W.; Swoboda, S.; Prohaska, T.; Glatzel, G. The role of calcium uptake from deep soils for spruce (Picea abies) and beech (Fagus sylvatica). *For. Ecol. Manag.* **2006**, *229*, 234–246. [CrossRef]

Article

Nitrogen Fertilization Increases Windstorm Damage in an Aggrading Forest

Christopher A. Walter [1,*], Zachariah K. Fowler [1], Mary Beth Adams [2], Mark B. Burnham [3], Brenden E. McNeil [4] and William T. Peterjohn [1]

[1] Department of Biology, West Virginia University, Morgantown, WV 26505, USA; zfowler@mail.wvu.edu (Z.K.F.); william.peterjohn@mail.wvu.edu (W.T.P.)

[2] Northern Research Station, USDA Forest Service (Emeritus), Morgantown, WV 26505, USA; Marybeth.adams@mail.wvu.edu

[3] Center for Advanced Bioenergy and Bioproducts Innovation, University of Illinois, Urbana, IL 61801, USA; mburnham@illinois.edu

[4] Department of Geology and Geography, West Virginia University, Morgantown, WV 26506, USA; brenden.mcneil@mail.wvu.edu

* Correspondence: cwalte12@mail.wvu.edu; Tel.: +1-304-293-5201

Abstract: Storms are the most significant disturbance events in temperate forests. Forests impacted by nitrogen deposition may face more severe storm damage as changes in soil and wood chemistry impact tree growth allocation, wood strength, and species composition. To examine these potential effects of nitrogen deposition, we measured tree damage from a windstorm in an aggrading forest that is part of a nitrogen fertilization experiment. We discovered that within the nitrogen fertilization treatment area there was significantly more basal area and stems damaged when compared to the reference treatment, and the nitrogen fertilization treatment had more snapped and severely damaged trees. Additionally, the effect of treatment and amount of damage to trees was different depending on tree species. If our results are indicative of the large and globally-distributed regions of temperate forests impacted by nitrogen deposition, then the increased windstorm disturbance risk posed by climate change could be more significant due to the effects of nitrogen deposition.

Keywords: disturbance; storm; damage; N deposition; temperate forest; acidification; acid rain; N saturation; windthrow

Citation: Walter, C.A.; Fowler, Z.K.; Adams, M.B.; Burnham, M.B.; McNeil, B.E.; Peterjohn, W.T. Nitrogen Fertilization Increases Windstorm Damage in an Aggrading Forest. *Forests* **2021**, *12*, 443. https://doi.org/10.3390/f12040443

Academic Editor: Leonor Calvo Galvan

Received: 19 March 2021
Accepted: 3 April 2021
Published: 6 April 2021

Publisher's Note: MDPI stays neutral with regard to jurisdictional claims in published maps and institutional affiliations.

1. Introduction

Storms are the most significant source of disturbance in temperate forests [1]. Storm disturbance causes ecological disruption through the creation of canopy openings that alter tree mortality and recruitment [2,3]. Additionally, the litter from storm disturbance can accelerate nutrient cycling [4] and cause forests to become major sources of atmospheric carbon [5]. Such ecological effects from storm disturbances can alter forest ecosystems for centuries [6], and even millennia [7]. At shorter time scales, storm disturbance also causes economic disruption. Costs to forest managers include reduced timber value from either a real or perceived decrease in wood quality, restoration of access to forest sites, reduced market value due to sudden supply increase in the storm-affected area, and either long-term storage to wait for the market to rebound or long-distance transport of timber from outside of the storm-affected area [8]. At longer time scales, storm disturbances may weaken trees and make them more susceptible to secondary disturbances (e.g., insects, disease, drought, fire, additional storms) [9,10], further increasing the ecological disturbance and lowering economic value.

The risk of forest storm disturbance has risen significantly with climate change [11,12]. Warming oceans have increased the frequency of large-magnitude storms [13,14]. Additionally, post-landfall storm power has increased, and it is more likely that storms will track further inland [15]. The economic effects of these extreme events can be staggering [16]. Hurricane Michael, which hit the Gulf Coast of the United States in October 2018, is an example of the threat storm disturbance poses to the timber industry. The hurricane damaged 1,133,120 ha of forest in the state of Florida, costing the timber industry $1.3 billion dollars—more than 5% of the state's annual timber industry value [17]. The neighboring states of Georgia and Alabama were also impacted by Hurricane Michael. Georgia had 121,406 ha of forest damaged, costing $763 million in timber revenue, and Alabama had 17,159 ha damaged, costing $32.7 million dollars to the timber industry [17]. The major destructive force in these severe storms is wind, and its impact on forests may depend on the presence of additional disturbance factors [18].

Adding to the increased wind disturbance risk posed by the growing frequency of severe storms, other global change factors may interact with storms to create even greater levels of disturbance [19–21]. Nitrogen (N) deposition is one such global change factor that causes physiological changes in trees that may leave them more susceptible to storm damage. N deposition has been shown to reduce belowground biomass allocation in forests [22,23], leaving trees with disproportionately more stems and leaves relative to roots. Trees fertilized with N also grow faster, creating more reaction wood (called compression wood in gymnosperms and tension wood in angiosperms) [24,25]. Reaction wood grows in reaction to tensile stress and has a greater concentration of cellulose than lignin [25–27]. It is structurally weaker and more brittle than regular grain wood, making it more susceptible to failure under force [28]. N deposition may also change the species composition of forests and select for trees that are more or less susceptible to storm damage [5,29]. Finally, N deposition could also lead to more susceptible trees by enhancing herbivory [30,31] and increasing pathogenic infection risk [32].

Evidence that N deposition may contribute to greater storm damage in forests is sparse. In a scrub mangrove forest, N fertilization led to a greater loss in leaf area index and slower recovery after two hurricanes damaged the forest in the same month [33]. The authors of the study suggested N fertilization led to increased aboveground biomass, which may have left mangrove trees more susceptible to intense wind [33]. In a tropical forest dominated by *Metrosideros polymorpha*, N and phosphorus (P) fertilization (P was more limiting to growth than was N at this site) caused greater damage from a hurricane [34]. However, the fertilized plots recovered from storm damage faster [34].

Perhaps the only evidence of an interaction between N deposition and storm damage in temperate forests is from Walter et al. 2019 [29]. In this study, the authors found damage from a severe snow storm (with relatively low wind speed) was lower in an N fertilized watershed and as N increased across a native N availability gradient [29]. Contrary to the increased damage expected from the horizontal forces of windstorms, the authors found that the more vertical vegetation structure linked to N enrichment provided an improved ability to shed heavy snowfall. Nevertheless, damage from snowstorms is not as common as from windstorms, and to our knowledge, no study within temperate deciduous forests has examined how N enrichment affects storm disturbance from wind.

Considering the lack of available studies documenting the interactive effects of N and storms in temperate forests, our objective was to examine whether enhanced N deposition led to greater windstorm damage in an aggrading temperate forest. To meet our objective, we surveyed trees in a long-term N fertilization experiment that experienced a severe windstorm. We hypothesized that the physiological and community composition effects of N fertilization would leave stands more susceptible to windstorm damage, causing damage to be greatest in areas undergoing N fertilization treatments.

2. Materials and Methods

2.1. Study Site

We assessed wind damage in the Long-Term Soil Productivity Experiment (LTSP) [35] at Fernow Experimental Forest (herein Fernow) in West Virginia, USA (39.0563° N, 79.6979° W). Fernow is a 1902 ha research forest that primarily contains Appalachian mixed mesophytic forest and receives 1430 mm yr^{-1} of precipitation on average. The LTSP experiment within Fernow was designed to test the effects of chronic N deposition in temperate forests. It consists of four treatments:

1. Whole-tree harvest (reference)
2. Whole-tree harvest + nitrogen (+N)
3. Whole-tree harvest + nitrogen + lime (+N+L)
4. An unharvested treatment (unused in this study)

All aboveground biomass was removed from the whole-tree harvested treatments in the winter of 1996–1997. The +N and +N+L treatments have been fertilized with ammonium sulfate at a rate of 35 kg N ha^{-1} yr^{-1} since 1997. The +N+L treatments have been additionally fertilized with dolomitic lime at a rate of 22.5 kg calcium ha^{-1} yr^{-1} since 1997. Each treatment is replicated four times and one replicate of each treatment was randomly assigned to a cell in each row of a 16-plot matrix at the beginning of the experiment. Each plot is 0.37 ha—containing a 0.2 ha quadrilateral area in the center where measurements are made, and a 7.6 m treated buffer that surrounds the treatment area on all sides. In addition to N fertilization, the LTSP also received an estimated 5.2 kg N ha^{-1} in 2011 from wet deposition of nitrate and ammonium (see Figure S1 for a time series of atmospheric wet deposition of N).

The inclusion of both +N and +N+L treatments in the LTSP experiment allows investigators to partition the dual ecosystem effects of N deposition: acidification and fertilization. The +N treatment both fertilizes and acidifies the soil, while the additional application of lime in the +N+L treatment attempts to reduce the acidification effect by adding base cations to soil. Therefore, ecosystem effects appearing in +N treatments and not +N+L treatments are more likely an effect of acidification. Effects appearing in both +N and +N+L treatments are attributed more to the fertilization effect of N.

In late December 2009, the LTSP experiment was damaged by a severe windstorm. There was very limited damage elsewhere in Fernow and extreme winds were not detected at nearby weather stations. Nevertheless, weather stations did indicate that thunderstorms crossed the region during late December, suggesting that the wind damage was from a severe downdraft or "microburst" that caused severe localized damage to just the mountainside containing the LTSP experiment. While we were unable to collect an instrumental estimate of wind speed, the damage to trees was extensive and the location of the LTSP experiment provided a globally rare opportunity to measure forest stand damage across a long-term N fertilization experiment.

Prior to the storm, the reference, +N, and +N+L treatments all had closed forest canopies. Fortunately, Fowler et al. 2014 completed a survey estimating forest composition in these treatments in summer 2009, several months prior to the storm [36]. Their survey determined that *Prunus pensylvanica*, *Liriodendron tulipifera*, *Betula lenta*, and *Prunus serotina* comprised 89% of the total basal area and 79% of the total stems when averaged across all treatments (see Table S1 in Supplementary Material for complete species composition). The pre-storm mean tree basal area across treatments was 19.7 m^3 ha^{-1} and the mean number of stems across treatments was 5691 ha^{-1}. Since species differ in physiology and response to N and lime fertilization, we formed species-specific hypotheses that were informed by published growth and allocation responses to N and lime (Table S2 in Supplementary Material for the initial species-specific damage hypotheses).

2.2. Damage Measurement

We measured damage from the December 2009 windstorm in summer 2011 in reference, +N, and +N+L treatment areas of the LTSP experiment. Based on regular visual assessments in the time between the storm and our survey, we did not note any other intervening disturbances that would have confounded ascribing damage to the 2009 event. To maximize our sampling effort, we took advantage of the composition survey by Fowler et al. 2014 [36] that had been completed just before the storm. Fowler et al. 2014 [36] had measured the basal area and identified all trees to species in six randomly selected subplots within each treatment plot. We returned to the same subplots in 2011 and identified each tree to species, measured the basal area, and visually categorized the damage of all damaged trees within the subplots. We categorized the damage of each tree in two stages: damage type and damage severity. Damage type consisted of the following categories:

- Bent—a tree that was bent by wind or a collision with another tree
- Tipup—a tree that had been uprooted and was leaning
- Snap—a tree that was snapped either at the main stem or at a major branch

Damage severity consisted of the following categories, based on angle relative to a vertical tree:

- Moderate—0 to 22.5 degrees from vertical
- Significant—23 to 45 degrees from vertical
- Extensive—45.5 to 67.5 degrees from vertical
- Prostrate—greater than 67.5 degrees from vertical

Damage severity could be assigned to any damage type, making all combinations of damage type and severity possible.

The 2009 composition survey provided the total number of stems and basal area, by species, in each of the six subplots within each treatment plot in the LTSP experiment. The 2011 damage measurement provided the same data, except it only included damaged trees, and also included the damage type and severity of each damaged tree. We combined of the 2009 composition survey dataset and the 2011 damage measurement dataset to calculate the response variables used in statistical analysis.

2.3. Statistical Analysis

To examine differences in damage across the LTSP experimental treatments, we used four main response variables: percentage of basal area damaged, percentage of stems damaged, damage type, and damage severity. To calculate both the percentage of basal area damaged and the percentage of stems damaged, we treated damage as a binary variable. Trees were either damaged (regardless of their damage type or severity) or undamaged. In each subplot, we divided the sum of basal area damaged (2011 data) by the total basal area (2009 data) and multiplied by 100 to calculate the percentage of basal area damaged. Similarly, in each subplot, we divided the number of damaged trees (2011 data) by the total number of trees (2009 data) and multiplied by 100 to calculate the percentage of stems damaged. To determine whether the percentage basal area or stems damaged differed by species, we calculated both variables in each subplot for *P. pensylvanica*, *L. tulipifera*, *B. lenta*, and *P. serotina*.

We calculated the damage type and severity as the percentage of total damaged stems in each category. In each subplot, we divided the number of trees in a damage category by the total number of damaged trees in the subplot and multiplied by 100. We used each category label as an individual response, which resulted in three response variables for damage type (bent, tipup, and snap) and four response variables for damage severity (moderate, significant, extensive, and prostrate). Using both of these categorical variables allowed us to determine whether treatments led to differences in either damage type or severity among the percentage of stems damaged.

To test whether the experimental treatments led to differences in the damage in any of the response variables examined, we used bootstrap hypothesis tests, following an approach from Walter et al. 2017 [37]. The following scheme outlines the process:

1. Sample 24 subplots in each treatment, with replacement
2. Calculate the mean of each response variable from the sample
   - Mean percentage of basal area damaged (five means: all species, *P. pensylvanica*, *L. tulipifera*, *B. lenta*, and *P. serotina*)
   - Mean percentage of stems damaged (five means: all species, *P. pensylvanica*, *L. tulipifera*, *B. lenta*, and *P. serotina*)

- Mean percentage of stems damaged by damage type (three means: bent, tipup, and snap)
- Mean percentage of stems damaged by damage severity (four means: moderate, significant, extensive, and prostrate)

3. Repeat 50,000 times
4. Calculate p-value as the fraction of times the bootstrap mean was different from the empirical mean of the comparison treatment

The *p*-value, testing whether the bootstrap mean of a treatment was greater than the empirical mean of a comparison treatment, was calculated as:

$$\frac{\sum_{i=1}^{n} [\overline{y}_{boot_i} < \hat{\mu}]}{n} \tag{1}$$

where n indicates the number of bootstrap sample iterations (50,000) and brackets indicate a Boolean true/false return of the inequality asking whether the bootstrap sample mean ($\overline{y}_{boot}$) of a treatment was less than the empirical mean of a comparison treatment ($\hat{\mu}$). Summation across all iterations (i) from 1 to n is simply the count of the number of times the inequality was true. In the cases where we tested whether the bootstrap sample mean of treatments was less than the empirical mean of a comparison treatment, the *p*-value was calculated as:

$$\frac{\sum_{i=1}^{n} [\overline{y}_{boot_i} > \hat{\mu}]}{n} \tag{2}$$

Testing differences in 17 response variables across three treatments resulted in 51 hypothesis tests. To estimate family-wise error in multiple comparisons, we calculated the false discovery rate among statistically significant ($p < 0.05$) test results using the Benjamini–Hochberg procedure [38]. This procedure estimates the fraction of false positive tests expected and has three major advantages over other methods (e.g., Bonferroni procedures): (i) its results are independent of test quantity, (ii) there are no additional arbitrary significance thresholds to select, and (iii) calculating and reporting false positive rate does not increase Type-II error probability. Out of 51 tests, we found 21 to be significant at $p < 0.05$, with a false positive rate of 11.7 percent (See Table S3 in Supplementary Material for a complete list of *p*-values and false positive rates). At this rate, we can expect about two or three (2.46) of the 21 positive tests to be false positives.

The LTSP experiment uses a randomized block design to examine the effect of controlled additions of fertilizer, but the blocking factor (slope position) [31] was not chosen to account for an unanticipated disturbance event. As a result, we used a bootstrap approach to maximize the statistical power of our datasets in a way that should improve our ability to assess the impact of a chance event on an established, long-term experiment. More specifically, bootstrap hypothesis testing, applied across sublots within the LTSP plots, allowed us to amplify the signal of damage differences among experimental treatments. We performed all analyses in R Statistical Software Version 4.0.2 for Mac [39], and all data and the R code used to perform the analysis are available on GitHub at https://github.com/waltscience/forest-wind-damage and archived on Zenodo [40].

3. Results

3.1. Basal Area and Stems Damaged

The 2009 windstorm damaged 29.7% of basal area, when averaged across all treatments and tree species. Across all species, the percentage of basal area damaged was greater in both the +N+L (33.4% damage; $p = 0.022$) and +N (33.2% damage; $p = 0.045$) treatments areas when compared to that of the reference treatment area (22.7% damage; Figure 1a). Treatments also affected the percentage of basal area damaged in individual species. The percentage of *P. serotina* basal area damaged was greater in both the +N+L (17.3% damage; $p < 0.001$) and +N (10.5% damage; $p = 0.031$) treatment areas, relative to that of the reference treatment area (1.2% damage; Figure 1a). *L. tulipifera* was damaged more in the +N treatment area (27.7% damage) than it was in the +N+L treatment area (10.9% damage; $p = 0.016$), while no significant difference was detected between either treatment and the reference treatment (17.6% damage).

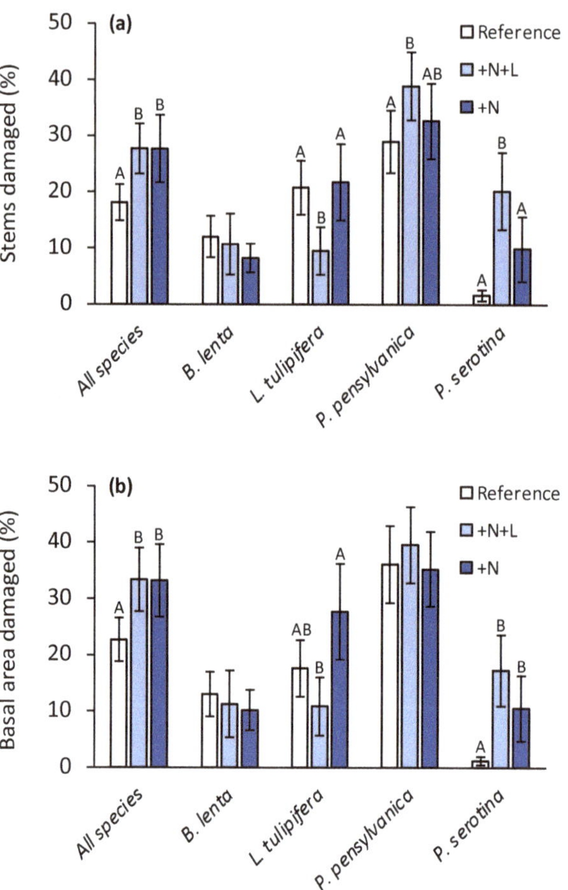

Figure 1. Tree damage from a windstorm across the reference, nitrogen fertilized (+N), and nitrogen and lime fertilized (+N+L) treatment areas in the Long-Term Soil Productivity Experiment (LTSP) Experiment: (**a**) Percentage of total basal area damaged across all tree species and within the four dominant tree species; (**b**) Percentage of total stems damaged across all tree species and within the four dominant tree species. Bars and error bars represent the empirical means and one standard error, respectively. Differences in capital letters within a species group denote significant differences ($p < 0.05$) between treatments, determined by bootstrap hypothesis tests.

The windstorm damaged 24.5% of individual tree stems, when averaged across all treatments and tree species. Across species, the percentage of stems damaged by the storm was greater in both +N+L (27.7% damage; $p = 0.012$) and +N treatment areas (27.7% damage; $p = 0.047$), when compared to that of the reference treatment area (18.2% damage; Figure 1b). Treatments also affected the percentage of stems damaged among individual tree species. The percentage of *P. pensylvanica* stems damaged was greater in the +N+L treatment area (73.1% damage) than it was in the reference treatment area (46.5% damage; $p = 0.045$), but the percentage of damaged stems in the +N treatment area (62.5% damage) was not statistically different from that in the +N+L treatment area or the reference treatment area. Similarly, the magnitude of damage among *P. serotina* stems was higher in the +N+L (9.5% damage) treatment area than it was in the +N (4.3% damage; *p* = 0.047) or reference treatment areas (8.1% damage; *p* < 0.001). Conversely, *L. tulipifera* stems were damaged less frequently in the +N+L treatment area (9.5% damage) than they were in either the +N (21.8% damage; *p* = 0.025) or reference treatment areas (20.8% damage; *p* = 0.008).

3.2. Damage Type and Severity

The majority of stem damage from the windstorm was from bent trees (55.9%), followed by tipup trees (14.9%), and snapped trees (10.9%), when averaged across all treatments. Comparing treatments, the percentage of snapped trees was greater in the +N treatment area (16.7%; *p* = 0.010) compared to that of the reference treatment area (6.1%). The percentage of snapped trees was also greater in the +N+L treatment area (10.2%; *p* = 0.021) than it was in the reference treatment area (Figure 2a).

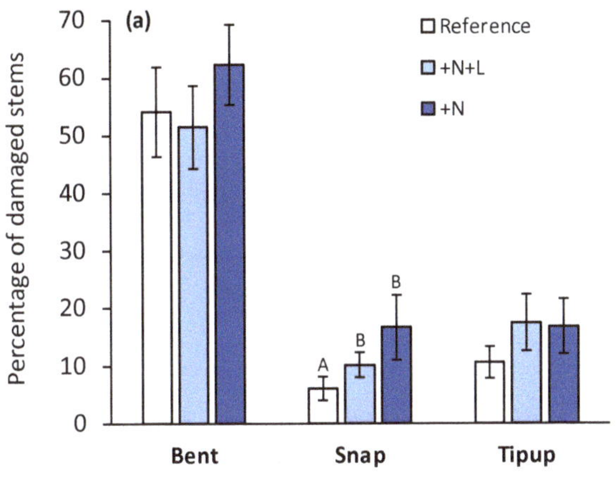

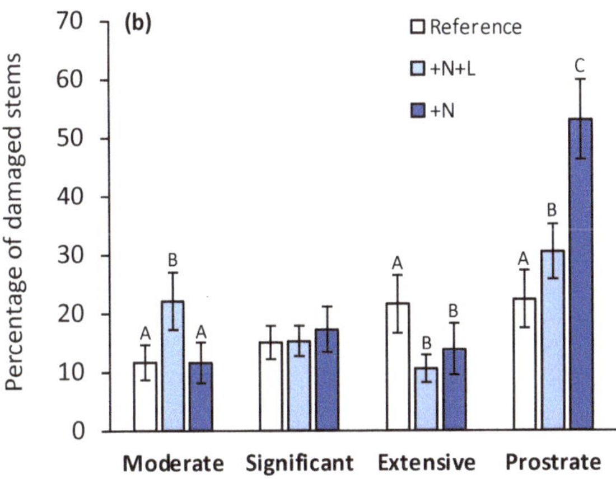

Figure 2. Stem damage from a windstorm across the reference, nitrogen fertilized (+N), and nitrogen and lime fertilized (+N+L) treatments in the LTSP Experiment: (**a**) Percentage of tree stems damaged by damage type; (**b**) Percentage of tree stems damaged by damage severity. Bars and error bars represent the empirical means and one standard error, respectively. Differences in capital letters within either a type or severity group denote significant differences (*p* < 0.05) between treatments, determined by bootstrap hypothesis tests.

When averaged across all treatments, the windstorm resulted in a greater frequency of severely damaged (prostrate) stems (35.3%) than the frequency of stems found in less severe damage categories: significant damage (15.9%), extensive damage (15.3%), and moderate damage (15.1%). Comparing treatments, there was a greater percentage of trees that experienced moderate damage in the +N+L treatment area (22.2%) when compared to that of either +N (11.6%; $p = 0.004$) or reference (11.7%; $p = 0.007$) treatment areas (Figure 2b). For trees that experienced extensive damage, there was a greater percentage found in the reference treatment area (21.6%) compared to that from either the +N+L (10.5%; $p < 0.001$) or +N (13.9%; $p = 0.048$) treatment areas. The percentage of damaged stems found in the prostrate category differed across all three treatment areas, with the +N treatment area having a greater frequency of prostrate trees (53%) than that of both the +N+L treatment area (30.5%; $p < 0.001$) and reference treatment area (22.4%; $p < 0.001$). The percentage of prostrate trees in +N+L treatment area was also greater than the percentage in the reference treatment area ($p = 0.038$).

4. Discussion

After 12 years of ammonium sulfate additions, we found that the extent of damage from a localized windstorm was enhanced by fertilizer additions. While numerous studies have reported forest disturbance levels from storms [41], few studies measured storm impacts in stands with experimental N treatments [29], and none, that we are aware of, attempted to partition the dual ecosystem effects of N deposition between fertilization (+N+L) and the combination of both fertilization and acidification (+N). Here, the fertilization-only treatment (+N+L) led to significantly more windstorm damage. However, we also found the combination of fertilization and acidification (+N) had a strikingly similar effect on stand damage, suggesting that greater stand damage resulted more from the fertilization effect of N, rather than acidification.

Across all tree species, both N fertilization and N and lime fertilization led to a 10.6% increase in basal area damage and a 9.5% increase in the number of stems damaged. N fertilization can increase both basal area (diameter growth) and decrease stem density (thinning) [42]. Taken together, the strong effect observed in both basal area and stem damage is evidence that N fertilization made stands more susceptible to a windstorm, irrespective of any secondary N effects on diameter growth or thinning. We suspect that N additions made the stands in our study more susceptible by changing patterns of biomass allocation and/or reductions in wood strength.

Multiple forest fertilization studies have found that trees tend to allocate more biomass aboveground with N fertilization [22,23,37]. This is consistent with evidence from a separate experiment at Fernow, where N fertilization has made trees 2.5 m taller, on average [29], and caused a 25% increase in aboveground net primary productivity with no detectable change in belowground productivity [43]. Thus, we suspect N fertilization led to a similar change between above and belowground allocation in the LTSP plots, possibly making trees more susceptible to wind damage because they become more "top heavy" from biomass being disproportionately allocated aboveground. Furthermore, tree height has widely been reported as being positively correlated with storm damage [44–46].

Beyond the potential effects of shifts in biomass allocation, N-fertilized trees may also be more susceptible to damage by producing weaker wood. Indeed, faster growing trees may have decreased wood strength due to the production of more tension wood [47]. Tension wood is created on the upper side of a tree stem under tension, and is characterized by an abundance of cellulose [27]. Tension wood is generally weaker than wood growing under normal tensile conditions [28] and N fertilization has been shown to increase production of tension wood in *Populus* [25,26]. Thus, while an increase in aboveground allocation in N fertilized treatments may help explain the overall increase in damage in N fertilized treatments (Figure 1), an increase in tension wood formation in N fertilized treatments could have led to the greater percentage of snapped and prostrate trees in the +N treatment (Figure 2).

When we examined the effects for different species, we found differences in damage between treatments for *P. serotina*, *P. pensylvanica*, and *L. tulipifera*, but the patterns varied depending on how storm damage was assessed: percent basal area or percentage of stems damaged. *P. serotina* has been observed to be particularly susceptible to wind damage because it has a relatively shallow rooting depth and grows taller than its neighbor trees [48]. The greatest percentage of *P. serotina* stems was damaged in the +N+L treatment area, with the addition of base cations. This was surprising, considering *P. serotina* growth is diminished on sites with greater soil base cation concentrations [49], and experimental lime fertilization slows *P. serotina* growth and increases mortality [50]. In the case of *P. serotina*, increased aboveground allocation may not explain the observed increase in the percentage of stems damaged. Instead, it is plausible that the addition of calcium (Ca) led to decreased wood strength in the +N+L treatment. Calcium nutrition has been shown to impact wood quality in other hardwood species [51], as such, it may be possible that lime fertilization reduced wood quality in *P. serotina*. Similar to *P. serotina*, there was a greater percentage of *P. pensylvanica* stems damaged in the +N+L treatment area, when compared to those damaged the reference area. We hypothesize these results may be because *P. pensylvanica* is a congener of *P. serotina*, although less is known about how base cations affect *P. pensylvanica* growth and mortality because of its short lifespan and noncommercial status.

In contrast to the *Prunus* results, a smaller percentage of *L. tulipifera* stems were damaged in +N+L treatment area when compared to those damaged in both the reference and +N treatment areas. Additionally, a smaller percentage of *L. tulipifera* basal area was damaged in the +N+L treatment area when compared to that in the +N treatment area (Figure 1). This was surprising because N fertilization in a nearby experiment at Fernow led to diminished growth in *L. tulipifera* [52]. Because of this, we expected lower *L. tulipifera* damage in both +N and +N+L treatment areas based on our hypothesis that increased aboveground allocation would increase wind damage. However, the main difference in *L. tulipifera* damage was between +N and +N+L treatments, which suggests the addition of lime had an effect in addition to altering biomass allocation. The addition of N fertilizer decreases *L. tulipifera* wood density [53], potentially by reducing the availability of Ca and magnesium (Mg) in soil [52]. If the process by which N fertilization reduced wood density was through diminished availability of soil Ca and Mg, the addition of Ca (such as in the +N+L treatment) could potentially increase wood density under N fertilization. The increased wood density may lead to greater wood strength and reduced damage from wind, as observed in the +N+L treatment area.

The species-specific responses in this study add evidence that allocation is not solely responsible for the observed differences in wind damage due to N fertilization. *P. serotina* was a tree we expected would shift allocation aboveground in response to N addition and acidification (+N treatment) because of its documented preference for soils with lower Ca and Ma concentrations. In response to greater aboveground allocation, we hypothesized *P. serotina* trees growing in the +N treatment would be more susceptible to wind damage. Instead, the number of damaged *P. serotina* stems was greatest with the addition of both N and lime (+N+L treatment). In a contrasting example, we expected *L. tulipifera* would allocate less aboveground with N addition and acidification (+N treatment), considering its observed decline in growth in response to N addition. In this instance, we hypothesized *L. tulipifera* trees growing in the +N treatment would be damaged less by the windstorm. However, we discovered a greater percent basal area and percentage stem damage in the +N treatment area, and less damage in the +N+L treatment. These results suggest that growth and allocation alone likely cannot fully explain the differences in damage we observed among species. Instead, there are likely many physiological factors (including wood strength) affected by N fertilization that determine the extent of forest damage due to wind.

While our experiment lacked the power to fully disentangle the intricate species composition and physiological interaction effects brought on by N fertilization, those interactions will likely remain quite difficult to elucidate experimentally [54]. The relatively scant number of long-term forest fertilization experiments, combined with the variable nature of severe storms, makes studying the interaction of the two a rarity. However, greater insight may be gained by studying storm disturbance across historic N deposition gradients (such as in the eastern United States) with available regional to continental-scale data products and monitoring programs, including the U. S. Forest Service's Forest Inventory and Analysis program (FIA), the Smithsonian Institution's Forest Global Earth Observatory (ForestGEO), and the Battelle Memorial Institute's National Ecological Observation Network (NEON).

Our study showed that N fertilization significantly increased tree damage from a windstorm in an aggrading temperate forest, and damage differed among tree species and N fertilization treatments. As global climate change increases the frequency and landfall power of large-magnitude storms, forests are already at an increased risk of disturbance. Based on our findings, temperate forests receiving current or historically high N deposition may be at an even greater risk of storm damage from wind.

Supplementary Materials: The following are available online at https://www.mdpi.com/article/10.3390/f12040443/s1, Figure S1: Total inorganic nitrogen wet deposition at Fernow, Table S1: Tree species composition by percentage of stems and percentage of basal area across three treatments in the LTSP experiment, Table S2: Species-specific hypotheses for storm damage based on published traits and responses to nitrogen and lime in fertilization and gradient studies, Table S3: Results for 51 bootstrap hypothesis tests across species group, damage type, and damage severity as either percentage of stems or basal area of trees damaged from the 2009 windstorm.

Author Contributions: Each author made significant and unique contributions to this paper: Conceptualization, C.A.W., M.B.A., Z.K.F., M.B.B., B.E.M., and W.T.P.; formal analysis, C.A.W.; investigation, C.A.W., M.B.A., Z.K.F., M.B.B., B.E.M., and W.T.P.; data curation, C.A.W.; writing—original draft preparation, C.A.W.; writing—review and editing, C.A.W., M.B.A., Z.K.F., M.B.B., B.E.M., and W.T.P.; supervision, M.B.A., B.E.M., and W.T.P.; funding acquisition, M.B.A., B.E.M., and W.T.P. All authors have read and agreed to the published version of the manuscript.

Funding: This research was funded by The National Science Foundation's Long-Term Research in Ecosystem Biology program, grant numbers DEB-0417678 and DEB-1019522, the U. S. Department of Agriculture Forest Service Northern Research Station, and the West Virginia University Eberly College of Arts and Sciences.

Institutional Review Board Statement: Not applicable.

Informed Consent Statement: Not applicable.

Data Availability Statement: The data and R code used to perform the analysis in this study are available on GitHub at https://github.com/waltscience/forest-wind-damage and archived on Zenodo at https://doi.org/10.5281/zenodo.4487873.

Acknowledgments: We thank Doug Owens and Chris Cassidy for building and maintaining the LTSP experiment, Dara Erazo Lilian Hill, Ty Heimerl, Justin Lego, and Devon Raiff for their help with field work and data entry, two anonymous reviewers for their feedback on our manuscript, and the United States Forest Service for their foresight and support of long-term forest ecology research.

Conflicts of Interest: The authors declare no conflict of interest and the funders had no role in the design of the study; in the collection, analyses, or interpretation of data; in the writing of the manuscript, or in the decision to publish the results.

References

1. Fischer, A.; Marshall, P.; Camp, A. Disturbances in deciduous temperate forest ecosystems of the northern hemisphere: Their effects on both recent and future forest development. *Biodivers. Conserv.* **2013**, *22*, 1863–1893. [CrossRef]
2. Batista, W.B.; Platt, W.J. Tree population responses to hurricane disturbance: Syndromes in a south-eastern USA old-growth forest. *J. Ecol.* **2003**, *91*, 197–212. [CrossRef]
3. Canham, C.D.; Papaik, M.J.; Latty, E.F. Interspecific variation in susceptibility to windthrow as a function of tree size and storm severity for northern temperate tree species. *Can. J. For. Res.* **2001**, *31*, 1–10. [CrossRef]
4. Houlton, B.Z.; Driscoll, C.T.; Fahey, T.J.; Likens, G.E.; Groffman, P.M.; Bernhardt, E.S.; Buso, D.C. Nitrogen Dynamics in Ice Storm-Damaged Forest Ecosystems: Implications for Nitrogen Limitation Theory. *Ecosystems* **2003**, *6*, 431–443. [CrossRef]

5. Busing, R.T.; White, R.D.; Harmon, M.E.; White, P.S. Hurricane disturbance in a temperate deciduous forest: Patch dynamics, tree mortality, and coarse woody detritus. *Plant Ecol.* **2008**, *201*, 351–363. [CrossRef]

6. Foster, D.R.; Knight, D.H.; Franklin, J.F. Landscape Patterns and Legacies Resulting from Large, Infrequent Forest Disturbances. *Ecosystems* **1998**, *1*, 497–510. [CrossRef]

7. Mitchell, S.J. Wind as a natural disturbance agent in forests: A synthesis. *Forestry* **2012**, *86*, 147–157. [CrossRef]

8. Hanewinkel, M.; Peyron, J.L. *The Economic Impact of Storms*; The European Forest Institute: Joensuu, Finland, 2013.

9. Paine, R.T.; Tegner, M.J.; Johnson, E.A. Compounded Perturbations Yield Ecological Surprises. *Ecosystems* **1998**, *1*, 535–545. [CrossRef]

10. Bigler, C.; Kulakowski, D.; Veblen, T.T. Multiple disturbance interactions and drought influence fire severity in rocky mountain subalpine forests. *Ecology* **2005**, *86*, 3018–3029. [CrossRef]

11. Reyer, C.P.O.; Bathgate, S.; Blennow, K.; Borges, J.; Bugmann, H.; Delzon, S.; Faias, S.P.; Garcia-Gonzalo, J.; Gardiner, B.; Gonzalez-Olabarria, J.R.; et al. Are forest disturbances amplifying or canceling out climate change-induced productivity changes in European forests? *Environ. Res. Lett.* **2017**, *12*, 034027. [CrossRef]

12. Seidl, R.; Thom, D.; Kautz, M.; Martin-Benito, D.; Peltoniemi, M.; Vacchiano, G.; Wild, J.; Ascoli, D.; Petr, M.; Honkaniemi, M.P.J.; et al. Forest disturbances under climate change. *Nat. Clim. Chang.* **2017**, *7*, 395–402. [CrossRef]

13. Zhao, M.; Held, I.M. An Analysis of the Effect of Global Warming on the Intensity of Atlantic Hurricanes Using a GCM with Statistical Refinement. *J. Clim.* **2010**, *23*, 6382–6393. [CrossRef]

14. Emanuel, K. Increasing destructiveness of tropical cyclones over the past 30 years. *Nat. Cell Biol.* **2005**, *436*, 686–688. [CrossRef]

15. Li, L.; Chakraborty, P. Slower decay of landfalling hurricanes in a warming world. *Nat. Cell Biol.* **2020**, *587*, 230–234. [CrossRef]

16. Smith, A.B. *2010–2019: A Landmark Decade of U.S. Billion Dollar Weather and Climate Disasters*; National Oceanic and Atmospheric Administratio: Washington, DC, USA, 2020. Available online: https://www.climate.gov/news-features/blogs/beyond-data/20 10-2019-landmark-decade-us-billion-dollar-weather-and-climate (accessed on 1 April 2021).

17. Brody, A.R. *Weathering the Storm: Assessing the Agricultural Impact of Hurricane Michael*; The Southern Office of The Council of State Governments: Atlanta, GA, USA, 2019.

18. Rogers, P. *Disturbance Ecology and Forest Management: A Review of the Literature*; USDA Forest Service Intermountain Research Station: Odgen, UT, USA, 1996.

19. Bradford, J.B.; Fraver, S.; Milo, A.M.; D'Amato, A.W.; Palik, B.; Shinneman, D.J. Effects of multiple interacting disturbances and salvage logging on forest carbon stocks. *For. Ecol. Manag.* **2012**, *267*, 209–214. [CrossRef]

20. Hermosilla, T.; Wulder, M.A.; White, J.C.; Coops, N.C. Prevalence of multiple forest disturbances and impact on vegetation regrowth from interannual Landsat time series (1985–2015). *Remote Sens. Environ.* **2019**, *233*, 111403. [CrossRef]

21. Cobb, R.C.; Metz, M.R. Tree Diseases as a Cause and Consequence of Interacting Forest Disturbances. *Forestry* **2017**, *8*, 147. [CrossRef]

22. Stober, C.; George, E.; Persson, H. *Carbon and Nitrogen Cycling in European Forest Ecosystem*; Shulze, E.D., Ed.; Springer-Verlag: Berlin, Germany, 2000; pp. 99–121.

23. Li, W.; Jin, C.; Guan, D.; Wang, Q.; Wang, A.; Yuan, F.; Wu, J. The effects of simulated nitrogen deposition on plant root traits: A meta-analysis. *Soil Biol. Biochem.* **2015**, *82*, 112–118. [CrossRef]

24. Sastry, C.B.R. Some Effects of Fertilizer Application on Wood Properties of Douglas Fir (Pseudotsuga menziesii). Master's Thesis, University of British Columbia, Vancouver, BC, Canada, 1967.

25. Pitre, F.E.; Pollet, B.; Lafarguette, F.; Cooke, J.E.K.; Mackay, J.J.; Lapierre, C. Effects of Increased Nitrogen Supply on the Lignification of Poplar Wood. *J. Agric. Food Chem.* **2007**, *55*, 10306–10314. [CrossRef] [PubMed]

26. Pitre, F.E.; Lafarguette, F.; Boyle, B.; Pavy, N.; Caron, S.; Dallaire, N.; Poulin, P.-L.; Ouellet, M.; Morency, M.-J.; Wiebe, N.; et al. High nitrogen fertilization and stem leaning have overlapping effects on wood formation in poplar but invoke largely distinct molecular pathways. *Tree Physiol.* **2010**, *30*, 1273–1289. [CrossRef]

27. Bentum, A.L.K.; Cote, W.A.; Day, A.C.; Timell, T.E. Distribution of lignin in normal and tension wood. *Wood Sci. Technol.* **1969**, *3*, 218–231. [CrossRef]

28. Wimmer, R.; Johansson, M. *The Biology of Reaction Wood*; Gardiner, B., Barnett, J., Saranp ää, P., Gril, J., Eds.; Springer-Verlag: Berlin, Germany, 2014; pp. 225–248.

29. Walter, C.A.; Burnham, M.B.; Adams, M.B.; McNeil, B.E.; Deel, L.N.; Peterjohn, W.T. Nitrogen Availability Decreases the Severity of Snow Storm Damage in a Temperate Forest. *For. Sci.* **2020**, *66*, 58–65. [CrossRef]

30. Massad, T.J.; Dyer, L.A. A meta-analysis of the effects of global environmental change on plant-herbivore interactions. *Arthropod Plant Interact.* **2010**, *4*, 181–188. [CrossRef]

31. Li, F.; Dudley, T.L.; Chen, B.; Chang, X.; Liang, L.; Peng, S. Responses of tree and insect herbivores to elevated nitrogen inputs: A meta-analysis. *Acta Oecol. Int. J. Ecol.* **2016**, *77*, 160–167. [CrossRef]

32. Hesterberg, G.A.; Jurgensen, M.F. The relation of forest fertilization to disease incidence. *For. Chron.* **1972**, *48*, 92–96. [CrossRef]

33. Feller, I.C.; Dangremond, E.M.; Devlin, D.J.; Lovelock, C.E.; Proffitt, C.E.; Rodriguez, W. Nutrient enrichment intensifies hurricane impact in scrub mangrove ecosystems in the Indian River Lagoon, Florida, USA. *Ecology* **2015**, *96*, 2960–2972. [CrossRef] [PubMed]

34. Herbert, D.A.; Fownes, J.H.; Vitousek, P.M. Hurricane Damage to a Hawaiian Forest: Nutrient Supply Rate Affects Resistance and Resilience. *Ecology* **1999**, *80*, 908–920. [CrossRef]

35. Adams, M.B.; Burger, J.; Zelazny, L.; Baumgras, J. *Description of the Fork Mountain Long-Term Soil Productivity Study: Site Characterization*; USDA Forest Service Northern Research Station: Newtown Square, PA, USA, 2004.

36. Fowler, Z.K. The Effects of Accelerated Soil Acidification on Aggrading Temperate Deciduous Forests: The Fernow Experimental Forest Long Term Soil Productivity (LTSP) Study at 13 Years. Ph.D. Thesis, West Virginia University, Morgantown, WV, USA, 2014.

37. Walter, C.A.; Adams, M.B.; Gilliam, F.S.; Peterjohn, W.T. Non-random species loss in a forest herbaceous layer following nitrogen addition. *Ecology* **2017**, *98*, 2322–2332. [CrossRef]

38. Benjamini, Y.; Hochberg, Y. Controlling the False Discovery Rate—A Practical and Powerful Approach to Multiple Testing. *J. R. Stat. Soc. Ser. B Methodol.* **1995**, *57*, 289–300. [CrossRef]

39. R-Core-Team. *R: A Language and Environment for Statistical Computing*; R Foundation for Statistical Computing: Vienna, Austria, 2020.

40. Walter, C.A.; Fowler, Z.K.; Adams, M.B.; Burnham, M.B.; McNeil, B.E.; Peterjohn, W.T. Tree damage from a 2009 windstorm in a temperate forest nitrogen fertilization experiment. *Zenodo* **2021**. [CrossRef]

41. Everham, E.M.; Brokaw, N.V.L. Forest damage and recovery from catastrophic wind. *Bot. Rev.* **1996**, *62*, 113–185. [CrossRef]

42. Harper, J.J. *Population Biology of Plants*; Academic Press: Cambridge, MA, USA, 1977; p. 892.

43. Eastman, B.A.; Adams, M.B.; Brzostek, E.R.; Burnham, M.B.; Carrara, J.E.; Kelly, C.; McNeil, B.E.; Walter, C.A.; Peterjohn, W.T. Altered plant carbon partitioning enhanced forest ecosystem carbon storage after 25 years of nitrogen additions. *New Phytol.* **2021**. [CrossRef]

44. Lohmander, P.; Helles, F. Windthrow probability as a function of stand characteristics and shelter. *Scand. J. For. Res.* **1987**, *2*, 227–238. [CrossRef]

45. Griess, V.C.; Knoke, T. Growth performance, windthrow, and insects: Meta-analyses of parameters influencing performance of mixed-species stands in boreal and northern temperate biomes. *Can. J. For. Res.* **2011**, *41*, 1141–1159. [CrossRef]

46. Lanquaye-Opoku, N.; Mitchell, S.J. Portability of stand-level empirical windthrow risk models. *For. Ecol. Manag.* **2005**, *216*, 134–148. [CrossRef]

47. Savidge, R.A. *Wood Quality and Its Biological Basis*; Barnett, J.R., Jeronimisi, G., Eds.; Blackwell Scientific: Oxford, UK, 2003.

48. Marquis, D.A. *Silvics of North America. Volume 2. Harwoods*; Burns, R.M., Honkala, B.H., Eds.; U.S. Department of Agriculture, Forest Service: Washington, DC, USA, 1990; pp. 594–604.

49. Long, R.P.; Horsley, S.B.; Hallett, R.A.; Bailey, S.W. Sugar maple growth in relation to nutrition and stress in the northeastern United States. *Ecol. Appl.* **2009**, *19*, 1454–1466. [CrossRef]

50. Long, R.P.; Horsley, S.B.; Hall, T.J. Long-term impact of liming on growth and vigor of northern hardwoods. *Can. J. For. Res.* **2011**, *41*, 1295–1307. [CrossRef]

51. Lautner, S.; Ehlting, B.; Windeisen, E.; Rennenberg, H.; Matyssek, R. Calcium nutrition has a significant influence on wood formation in poplar. *New Phytol.* **2007**, *173*, 743–752. [CrossRef]

52. Jensen, N.K.; Holzmueller, E.J.; Edwards, P.J.; Gundy, M.T.-V.; Dewalle, D.R.; Williard, K.W.J. Tree Response to Experimental Watershed Acidification. *Water Air Soil Pollut.* **2014**, *225*. [CrossRef]

53. Ross, D.; Buckner, E.; Core, H.; Woods, F. Nitrogen Fertilization Decreases Yellow-Poplar Wood Density. *South. J. Appl. For.* **1979**, *3*, 119–122. [CrossRef]

54. Xi, W.; Peet, R.K. *Recent Hurricane Research–Climate, Dynamics, and Societal Impacts*; Lupo, A., Ed.; Intech Publishers: London, UK, 2011; pp. 503–534.

MDPI

Article

Nitrogen Fertilization, Stand Age, and Overstory Tree Species Impact the Herbaceous Layer in a Central Appalachian Hardwood Forest

Lacey J. Smith [†] and Kirsten Stephan *

Division of Forestry and Natural Resources, West Virginia University, Morgantown, WV 26506, USA; LJSmith1@mix.wvu.edu
* Correspondence: kirsten.stephan@mail.wvu.edu
† Present Affiliation: West Virginia Division of Natural Resources, Farmington, WV 26571, USA.

Abstract: Research Highlights: Herb-layer community composition, abundance, species richness, and Shannon–Wiener diversity index are shaped by nitrogen fertilization, disturbance history, and the overstory tree species in its immediate vicinity. Background and Objectives: While the herbaceous layer in deciduous forests is increasingly recognized for its importance in various aspects of forest ecosystem function, this study sought to describe the factors impacting the herbaceous layer. Specifically, this study's objective was to quantify and compare herb-layer species composition, cover, and other community indices in watersheds with (a) different levels of N deposition, (b) different stand ages due to differing disturbance histories, and (c) different watershed aspects. This study also tested the hypothesis that herb-layer characteristics vary beneath tree species with contrasting nutrient dynamics (i.e., red and sugar maple). Materials and Methods: At the Fernow Experimental Forest in West Virginia (USA), the cover of all herb-layer species was recorded directly under nine red maple and nine sugar maple trees in each of four watersheds (WS): long-term fertilized WS3 and unfertilized WS7, both with a stand age of about 50 years, and two unmanaged watersheds with 110-year-old stands and opposite watershed aspects (south-facing WS10, north-facing WS13). Community composition and plot-level indices of diversity were evaluated with multivariate analysis and ANOVA for watershed-level differences, effects of the maple species, and other environmental factors. Results: In the fertilized watershed (WS3), herb-layer diversity indices were lower than in the unfertilized watershed of the same stand age (WS7). In the unfertilized watershed with the 50-year-old stand (WS7), herb-layer diversity indices were higher than in the watershed with the 110-year-old stand of the same watershed aspect (WS13). WS10 and WS13 had similar herb-layer characteristics despite opposite watershed aspects. The presence of sugar maple corresponded to higher cover and diversity indices of the herb-layer in some of the watersheds. Conclusions: Despite the limitations of a case study, these findings bear relevance to future forest management since the forest herb layer plays important roles in deciduous forests through its influence on nutrient cycling, productivity, and overstory regeneration.

Keywords: understory plant community; plant cover; species richness; diversity; nitrogen deposition; maple; Fernow Experimental Forest

check for
updates

Citation: Smith, L.J.; Stephan, K. Nitrogen Fertilization, Stand Age, and Overstory Tree Species Impact the Herbaceous Layer in a Central Appalachian Hardwood Forest. *Forests* **2021**, *12*, 829. https://doi.org/10.3390/f12070829

Academic Editor: Frank S. Gilliam

Received: 1 May 2021
Accepted: 18 June 2021
Published: 24 June 2021

Publisher's Note: MDPI stays neutral with regard to jurisdictional claims in published maps and institutional affiliations.

1. Introduction

A growing number of studies underscores the importance of the herbaceous layer in forests despite its small (<1%) biomass relative to total aboveground biomass [1,2]. The herb layer is the most diverse stratum in the forest, containing between 75% and 91% of all plant species [1]. Plant species and functional richness may impact many ecosystem functions [3,4]. In grasslands, the source of much biodiversity research, higher plant species diversity significantly increased ecosystem productivity, nutrient use, and nutrient retention [5]. In forests, the vernal dam hypothesis by Muller and Bormann [6] highlights

95

the role of spring ephemeral herbaceous plants in nutrient retention. During the growing season, the herb layer can contribute up to 20% of the foliar litter to the forest floor [1]. Forest herb litter, with its low C:N ratio and high C quality, decomposes on average more than twice as fast as tree litter [2,7]. This provides a rapid pathway for the recycling of nutrients as demonstrated by high herbaceous biomass being associated with high rates of nitrogen mineralization, nitrogen availability, tree litterfall mass, and total tree litterfall N in an eastern deciduous forest [8]. The understory also influences soil microbial abundance. Removing the understory significantly reduced the amount of phospholipid fatty acids (PLFAs, used to estimate microbial biomass), the fungi to bacteria ratio, and N availability in the soil [9]. The herb layer may also shape the future overstory by inhibiting tree seedlings. Dense layers of ferns have been shown to limit maple (*Acer* L.) and black cherry (*Prunus serotina* Ehrh.) seedlings via foliage shade [10,11].

A number of abiotic and biotic factors influence the herb-layer within a forest [12]. Variability in solar radiation received, with its impact on the heat and water balance, leads to microclimates [13–15] influencing microbial activity [16], decomposition rates [17], and N availability [18]. Forest trees have a direct effect on the availability of resources for species in the herb layer. Obviously, tree crowns impact the amount and quality of light reaching the forest floor, and tree fine roots can decrease nutrients and moisture available to the herb layer [19]. Stemflow could be important in establishing soil moisture and mineral gradients around the tree base and, thus, be a determining factor of herb distribution by affecting and establishing microhabitats underneath the overstory [20,21]. However, can trees of different species affect the herb layer in distinguishable ways?

In a northern hardwood forest, tree species identity correlated strongly with soil chemistry [22], via litter-soil feedback supported by localized litter dispersal [23]. Mycorrhizal type has also been shown to correlate with nutrient dynamics [24] in that trees with arbuscular mycorrhizal fungi (AM) had higher levels of N and P in their leaves and faster litter decomposition compared to trees with ectomycorrhizal fungi (ECM) [24,25]. Associations between tree species and soil nitrate availability across spatial scales were found at the Fernow Experimental Forest (FEF) in the Appalachian Mountains of West Virginia [18]. At the scale of individual trees, small plots, and entire watersheds, sugar maple (*Acer saccharum* Marsh.) (AM), tulip-poplar (*Liriodendron tulipifera* L.) (AM), and black cherry (AM) strongly correlated with locations of higher soil nitrate availability, whereas red maple (*Acer rubrum* L.) (AM), American beech (*Fagus grandifolia* Ehrh.) (ECM), and chestnut oak (*Quercus montana* Willd.) (ECM) strongly correlated with locations of lower soil nitrate availability. At the scale of individual trees, the study also found lower soil C:N ratios and higher soil pH values around trees associated with higher soil N availability than trees associated with lower soil N availability. Interestingly, while sugar maple and red maple could be predicted to affect their soil environment in a similar way due to both having AM fungi, Peterjohn et al. [18] found these species associated with soils of differing N availability. While the study of Peterjohn et al. [18] could not determine whether tree species cause or reflect patterns of soil nitrate availability, their results supported the hypothesis that the nature of leaf litter alters soil C:N ratios in ways that influence rates of nitrification [26,27], which in turn might influence the herb layer.

At a broader scale, anthropogenic influences such as atmospheric N deposition or disturbance may directly (and perhaps indirectly via an influence on tree species composition) impact the herbaceous layer. Excess nitrogen from anthropogenic activities has led to an increase in nitrogen deposition in the eastern U.S. forests [28]. An increase in atmospheric nitrogen deposition may reduce plant diversity in forests, especially in the herb layer [12,29]. As the herb layer is sensitive to nutrient availability, nitrogen additions can create a competitive environment supporting the survival and growth of nitrophilic species while decreasing species richness [2,30,31]. The former was demonstrated at the FEF; after 25 years of nitrogen fertilization in a watershed, the cover of nitrophilic *Rubus* spp. (mostly Allegheny blackberry, *R. allegheniensis* Porter) had increased from 1 to 19% of total herb cover. Walter et al. [32] concluded that the increase in *Rubus* cover was consistent with the

N homogeneity hypothesis [29], which states that, as an ecosystem shifts from N limitation to N saturation due to the homogeneous supply of soil nutrients, species richness decreases due to the exclusion of N-efficient species by nitrophilic species [33]. While excess N may be detrimental to the forest herb-layer community, disturbances such as forest harvesting lead to an initial increase in herb-layer abundance and richness if resources (e.g., light, water, nutrients) become more available post-harvest [34–36]. Even so, studies describing short- to long-term effects of forest harvest on the herb layer have reported variable results, including higher, equal, or lower herb layer richness or diversity after harvesting relative to their controls [37–39].

With emerging and persisting perturbations impacting forest ecosystems, e.g., climate change, introduced pests, atmospheric N deposition, and management practices, a better understanding of all components of the ecosystem, i.e., including the herbaceous layer, is required to inform management aiming at improving resistance or resilience of forest ecosystems. To alleviate the relative paucity of quantitative data of the herbaceous layer in contemporary forests in general and in response to perturbations, this study's objectives were to quantify and compare herb-layer characteristics (species composition, cover, richness, diversity, evenness) in watersheds with (a) different levels of N deposition, (b) different stand ages due to past harvests, and (c) different watershed aspects (N vs. S). We expected fertilization, older stand age, and a south-facing watershed aspect to negatively impact the herb layer relative to no fertilization, younger stand age, and a north-facing watershed aspect, respectively. This study also tested the hypothesis that herb-layer characteristics vary beneath trees species with contrasting nutrient dynamics (i.e., red and sugar maple), such that the herb layer benefits from the vicinity of sugar maple.

2. Materials and Methods

2.1. Study Site

This study was conducted at the Fernow Experimental Forest (FEF), located in Tucker County in north-central West Virginia (Figure 1), in 2018 and 2019. The FEF lies within an area classified as the Allegheny Mountain Section of the Central Appalachian Broadleaf Forest [40]. The growing season extends from May through October, with tree leaves emerging in late April and being fully developed by mid-June. Leaves begin to fall in late August to early September. The mean annual precipitation is approximately 1460 mm with most precipitation occurring during the growing season (March through August) [41,42]. During the sampling periods (April–August) of 2018 and 2019, total precipitation was 914 mm and 842 mm, respectively. These amounts are in the upper 20th percentile of amounts recorded since 1990, averaging 708 mm for this 30-year period. Mean monthly air temperatures range from $-2.8\,^{\circ}\text{C}$ in January to $20.4\,^{\circ}\text{C}$ in July [43]. In 2018 and 2019, the mean July temperatures were $20\,^{\circ}\text{C}$ and $20.5\,^{\circ}\text{C}$, respectively. The most common soil at FEF is Calvin channery silt loam (loamy-skeletal, mixed, active, mesic Typic Dystrudept) [43] derived from acidic sandstone parent material. In the top 10 cm of mineral soil, soils comprised about 67% sand, 22% silt, 11% sand, and 14% organic matter [33]. In the top 5 cm of the mineral soil, average pH ($CaCl_2$) is 3.7 (range 3.0–4.4), % C is 5.5 (range 1.9–12.5), % N is 0.35 (range 0.15–1.0) [44]. Dominant overstory species at the FEF are sugar maple, sweet birch (*Betula lenta* L.), American beech, tulip-poplar, black cherry, and northern red oak (*Quercus rubra* L.) [45].

Four experimental watersheds (WS) in close proximity to each other (WS3, WS7, WS10, WS13) were included in this study (Figure 1). These watersheds support closed-canopy forest and are similar with respect to elevation, soil series, geology, climate, and most of their disturbance history. Watersheds were heavily logged around 1910 (along with most forests of West Virginia) [46] and then left to regenerate naturally [47]. American chestnut (*Castanea dentata* (Marshall) Borkh.), a dominant species at the time, was harvested from the watersheds in this study in the 1940's, after the trees had died from the introduced chestnut blight [46]. Watersheds differed in their more recent anthropogenic disturbance history (detailed below) and watershed aspect (Table 1), providing the foundation for this study.

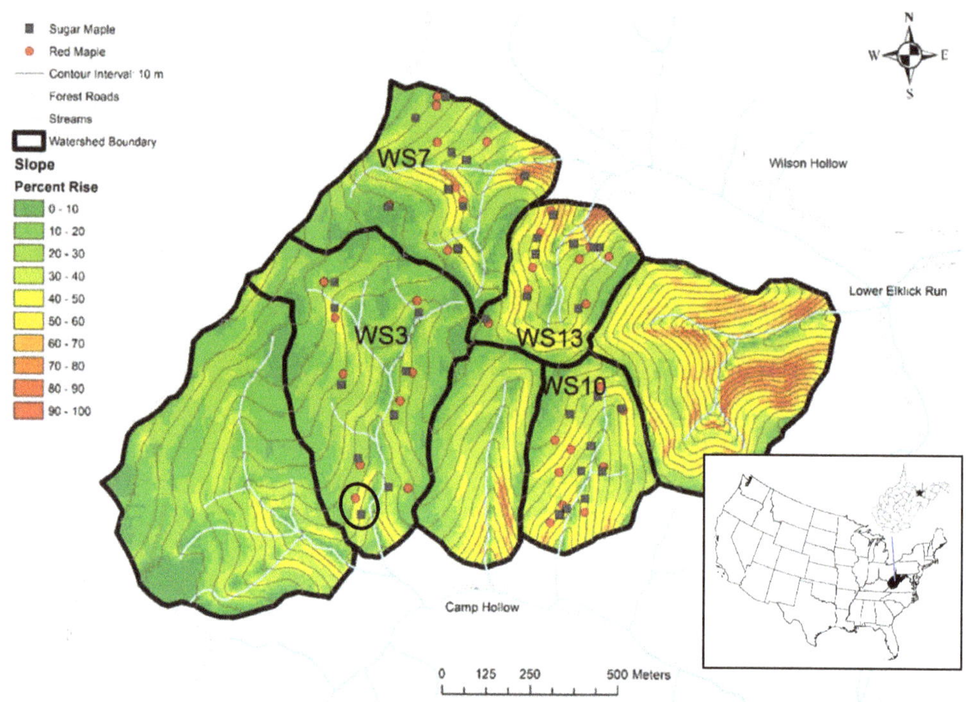

Figure 1. Location of the Fernow Experimental Forest, West Virginia, USA (insert) and watersheds used in this study (WS3, WS7, WS10, WS13). Location of study plots are marked by the tree at the plot center (square—sugar maple, circle—red maple). The oval highlights a site consisting of a plot pair.

Table 1. Characteristics of the watersheds in this study, located at the Fernow Experimental Forest, West Virginia, USA.

Watershed ID	Location *	Area (ha)	Elevation (m)	Average Slope (%) (Min–Max)	Dominant Tree Species **	Aspect	Stand Age (year) in 2020	Fertilization Treatment
WS3	39.05413 N 79.68625 W	34.3	730–870	20.6 (0–60)	Black cherry, red maple, sweet birch northern red oak	S	~50	Fertilized
WS7	39.06388 N 79.68029 W	24.2	705–860	25.8 (0–90)	Tulip-poplar, black cherry, sweet birch, red maple	E	~50	Not fertilized
WS10	39.05411 N 79.68029 W	15.2	700–805	33.4 (0–70)	Chestnut oak, northern red oak, red maple, blackgum	S	~110	Not fertilized
WS13	39.06280 N 79.67917 W	14.2	695–810	35.2 (0–100)	Northern red oak, sugar maple, red maple, tulip-poplar	N	~110	Not fertilized

* Lowest point in the watershed, i.e., the location of the stream weir. ** Tree species are listed in order of descending dominance (see text below).

Watershed 3 (WS3) had received long-term fertilization/acidification treatments to study the effects of atmospheric N deposition. Since 1989, granular ammonium sulfate has been applied aerially three times a year at a rate of 7.1 kg N ha^{-1} in March and November, and 21.2 kg N ha^{-1} in July [40]. Partial cuts were made in 1958, 1963, and 1968, respectively removing 14%, 9%, and 6% of trees with a diameter at breast height (DBH) $\geq$ 12.7 cm. The watershed was subsequently clearcut in 1969–1970, removing all trees with a DBH $\geq$ 12.7 cm; all saplings with a DBH between 2.5 cm and 12.5 cm were sprayed with herbicide [46]. A 3-ha riparian/protection buffer strip (approximately 40 m wide and 730 m long) was initially left along the perennial stream to help protect water quality; it was cut in 1972 [41,48]. In 2003, the dominant tree species (as % basal area) in this watershed were black cherry (51), red maple (11.5), northern red oak (5.1), and sweet birch (5.1); sugar maple made up 1.3% of the total basal area in WS3 [49].

To assess fertilization/acidification effects in WS3, watershed 7 (WS7) was used as an unfertilized reference watershed due to similar disturbance history. The upper half of WS7 was clearcut from 1963–1964; the lower half was clearcut from 1966–1967. Following the clearcuts, the watershed was then treated annually with herbicide till 1969 [43]. In 2003, the dominant tree species (as % basal area) for this watershed were tulip-poplar (26.2), sweet birch (20.5), black cherry (20.5), red maple (8.2), and sugar maple (4.9) [45].

Watershed 10 (WS10) and Watershed 13 (WS13) served as "unmanaged" watersheds although a final, partial cut took place in WS13 in the early 1950s [47]. In 2000, the dominant tree species in WS10 (as % basal area) were chestnut oak (24), northern red oak (22) red maple (19), blackgum (*Nyssa sylvatica* Marshall, 8), and white oak (*Quercus alba* L., 6); sugar maple made up 2% of the total basal area in WS10. In WS13 dominant species were northern red oak (30), sugar maple (22), red maple (13), tulip-poplar (7), and American beech (7) [50]. WS10 and WS13 differ in watershed aspect (S vs. N, Table 1).

2.2. Experimental Design

To collect herb-layer data, a total of 18 plots were established in each of the four watersheds. Two adjacent plots, with one centering around a stem of sugar maple and the other around a stem of red maple, represent a site (Figures 1 and 2). Prior to selecting maples for this study, a total of 151 maple trees had been located, if possible, upslope from the riparian area and downslope from the watershed boundary (to minimize variability due to a moisture gradient). In each watershed, at least twice as many trees per species were initially located than were used in this study. Tree locations were recorded with GPS and mapped. From this map, sites were identified if a red and sugar maple tree were reasonably close, and of those sites, four or five were randomly selected on either side of the main stream (totaling nine sites per watershed). The distance between plot centers of a red and sugar maple pair averaged 32.5 m (range: 6.2–83.1 m) while the average distance between neighboring plot pairs (closest plot centers) was 112.6 m (range: 16–257.9 m) (Figure 1). Selected plot-center trees were vigorous (i.e., without signs of disease or injury) and had a DBH $\geq$ 10 cm (average DBH 20 cm, range 10 cm–41 cm).

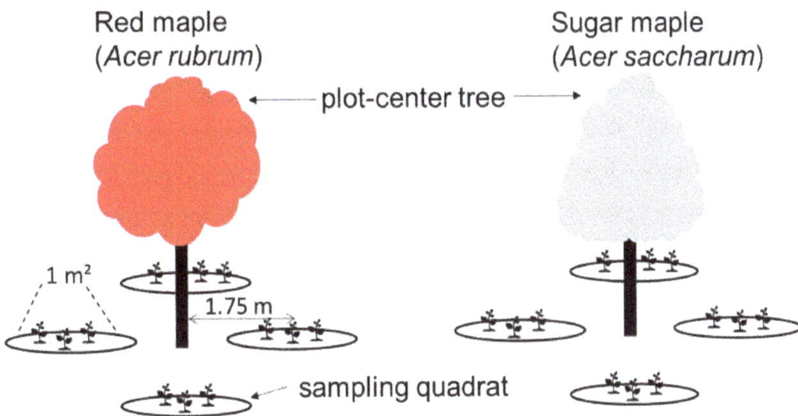

Figure 2. Experimental design of plot pairs at each site (replicated nine times in each watershed) for collecting herb-layer composition and cover.

2.3. Data Collection and Analysis

Cover of herb-layer species (<1 m tall) were assessed in four circular 1-m^2 sampling quadrats established at the four cardinal directions (N, S, E, W) from the plot-center tree (Figure 2). Sampling quadrats represented subsamples of the plot; therefore, cover was averaged across the four quadrats per plot prior to analysis. The center of the sampling quadrats was 1.75 m away from the base of the plot-center maple, roughly halfway between the stem and the edge of the crown to avoid stem flow and canopy drip. Each quadrat location was marked with three stake flags on the perimeter so that sampling quadrats could be placed in exactly the same spot during the different sampling campaigns. Herb-layer species composition and cover were collected in early summer 2018 (16 June–5 July), spring 2019 (3 May–6 May), and summer 2019 (29 June–29 July). Environmental variables that may influence the herb layer (slope angle, aspect, slope position, canopy cover, DBH of plot-center maple, DBH of neighbor trees) were collected from June through August 2018; as an exception, leaf area index (LAI) was collected on August 6 and September 5, 2019.

Within each sampling quadrat, herb-layer composition was determined by identifying plants to species level, with exceptions for grasses and sedges (identified as graminoids), and *Rubus* L., *Viola* L., and *Anemone* L. (identified to genus level). These taxonomically difficult groups were not identified to species level to limit misidentification. For example, at the FEF, the vast majority of *Rubus* individuals are *R. allegheniensis* (Allegheny blackberry), but *R. idaeus* L. (common red raspberry) also occurs at this location. *Rubus* species can hybridize and are difficult to identify without fruit or flowers [32]. These exceptions might have resulted in an underestimation of species richness but had little influence on the other diversity indices since graminoids and Anemone had very low abundances and *Rubus* likely was *R. allegheniensis*.

For each taxon, cover was measured as leaf area using the hand-area (HA) method [51]. In brief, the HA method compares the area of a hand with the area of the leaves of all plants of a given species. The observer places a hand (equivalent to 1% of 1 m^2), palm down and fingers closed, directly above the leaves or leaflets of the species of interest and then determines the size of the leaf in relation to their hand, either as individual or group, until all leaf surfaces are observed within the quadrat. To improve the accuracy and precision, (a) observer hands were calibrated to 1 dm^2 by folding under the thumb and or fingertips depending on the actual size of the observer's hand and (b) two observers independently recorded cover, with the average of the two estimates being recorded [51]. Cover was estimated with a precision of 0.01% (i.e., 1 cm^2 leaf area in the 1-m^2 quadrat).

Species richness (S), Shannon–Wiener diversity index (or Shannon index, H), and Pielou's evenness (J) were calculated at the plot level. Species richness is the number of species per unit of area, i.e., in this study, S is the total number of species found across the four sampling quadrats per plot. Shannon index is commonly used to characterize species diversity, which accounts for both abundance and evenness of species present [52]. H will increase with increasing species richness and with increasingly equitable contributions of the species to the community. Pielou's evenness is another measure of diversity. Values for J are the ratio of actual H to maximally possible H (if all species were present in equal proportion) [52,53].

The Shannon index is calculated as

$$H = -\sum_{i=1}^{s} Pi \ln Pi \tag{1}$$

where Pi is the relative abundance of each herbaceous species in the plot and where s is the number of species [54].

Pielou's evenness is calculated as

$$J = \frac{H}{H_{max}} = H / \ln S \tag{2}$$

where H is the Shannon index and S is the total number of species [53].

To characterize the abiotic environment of the herb layer, several variables relating to light availability, soil moisture, and temperature were measured in each plot. To quantify the light environment above the herb layer, percent canopy cover was measured with a densiometer over each sampling quadrat while facing the plot-center tree. LAI was measured in the same locations (albeit in 2019) using an Accupar LP-80 PAR/LAI ceptometer (Meter Group, Inc., Pullman, WA, USA) (details in [55]). The four respective densiometer and LAI measurements were then averaged for each plot. Slope (%) was measured using a clinometer and aspect (°) was determined with a compass; both measurements were taken at the plot center facing downhill. Slope position, assumed to correlate with soil moisture, was measured in ArcMap as the slope distance from the plot center to perennial stream. To align aspect with ecologically relevant effects of radiation (e.g., heat and moisture balance), and to be able to analyze it as a continuous variable, aspect was transformed into a linear scale that ranges from 0 to 2, with 0 being the relatively warmer and drier southwest and 2 being the relatively cooler and moister northeast, using the formula:

$$A' = \sin(A + 45) + 1 \tag{3}$$

where A' is the transformed aspect code and A is the aspect defined as the direction of the prevailing slope [56].

To assess the potential influence of neighboring trees on herb-layer characteristics, DBH of the five closest neighboring trees (>2 m height) was measured. For each neighbor tree, ectomycorrhizal (ECM) or arbuscular mycorrhizal (AM) association was determined (based on [57,58]), and basal area (BA, cm^2) was summed for each association. The average distance of the farthest tree from the plot center was 5.7 m (range 2.9–11.6 m).

2.4. Statistical Analysis

The relationships between herb-layer composition, watersheds, overstory maple, and environmental variables in summer 2018, were analyzed with a constrained ordination (i.e., direct gradient analysis including species and environmental data) in Canoco 5.1 [59]. In a constrained ordination, only the species composition variability is shown that can be explained by environmental variables. Specifically, canonical correspondence analysis (CCA) was performed with forward selection of significant predictor variables. Most of the environmental variables were not highly correlated with each other (VIF < 3.3). However, DBH of the largest neighbor tree and BA of ECM trees had a VIF of 11.3 and 7.9, respectively. These variables were highly correlated with each other (0.83) and moderately correlated with WS13 (0.61 for large neighbor DBH and 0.65 for BA of ECM trees).

In the analysis of community-level indices (cover, S, H, J), analysis of variance (ANOVA) was conducted separately for watershed pairs. Watershed pairs were WS3-WS7 (treatment: fertilization level; same, "younger" stand age); WS7-WS13 (treatment: stand age; similar aspect); and WS10-WS13 (treatment: watershed aspect; same, "older" stand age). The analysis of watershed pairs reduced the potential for confounding the intended watershed-level treatment (e.g., fertilization) with other differences between watersheds (e.g., stand age). Residuals of response variables were checked for normality; response variables did not require transformation. After detecting extreme observations in some predictor variables (distance-to-stream, BA of ECM trees, BA of AM trees) when conducting influence diagnostics in regression analyses, these variables were square-root-transformed to reduce the influence of high-leverage extreme observations in model predictions. No significant multicollinearity was detected between the continuous environmental variables (VIF < 10). Using JMP and SAS (JMP®, Version Pro 12.2, SAS Institute Inc., Cary, NC, USA, Copyright ©2015; SAS®, Version 9.3, SAS Institute Inc., Cary, NC, USA, Copyright ©2002–2010), analyses were conducted separately for the spring and summer datasets. For the spring dataset, repeated measures analyses of variance were undertaken to account for the spatial correlation between sugar maple/red maple plot pairs within a site (Figure 1) (SAS code: Repeated Tree/Subject = Site * WS). Datasets from summers 2018 and 2019 were jointly analyzed in a doubly repeated ANOVA accounting for the spatial relationships of plot pairs and the repeated measurement in time (SAS code: Random Tree/Subject = Site * WS; Repeated Year/Subject = Tree * Site * WS). Models evaluated the effects of watershed-level treatment (factor Watershed, WS), plot-center overstory maple species (factor Maple, M), their interaction (WS × M), and variables describing the physical and biotic environment on each herb-layer characteristic (cover, S, H, J) (Table 2). Thus, 12 models (three watershed pairs × four herb-layer indices) were run for the summer and spring data, respectively. A reduced model was created from the full model by removing predictor variables (other than WS, M, WS × M) with high p-values (i.e., $p > 0.4$); applying a backward elimination regression was not possible due to not all predictor variables being continuous variables. Final model selection (i.e., choosing between the full versus reduced model) was determined by the lowest AIC value. Graphs and tables show untransformed raw data.

Table 2. Dependent and independent (predictor) variables used in ANOVA for summer and spring herb-layer datasets. The four watersheds in this study were grouped into pairs to avoid confounding the treatment effect of interest with other watershed-level differences. In total, 12 models were run for the summer and spring dataset, respectively.

Dependent Variable	Predictor	Description
	Watershed (WS)	Watershed-level treatment effect; 2 factor levels; factor levels (treatments) varied by watershed pair: • WS3 vs. WS7: fertilized vs. not fertilized • WS7 vs. WS10: 50-year old stand vs. 110-year-old stand • WS13 vs. WS10: northerly vs. southerly watershed aspect
Cover Species richness (S) Shannon index (H) Evenness (J)	Plot-center maple (M)	2 factor levels: red maple, sugar maple
	WS × M	Watershed (treatment)—maple species interaction
	Slope	Plot-level slope in %
	Aspect code	Plot-level aspect code A'
	Distance to stream	Slope distance from plot center to perennial stream
	DBH	DBH of plot-center tree
	Large neighbor DBH	DBH of the largest of the five nearest neighbor trees to the plot center
	Canopy cover	Percent of sky covered by canopy
	Large Neighbor Myc.	Mycorrhizal association of the largest of the five nearest neighbor trees to the plot center; 3 factor levels: AM, ECM, both AM and ECM
	Leaf Area Index	LAI, total leaf area (m^2) found over 1 m^2 of ground
	BA of ECM trees	Basal area of ectomycorrhizal trees among the five nearest neighbors of the plot center (BA-ECM)
	BA of AM trees	Basal area of arbuscular mycorrhizal trees among the five nearest neighbors of the plot center (BA-AM)

Within each watershed pair, pairwise comparisons of herb-layer indices were conducted (a) between the two overstory maple species per watershed and (b) between the two watersheds under a given overstory tree species. There was no adjustment made for multiple comparisons, but the number of comparisons were minimized (to four, using the Slice function in SAS) to reduce false positives (Type I error). In all statistical analyses, significance criterion alpha was 0.05 and a statistical trend was declared when $p \leq 0.1$.

To determine associations between individual herb-layer plant species with watershed and overstory maple species, ANOVA was conducted separately for 22 frequent herb-layer species in each of the three watershed pairs. Since the probability of a Type I error increases with the number of tests conducted (familywise error rate), the Benjamini–Hochberg method was applied to control the false discover rate (i.e., a false positive or a Type I error). The concept is similar to the Bonferroni adjustment for multiple comparisons, yet less conservative [60]. To perform the Benjamini–Hochberg method, individual p-values from ANOVAs are ranked from smallest to largest. The smallest p-value receives a rank (i) of 1, the next larger p-value receives the rank of 2, etc. Next, each p-value is compared to the Benjamini–Hochberg critical value (i/m)Q, where i is the rank, m is the total number of tests (total number of individual *p*-values ranked), and Q is the false discovery rate selected by the researcher [60]. In this study, the false discovery rate was set at 0.10. For the cover of a plant species to vary significantly by WS, M, or WS × M, the *p*-value must be smaller than the Benjamini–Hochberg critical value. For example, with 22 species and two main effects (WS, M) and an interaction effect, there is a family of 66 analyses (i.e., m = 66). For the smallest *p*-value obtained by ANOVA to be considered significant, it would have to be smaller than $1/66 \times 0.1 = 0.0015$; the *p*-value at rank two would have to be smaller than $2/66 \times 0.1 = 0.003$, etc. The largest *p*-values that is smaller than the Benjamini–Hochberg critical value, and all of the *p*-values smaller than it, are considered significant [60].

While the analyses of the overstory maple species effect on herb-layer indices and the cover of individual species are truly replicated, the analyses of differences between watersheds are pseudo-replicated, i.e., no inference can be made beyond the watershed pairs used in this case study.

3. Results

3.1. Community Composition

3.1.1. Watershed-Level Herb-Layer Community Composition

Among all four watersheds, a total of 63, 57, and 64 taxa (including three genera and not counting graminoids) were recorded during the summer 2018, spring 2019, and summer 2019 sampling periods (Table 3, Table S1), respectively. Except for multiflora rose (*Rosa multiflora* Thunb. ex Murr.) and Japanese barberry (*Berberis thunbergii* DC.), all herb-layer species were native species. Species numbers were consistently highest in WS7 (younger stand, unfertilized), lowest in WS3 (younger stand, fertilized), and intermediate in WS10 and WS13 (older stands) (Table 3, Figure 3A). Herb-layer indices were similar between sampling periods, likely due to most species being perennials and resampling the same quadrats. Cover was expectedly lower in spring than in summer (Table 3). In all watersheds, most herb-layer species were tree seedlings or herbs (Figure 3A) with similar proportions of species in each plant type (fern, herb shrub/vine, tree seedling) between the watersheds (Table S2). However, the absolute number of herb and tree species (as seedlings) was lower in WS3 (younger stand, fertilized) than in the other watersheds (Figure 3A). Watersheds also differed in the proportion of total cover in different plant types. The younger stands (WS3, WS7) had a higher proportion of cover in ferns than the older stands (WS10, WS13). Comparing the two younger stands, half of the cover in fertilized WS3 was due to shrubs/vines and less than 3% of all cover was due to herbs, whereas the contribution to cover from these two plant types in WS7 was about equal (Figure 3B). Blackberry, intermediate shield fern (*Dryopteris intermedia* (Muhl. ex. Willd.) Grey), and hay-scented fern (*Dennstaedtia puntilobula* (Michx.) Moore) made up two thirds of all cover in WS3 (Figure 4), which is reflected in a lower Shannon index and evenness than in the other watersheds (Table 3). The 14 species with the highest relative cover in the study area contributed 97%, 89% 85%, and 93% to the total cover in WS3, WS7, WS10, and WS13, respectively (Figure 4).

Table 3. Herb-layer cover (in m^2 of leaf area per 100 m^2 of ground), species richness (S), Shannon diversity (H), and Pielou's evenness (J) at the watershed scale in each of the three sampling campaigns. Data are from the Fernow Experimental Forest, West Virginia, USA.

	Cover			S			H			J		
	July 2018	May 2019	July 2019	July 2018	May 2019	July 2019	July 2018	May 2019	July 2019	July 2018	May 2019	July 2019
WS3	32.6	12.4	32.5	30	32	31	2.0	2.2	1.8	0.59	0.62	0.53
WS7	36.4	18.6	31.7	48	46	46	2.7	2.8	2.6	0.69	0.74	0.69
WS10	16.8	10.0	11.3	43	42	45	2.8	3.1	2.9	0.74	0.82	0.77
WS13	19.5	10.9	15.4	36	35	40	2.5	2.6	2.5	0.69	0.72	0.68

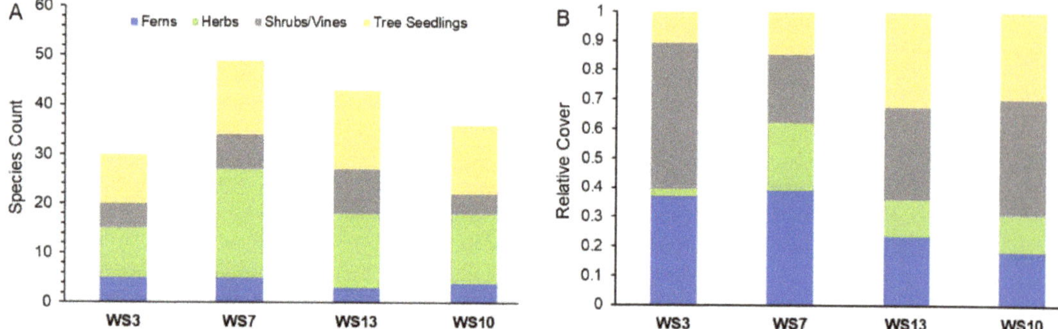

Figure 3. Species count (**A**) and the proportion of overall cover (**B**) by herb-layer plant type at the watershed level for the Summer 2018 sampling. Note that watersheds (WS) are arranged on the *x*-axis so that those analyzed as a pair (Table 2) are adjacent to each other. Data are from the Fernow Experimental Forest, West Virginia, USA.

3.1.2. Plot-Level Herb-Layer Community Composition

At the plot level, multivariate analysis (CCA) with forward selection of significant predictor variables and factors levels revealed that factor Watershed, factor Maple (red vs. sugar maple), and four environmental variables (LAI, basal area of ectomycorrhizal neighbor trees—BA-ECM, aspect code, distance to stream) were significant ($p \leq 0.1$) in explaining variability in herb-layer species composition. Of the total variability in species composition, the selected predictor variables together could explain 19.5%. While this number is low, it is not unusual because species data are inherently noisy [61]. Individually, factor Watershed explained 10.3%, the group of four environmental variables explained 8.5%, and factor Maple explained 2.2% of the total variability in the species data.

While influencing species composition, the physical attributes of the plots (e.g., slope angle, plot-level aspect, and slope position [i.e., distance to stream]) were not of primary interest and were subsequently analyzed as covariates in a partial CCA. After removing the compositional variability explained by physical plot attributes, the remaining explanatory factors/variables (Watershed, Maple, LAI, BA-ECM) together accounted for 16% of the total variation in the species data. Conversely, the group of physical plot attributes explained 6.8% of variation in herb-layer species data when Watershed, Maple, LAI, and BA-ECM were used as covariates.

Most (76%) of the variation in species composition (explained by Watershed, Maple, LAI, and BA-ECM) was represented by the first three canonical axis (Figure 5A,B). Species composition in WS3 (fertilized, younger stand) was most dissimilar from WS10 and WS13 (unfertilized older stands); unfertilized WS7 had a species composition similar to fertilized WS3 of the same stand age and WS13 of older stand age but same fertilization status (Figure 5A). The proximity of species symbols for blackberry, intermediate shield fern, and hay-scented fern to the symbols for WS3 and WS7 indicated the species' higher relative abundance in these watersheds (with younger stand age). Star chickweed (*Stellaria pubera* Michx.), violets, yellow fairybells (*Prosartes lanuginosa* (Michx.) D.Don) and jack-in-the-pulpit (*Arisaema triphyllum* (L.) Schott) had higher relative abundances in unfertilized WS7 than fertilized WS3. Tree seedlings had a higher relative abundance in plots of WS10 and WS13 (older stand age) than in WS3 and WS7 (younger stand age). The somewhat different species compositions under red maple compared to sugar maple is shown by the separation of these factor levels along Axes 2 and 3 (Figure 5A,B). Plots of WS10 and WS13 (older stands) were separated from the plots of WS3 and WS7 (younger stands) along Axis 1 that correlated with the basal area of EM-associated trees among the five nearest neighbors of the plot-center maple (Figure 5A). Among the younger stands, plots of fertilized WS3 were separated from plots of unfertilized WS7 along Axis 2 that mainly correlated with LAI (Figure 5A). BA of ECM trees was highly (0.83) correlated with the DBH of the largest neighboring trees, a variable that therefore likely did not come up as significant in the forward selection. Factor level WS13 was moderately positively correlated with BA-ECM (0.56) and DBH of the largest tree neighbor (0.6), and factor level WS3 was moderately positively correlated with LAI (0.45). This suggests partial mechanisms for watershed-level differences. Interestingly, red maple seedlings (found in nearly all plots, Table S1) had higher cover under sugar maple than under red maple (Figure 5B) and no seedlings of sugar maple were found in plots in the fertilized watershed.

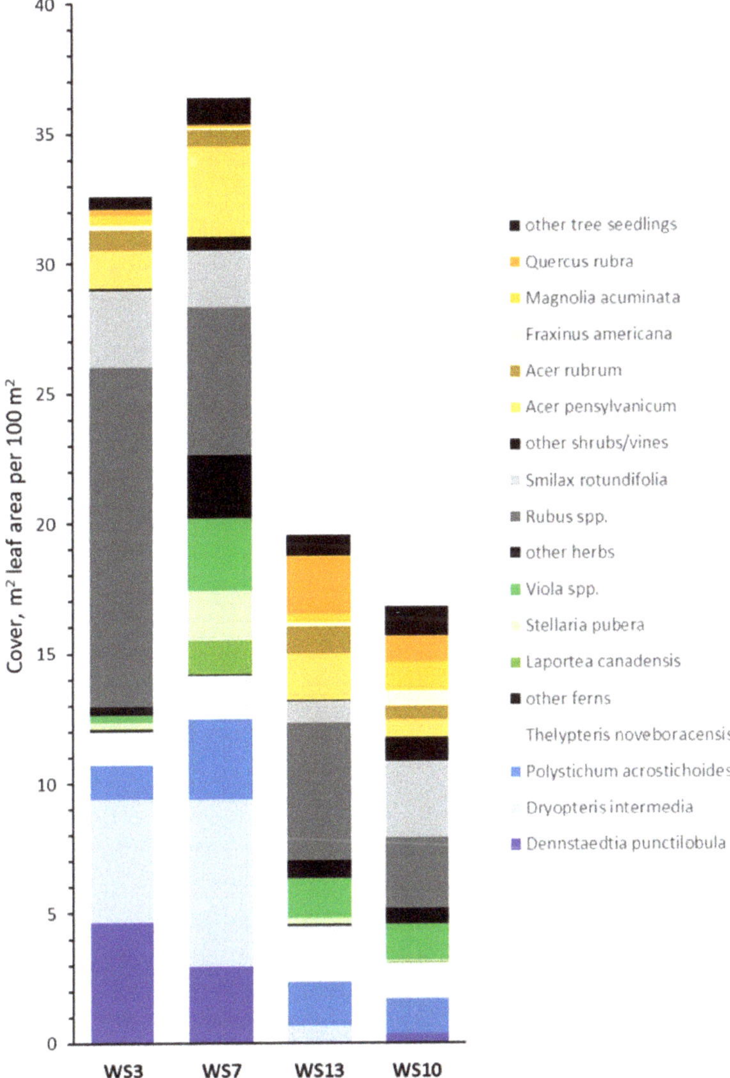

Figure 4. Watershed-level plant cover for the most important species in the summer of 2018. The 14 species shown contributed 97%, 89% 85%, and 93% to the total cover in WS3, WS7, WS10, and WS13, respectively. Note that watersheds (WS) are arranged on the *x*-axis so that those analyzed as a pair (Table 2) are adjacent to each other. Data are from the Fernow Experimental Forest, West Virginia, USA.

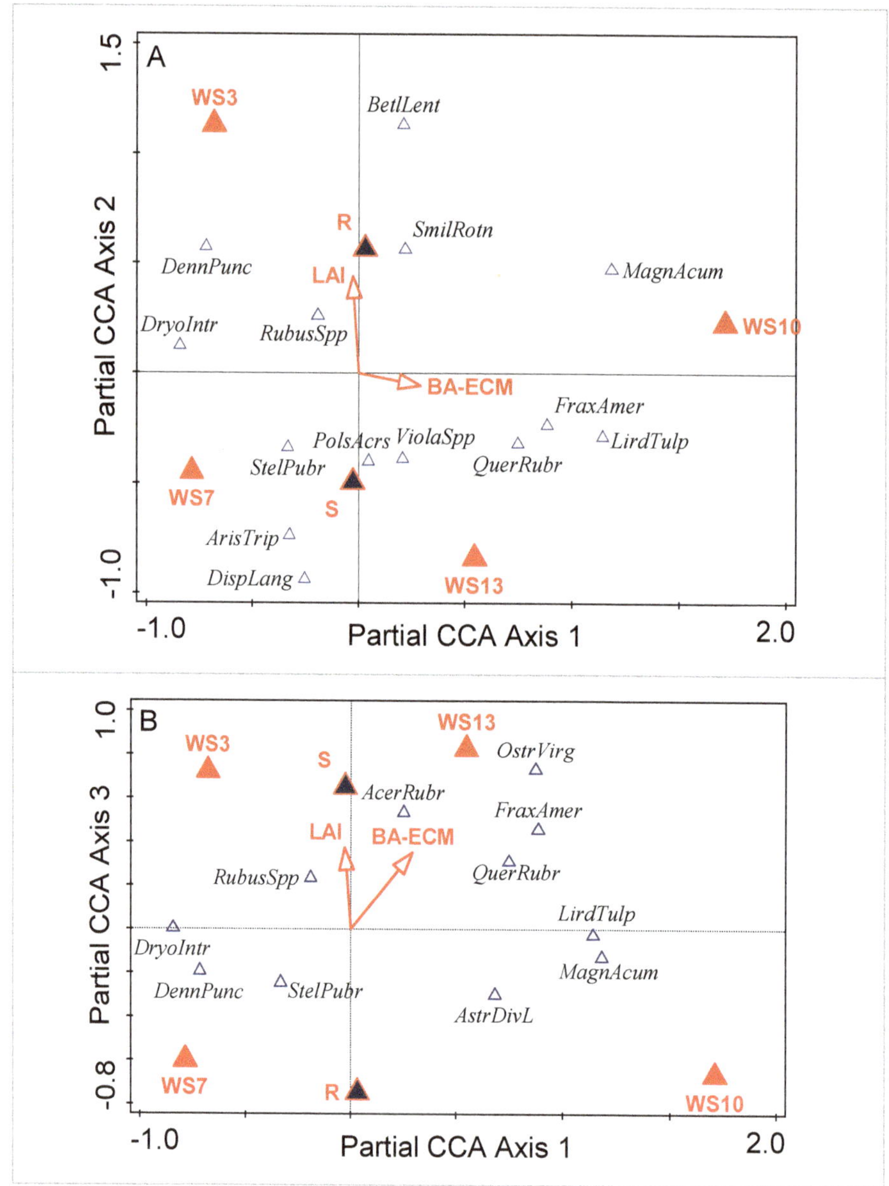

Figure 5. Ordination diagrams of a partial canonical correspondence analysis (CCA), plotting (**A**) Axis 1 against Axis 2 and (**B**) Axis 1 against Axis 3, of summer 2018 herb-layer species in relation to environmental factors and variables at Fernow Experimental Forest, West Virginia, USA. Covariates in this analysis were physical attributes of the plots (e.g., slope angle, aspect, and slope position [i.e., distance to stream]). Shown are only species that occurred in at least seven of the 72. plots and that have >5% correlation with each axis. Species are plotted at their optimum positions along each of the ordination axes with their abundances decreasing symmetrically in all directions. Symbols of environmental variables (solid triangles for each watershed (WS) and the two maple species—red maple (R), sugar maple [S]) represent a level of factors Watershed and plot-center Maple, respectively, i.e., each symbol can be interpreted as representing a group of plots (e.g., R represents all plots under red maple, WS10 represents all plots of WS10) [62].

3.2. Herb-Layer Indices: Cover, S, H, and J

The importance of factors watershed, maple, and environmental variables was also reflected in herb-layer community-level indices (cover, S, H, J). In the ANOVAs of summer herb-layer cover, S, H, and J in watershed pairs, factor Watershed (WS), in the sense of "treatment", affected the herb layer in watershed pairs WS3 vs. WS7 (fertilized vs. unfertilized) and WS7 vs. WS13 (younger vs. older stand age), but not in watershed pair WS10 vs. WS13 (south-facing vs. north-facing watershed aspect). Factor maple (M) was not statistically significant as a main effect in any of the models, but there was a statistically significant effect or a trend of the WS × M interaction in five of the 12 models (Table S3a). In these models, herb-layer cover, species richness, and/or Shannon–Wiener diversity were lower under red than sugar maples in one of the watersheds in the pair (WS3, WS13) but not the other (WS7, WS10).

3.2.1. WS3 vs. WS7: Fertilized vs. Unfertilized 50-Year-Old Stands

In the comparison of the fertilized and unfertilized watersheds (both with 50-year-old stands), factor Watershed and/or WS × M interactions were statistically significant predictors for all herb-layer indices except cover. Species richness, diversity, and evenness overall were lower in fertilized WS3 than unfertilized WS7 (WS: $p < 0.001$, $p < 0.001$, $p = 0.02$, respectively) (Figure 6B–D, Table S3a). Species richness and diversity tended to be greater under sugar maples than red maples in fertilized WS3 with an opposite pattern or no difference between maples in the unfertilized WS7 (WS × M: $p = 0.001$, $p = 0.09$, respectively) (Figure 6B,C).

Analyses of individual herb-layer species showed that cover of three species varied with watershed or overstory maple species. Violets (*Viola* spp.) and jack-in-the-pulpit (*Arisaema triphyllum* L.) cover was significantly lower in fertilized WS3 than in unfertilized WS7 (WS: $p = 0.001$ and $p = 0.003$, respectively; Table S3). The cover of New York fern (*Thelypteris noveboracensis* L.) was greater beneath sugar maples than red maples in fertilized WS3 with an opposite pattern in WS7 (WS × M $p = 0.002$; Table S4).

Various environmental factors explained significant variation in herb-layer indices (Table S3a). S, H, and J decreased with increasing slope ($p = 0.01$, $p = 0.004$, $p = 0.03$, respectively), increased with increasing aspect code (i.e., aspect changing from southwest toward northeast) ($p = 0.04$, $p = 0.02$, $p = 0.02$, respectively), and decreased with increasing distance from the stream ($p = 0.01$, $p = 0.01$, $p = 0.05$, respectively) (Table S3a).

3.2.2. WS7 vs. WS13: Stand Age of 50 Years vs. 110 Years

In comparison of the two watersheds with different stand ages, factor Watershed was a statistically significant predictor for all community indices but species richness. Herb-layer cover was greater in WS7 (younger stand) than in WS13 (older stand) (WS $p = 0.01$). Cover tended to be higher under sugar maple than under red maple in the older stand, with an opposite pattern in the younger stand (WS × M $p = 0.07$) (Figure 6E, Table S3a). Diversity and evenness were higher in the younger stand (WS $p = 0.01$ and $p = 0.02$, respectively) than the older stand and did not vary by overstory maple species (WS × M $p > 0.05$) (Figure 6G,H; Table S3a).

Considering individual understory species, the cover of red maple seedlings was greater beneath sugar maples than red maples in both watersheds (M $p = 0.0003$; Table S4).

Four environmental variables showed a trend (i.e., $p < 0.1$) in explaining herb-layer indices, but none of the variables was consistently significant across multiple indices (Table S3a).

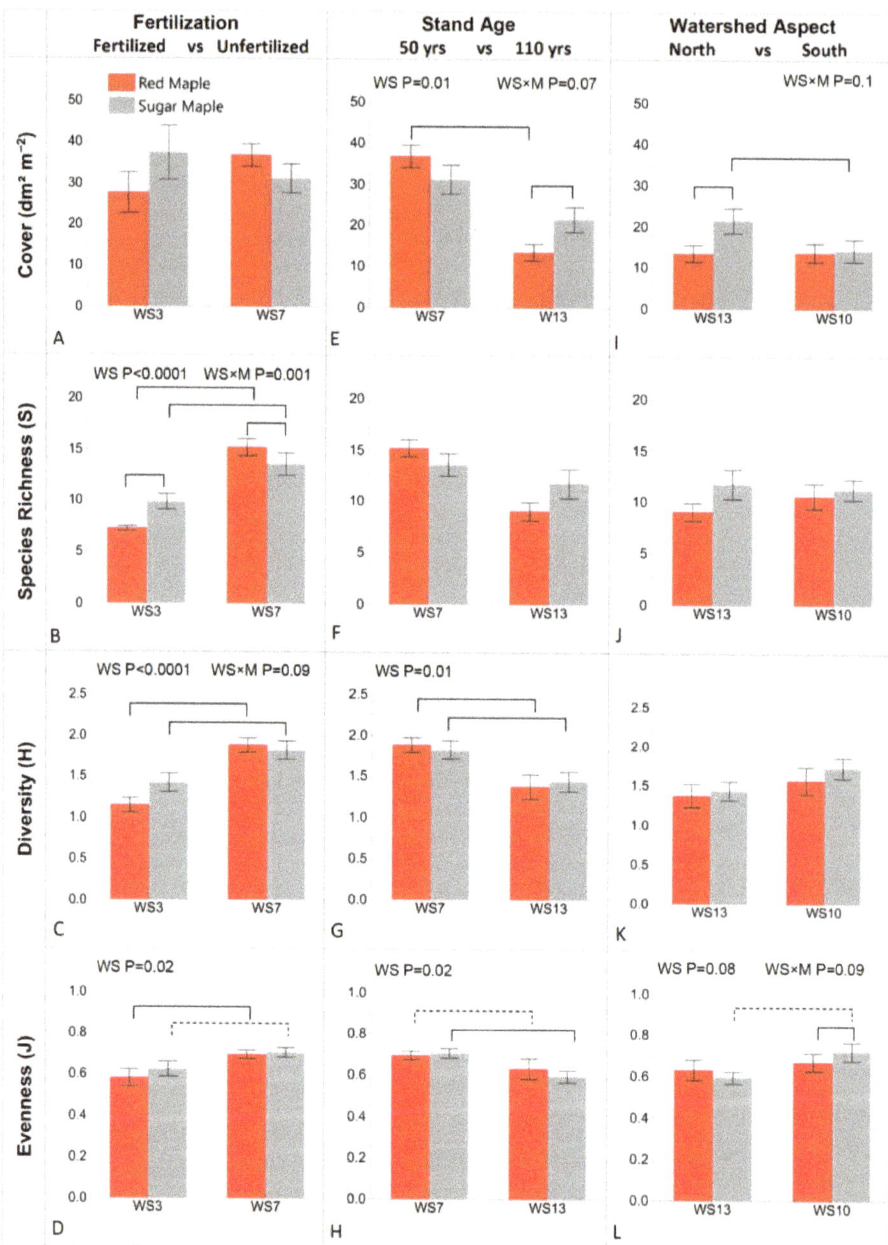

Figure 6. Herb-layer indices in red and sugar maple (M) plots in each of the studied watersheds (WS) at the Fernow Experimental Forest, West Virginia, USA, in the summers of 2018 and 2019. Watershed comparisons are: fertilized (WS3) vs. unfertilized (WS7) younger stands (**A–D**), younger (WS7) vs. older (WS13) unfertilized stands (**E–H**), and older unfertilized stands with northerly aspect (WS13) vs. southerly aspect (WS10) (**I–L**). If a horizontal bracket is connecting two means, the pairwise comparison (SAS slice effect) was significant at $p \leq 0.05$ (solid bracket) or $p \leq 0.1$ (dashed bracket). Error bars represent 1 SE based on nine plots per maple species per watershed, showing spatial variability within a watershed after respective plot-level values from 2018 and 2019 had been averaged. Statistical results are excerpts from the final ANOVA model (Table S3a), showing only factors WS, M, and WS $\times$ M if $p \leq 0.1$.

3.2.3. WS13 vs. WS10: Northerly vs. Southerly Watershed Aspect in 110-Year-Old Stands

In comparison of the two watersheds with the same older stand age but varying watershed aspect, factor Watershed, factor Maple, or their interaction (WS $\times$ M) were not statistically significant predictors for any of the herb-layer characteristics. There was a statistical trend of herb-layer cover under sugar maples being greater than under red maples in the north-facing watershed (WS13), while there was no difference in cover under the different maple species in the south-facing watershed (WS10) (WS $\times$ M $p = 0.1$) (Table S3a, Figure 6I). Evenness tended to be greater under sugar than red maples in WS10, with an opposite pattern in WS13 ($p = 0.09$) (Figure 6L). Species richness and diversity did not differ between watersheds (Figure 6J,K) (Table S3a).

Considering individual understory species, the cover of red maple seedlings was greater beneath sugar maples than red maples (M $p = 0.0003$) (Table S4). Environmental factors explaining significant variation (or a trend) in more than one herb-layer characteristic were plot-level aspect code and DBH of the plot-center maple. Herb cover increased and evenness decreased with increasing aspect code ($p = 0.09$ and $p = 0.01$). Diversity and evenness decreased with increasing DBH of the plot-center maple ($p = 0.09$ and $p = 0.03$) (Table S3a).

For all watershed pairs, spring herb-layer indices (Table S3b, Figure S1) were generally similar to summer herb-layer indices with the exception of expectedly lower cover.

4. Discussion

Plant communities are complex, which often results in noisy data sets. Nevertheless, this study showed that herb-layer composition and community indices can be affected by numerous factors, including abiotic environmental factors such as plot-level aspect and slope (expected results), overstory tree species (a novel finding), and anthropogenic activities, such as fertilization and land use history (i.e., stand age).

4.1. Influence of Abiotic Environmental Factors

The influence of abiotic factors that affect plant resources (e.g., soil moisture) has long been established in shaping plant communities. Therefore, it was not surprising to most frequently find a positive relationship of aspect code (i.e., increasing moisture availability) and negative relationships between slope steepness or plot distance to stream and the herb-layer indices (cover, S, H, or J) in the mountainous terrain of the FEF (Table S3a,b). A partial CCA indicated a relatively smaller importance of these abiotic variables compared to the group of predictor variables containing watershed-level treatment (fertilization, stand age, watershed aspect) and biotic variables (plot-center maple species, LAI, and basal area of ECM trees) in this study.

4.2. Herb-Layer Responses to Overstory Red Maple vs. Sugar Maple

Linkages between the overstory and the herb layer have been suggested to exist at spatial scales smaller than the landscape scale and may arise from parallel responses of strata to similar environmental gradients (e.g., soil pH, soil fertility, light) [19]. At the tree scale, this study revealed that the identity of the overstory tree species has a small but significant influence on the composition of the herb layer growing underneath (Figure 5). This response was more difficult to detect when the herb-layer composition was summarized into indices. There was no consistent response of the herb layer to the plot-center maple species across all watersheds; however, in summer (but not spring) the herb-layer response differed by maple species depending on the watershed. In five of the 12 models (four indices and three watershed pairs) there was a significant effect or trend of the WS $\times$ M interaction, indicating that the herb layer in summer benefitted from being under sugar maple relative to red maple in fertilized WS3 (50-year-old stand) and unfertilized WS13 (110-year-old north-facing stand), but not in unfertilized WS7 (50-year-old stand) and unfertilized WS10 (110-year-old south-facing stand) (Figure 6).

With none of the abiotic or biotic environmental variables measured in this study differing between red and sugar maple plots (Figure 7), mechanisms of a sugar maple effect on the herb layer may be located belowground. Since this study was not designed to establish causation of how individual tree species influence the herb layer (a heretofore unknown linkage), we can only speculate that differences in nutrient cycling below red and sugar maple may be a contributing factor (Figure 8). A positive feedback between litter quality and soil nutrient concentrations has been widely described (e.g., [63]). Soils beneath sugar maple have significantly less forest floor biomass (due to more rapid decomposition), a lower C:N ratio in the mineral soil layer [22,64], and significantly more calcium than soils under red maple [64,65]. At the FEF, sugar maple is associated with soils that have higher rates of nitrification and nitrate production than red maple [18]. With high mobility, nitrate may be more easily accessible to understory plants, but it is also susceptible to leaching [66]. Nitrification slightly decreases the pH of the soil, resulting in increased concentrations in soil solution and, thus, mobility of cations, such as calcium [67], which may benefit the herb layer under sugar maple (Figure 8).

Mycorrhizal fungi play an important role in the nutrition of forest trees and many herb layer species [68]. Mycorrhizae may, in fact, tighten the plant litter-soil feedback loop by facilitating more direct plant access to litter nutrients that otherwise would cycle first through the soil organic matter pool [69]. Since maples and most forest herbs are both associated with AM fungi [57,70], the linkage between these trees and their surrounding herb layer maybe a direct one via hyphal connections. While the fungal benefit from connections to multiple plant species lies in increased access to carbon, the benefits to the connected plants are not yet clear. It may be possible that "cheater" plants obtain their mineral nutrients from fungi at a lower carbon expense than other plants that provide relatively more carbon to the shared mycelium [71]. On native soil, red maple compared to sugar maple has been shown to have higher levels of mycorrhizal colonization [72]; whether this has any relevance to the herb layer is an avenue for exploration.

This study indicates that a sugar maple effect on the herb layer may not manifest under some conditions such as those in the younger unfertilized stand (WS7) and the older unfertilized stand with a southerly watershed aspect (WS10). It is possible, that in these watersheds a sugar maple effect is masked by other factors (e.g., stand age, watershed aspect). These factors may influence the herb layer via soil nutrient cycling (as reflected in differing stream water nitrate concentrations, Table 4). Under conditions of high external N inputs in WS3, sugar maple may be able to buffer—at the tree scale via litter-soil feedback—against potential nutrient imbalances at the watershed scale following excessive nitrate and calcium leaching (Table 4, [73]).

Table 4. Average stream water nutrient concentrations at the Fernow Experimental Forest (West Virginia, USA) over a 25-year period (1991–2015) after fertilization began in 1989. The linear regression between stream water nitrate and calcium concentrations yielded an $R^2 = 0.99$ ($n = 4$). SD—standard deviation (based on $n = 25$ years).

Watershed ID	Stream Water Ca^{2+} (mg/L) *	SD	Stream Water NO_3^- (mg/L) *	SD	Treatment	Stand Age in 2020 (year)
WS3	2.28	0.21	8.55	1.14	Fertilization	~50
WS7	2.07	0.14	4.64	0.67	No fertilization	~50
WS13	1.76	0.18	1.96	0.63	No fertilization, north aspect	~110
WS10	1.60	0.20	0.83	0.32	No fertilization, south aspect	~110

* Values are based on weekly grab samples taken upstream of the weirs. Stream water chemical analyses were conducted at the USDA Forest Service's Timber and Watershed Laboratory in Parsons, WV, using EPA—approved protocols. Sampling and analysis methods are described in Adams et al. [40]; data available at [74]. For this table, weekly values were first averaged for each year and then averaged over the 25-year period.

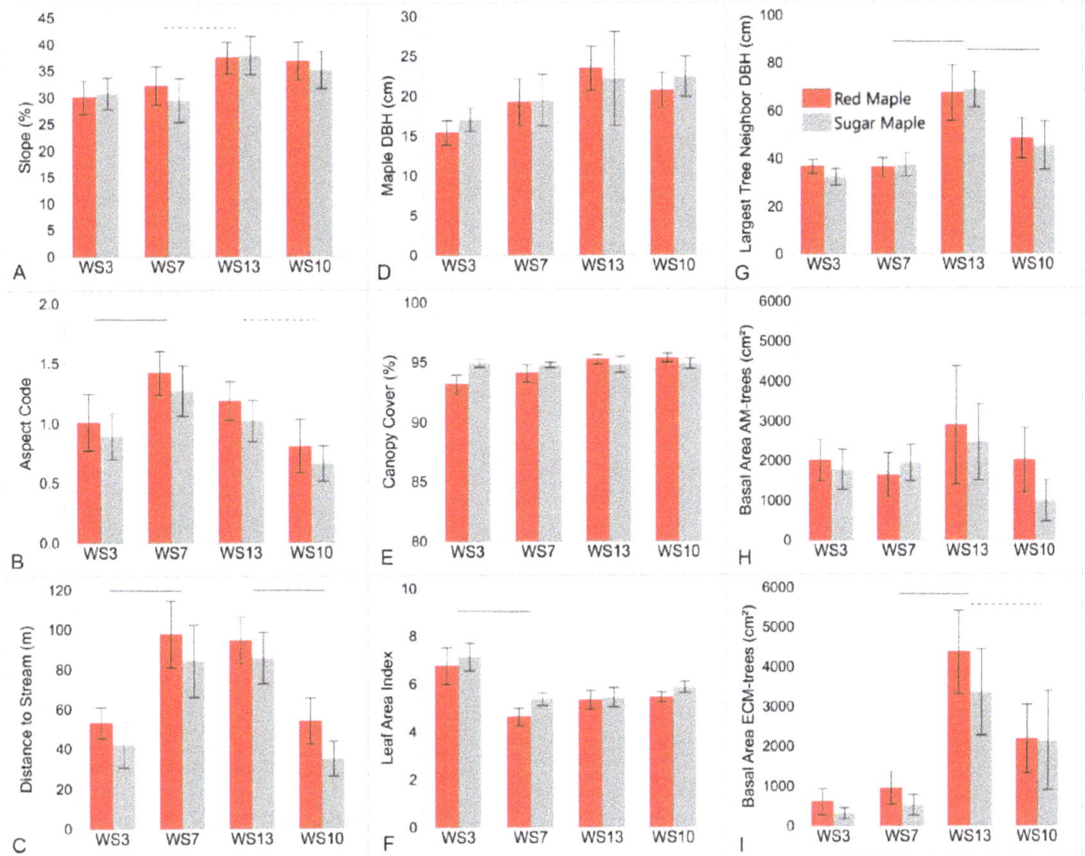

Figure 7. Abiotic and biotic factors potentially influencing study plots under red and sugar maples in each of the watersheds (WS) at the Fernow Experimental Forest, West Virginia, USA. (**A**) Slope measured at each plot, (**B**) plot-level aspect code ranging from 0 to 2, with 0 = southwest (relatively warm, dry) and 2 = northeast (relatively cool, moist), (**C**) slope distance between plot-center and the closest perennial stream, (**D**) diameter at breast height (DBH) of plot-center maples, (**E**) canopy cover measured as crown closure (in July 2018), (**F**) leaf area index (LAI) in Aug/Sep 2019, (**G**) DBH of the largest of the five trees closest to the plot-center maple, and basal area of arbuscular mycorrhizal (**H**) or ectomycorrhizal (**I**) trees among the five trees closest to the plot-center maple. Error bars represent 1 SE (*n* = 9). Based on two-way ANOVA containing all four watersheds, followed by least-squares-means contrasts, statistically significant differences between watersheds of watershed pairs are shown by solid lines (*p* < 0.05) and dashed lines (*p* < 0.1); factor Maple and Watershed × Maple interactions were not statistically significant for any of the environmental variables.

4.3. Herb-Layer Responses to Fertilization: WS3 vs. WS7

In this study, the herb-layer diversity indices (S, H, J) were consistently negatively affected by long-term N amendments to WS3 compared to unfertilized WS7. As the applied fertilizer was ammonium sulfate, changes in the understory may reflect direct responses to NH_4^+ or concomitant changes in soil properties (i.e., soil acidification, calcium loss; [73]), and indirect responses via competitive exclusion from nitrophilic species [33] or high LAI (Figure 7F). Interestingly, the tree canopy in fertilized WS3 was taller and more open than in unfertilized WS7 [75], which may have ameliorated potential LAI effects on the herb layer in fertilized WS3. Additionally, the average aspect code indicated a somewhat moister microclimate in unfertilized WS7 than fertilized WS3, and plot-center maples were located closer to the stream in fertilized WS3 than unfertilized WS7 (Figure 7B,C). These differences could, in part, be confounded with a fertilization effect. Alternatively, in WS3, the lower slope positions, where maples were found, may have been in compensation for the overall southerly watershed aspect (relative to WS7) without affecting the herb-layer indices measured in this study.

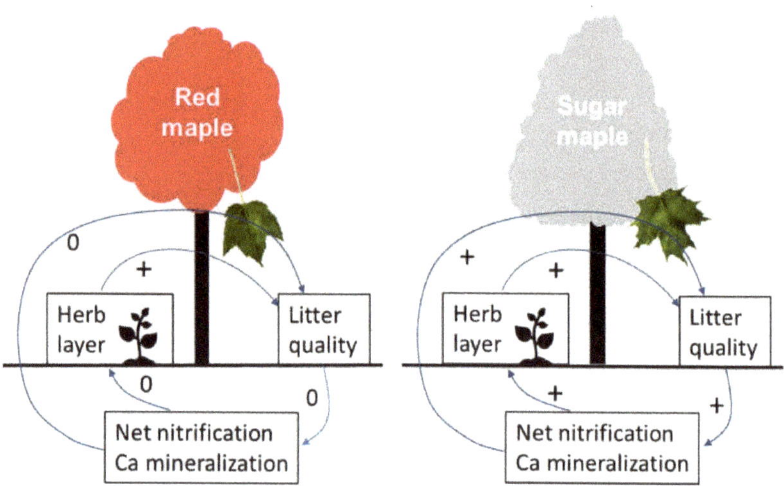

Figure 8. Conceptual diagram of how sugar maple and red maple may affect herb-layer cover through neutrally (0) or positively (+) influencing soil nitrate and calcium availability.

The lower values for S, H, and J in the fertilized than unfertilized watershed are predicted by the N homogenization hypothesis [29], stating that, as an ecosystem shifts from N limitation to N saturation, species richness decreases due to the displacement of N-efficient species by nitrophilic species. In this study, as in the study of Gilliam et al. [33], N-efficient species were displaced by species such as *Rubus* spp., altering community composition and decreasing biodiversity. The N homogenization hypothesis also states that the response time of the herbaceous layer to fertilization will depend on ambient N. For example, an environment with low N would react more quickly to additional N than an environment with high N. As sugar maple is known to be associated with higher soil N availability than red maple at the FEF [18], the vicinity of sugar maple could potentially delay the negative effects of fertilization on understory richness, diversity, and evenness. However, since fertilization began almost 30 years prior to this study, the sugar maple effect observed in this study is more likely due contemporaneous effects, e.g., sugar maple locally buffering against nutrient imbalances (Figure 8).

This study provided additional data to the research by Gilliam et al. [33] at the FEF comparing herb-layer characteristics during the first 25 years of N fertilization in WS3 to a different unfertilized watershed (WS4) with similar land use history as WS10/WS13. In their study, there was an increase in herb-layer cover in the fertilized watershed with a younger stand (WS3) in comparison to the unfertilized watershed with an older stand (WS4). In our study, the unfertilized watershed with a younger stand (WS7) also had a higher cover than the unfertilized watershed with the older stand (WS13) (Figure 6E). This indicates that the findings of Gilliam et al. [33] may also, at least in part, represent a stand age effect rather than solely a fertilization effect and, thus, highlights the need for careful selection of reference watersheds in case studies.

4.4. Herb-Layer Responses to Land Use History (Stand Age): WS7 vs. WS13

Herb cover and diversity indices (S, H, J) were consistently greater in the younger watershed (WS7) compared to the older watershed (WS13). The results of this study are in agreement with Bormann and Likens [76], who noted greater diversity in recently disturbed stands (25–75 years since disturbance) relative to mature stands at the Hubbard Brook Experimental Forest in New Hampshire. In a compilation of several studies on the effects of clearcutting on herb-layer richness and diversity (S, H) in northeastern US forests, effects varied between studies [77]. In 2-year to 26-year-old clearcuts, in comparison to mature second growth forest (55 to >150 years old), S and H were either higher than, or the same as, in clearcut sites than their reference sites. In contrast, S and cover of spring flora in secondary forests (stand age 45–87 years) of the southern Appalachian Mountains were lower than in old-growth reference stand [39].

Changes in species richness are influenced by stand development during succession after harvest, where initial high species richness is due to colonization of remnant shade-tolerant communities by shade intolerant herbs, followed by a decrease during the stem exclusions stage of stand development due to low light levels below the canopy, followed by an increase in the old-growth stage due to increases in canopy horizontal and vertical heterogeneity [77]. However, site conditions, such as soil fertility, may exert an overriding control on successional patterns of species diversity [77,78].

In this study, the two watersheds with different stand ages appeared to vary in soil nutrient resources. Stream water nitrate and calcium concentrations were higher in the younger stand (WS7) than in the older stand (WS13) (Table 4), which may reflect the high availability of these ions in soil solution prior to leaching. Higher nitrate concentrations in soil water 45 cm below the soil surface were found in the study plots of younger WS7 than older WS13 [55]. Since both watersheds are similar in their physical characteristics (Figure 7A–C), higher soil nutrient levels in WS7 than WS13 may be the result of the more recent disturbance. Mechanistically, this may be due to early successional tree species (black cherry, tulip-poplar) still dominating in WS7 50 years after harvest, whereas WS13 is dominated by oak (Table 1). In this study, neighboring trees, growing on average 4 m away from the plot-center maple (range 0.5–11.6 m), could influence the litter quality in the study plots. Surrounding the plot-center maples in WS13 were ECM trees (mostly northern red oak, Table 1) with a basal area about four times higher than ECM trees surrounding plots in WS3 (Figure 7I). Trees associated with ectomycorrhizal fungi produce litter that breaks down more slowly due to low N and P and high lignin content compared to arbuscular mycorrhizal trees [17,24,79]. The resulting feedback between low-quality litter and soil may affect the herbaceous layer by impacting nutrient access [69]. Further, while canopy cover and light levels above the herb layer (Figure 7E,F) were similar between the younger and the older stand, canopy structure might differ and might exert influence over the herb layer [80].

4.5. Herb-Layer Responses to Watershed Aspect: WS13 vs. WS10

For the two unfertilized watersheds with 110-year-old stands, there was no main effect ($p \leq 0.5$) of WS, i.e., watershed aspect did not predict herb-layer indices. While the average aspect code and distance to stream differed between these two watersheds, they were only significant predictors for evenness, but not richness or Shannon–Wiener diversity index. This supports the interpretation of differences in richness and diversity between fertilized WS3 and unfertilized WS7 as fertilization effect rather than these watersheds' differences in aspect code and plot distance to stream. Thus, while pseudo-replicated, the watershed-level differences of interest (i.e., fertilization and stand age) likely cause the differences in herb-layer indices in the comparisons of WS3 vs. WS7 and WS7 vs. WS13, respectively.

5. Conclusions

This study demonstrated (1) the possible existence of a sugar maple effect, i.e., sugar maple having a positive effect on herb-layer cover, (2) that long-term N enrichment can reduce ecosystem biodiversity by favoring nitrophilic plant species, (3) that herb-layer characteristics can be influenced by stand age, in that lower litter quality in later-successional stands (dominated by ECM trees, i.e., oaks) may lead to lower herb-layer abundance and diversity than in earlier successional stands dominated by AM trees, and (4) that watershed aspect did not influence herb-layer indices. While lower herb-layer cover and diversity indices in older relative to younger stands may be part of the natural successional trajectory in forests recovering from disturbance, lower herb-layer cover and diversity in the N-fertilized watershed indicates that anthropogenic activity may have fundamentally altered the overall structure and function of the eastern deciduous forest over the past decades of high atmospheric N deposition. Looking into the future, as maple species are shifting in abundance in the eastern United States [81] and atmospheric N deposition is decreasing in the eastern United States [82], concomitant change in the understory can be expected along with changes in ecosystem function due to feedback between diversity and productivity. Given the urgent need to adapt forest management to meet the challenges of climate change and other anthropogenic influences, this study justifies further examination of tree-herb layer interactions for a wider range of tree species, N-deposition levels, and stand ages in future studies.

Supplementary Materials: The following are available online at https://www.mdpi.com/article/10.3390/f12070829/s1; Table S1. Species found in each sampling campaign (Su—summer, Sp—spring) in studied watersheds (WS3, WS7, WS10, WS13) at the Fernow Experimental Forest, West Virginia, USA; Table S2. Proportion of the total number of species in different plant types at the watershed (WS) level at the Fernow Experimental Forest, West Virginia, USA; Table S3. Results of the statistical analyses (repeated measures ANOVA) of differences in herb-layer indices between watersheds (WS) at the Fernow Experimental Forest, West Virginia, USA, in (a) the summers of 2018 and 2019 and (b) the spring of 2019.; Table S4. ANOVA results testing the effect of watershed (WS), plot-center maple species (M) and their interaction (WS × M) for 22 individual herb-layer species found at the Fernow Experimental Forest, West Virginia, USA. The Benjamini-Hochberg method was applied to account for the familywise error rate; Figure S1. Spring 2019 plot-level herb-layer indices in red maple and sugar maple (M) plots in each of the studied watersheds (WS) at the Fernow Experimental Forest, West Virginia, USA.

Author Contributions: Conceptualization, K.S.; methodology, K.S.; formal analysis, L.J.S. and K.S.; investigation, L.J.S.; writing—original draft preparation, L.J.S. and K.S.; writing—review and editing, K.S.; visualization, L.J.S. and K.S.; supervision, K.S.; project administration, K.S.; funding acquisition, K.S. All authors have read and agreed to the published version of the manuscript.

Funding: This work was supported by the USDA National Institute of Food and Agriculture, McIntire Stennis Cooperative Research Program, project # WVA00129, the West Virginia Agricultural and Forestry Experiment Station, and a joint venture agreement (# 17-JV-11242303-065) with the USDA Forest Service, Northern Research Station.

Data Availability Statement: The data presented in this study are available on request from the corresponding author and at http://www.as.wvu.edu/fernow/data.html, accessed on 1 May 2021, Supplementary Material.

Acknowledgments: We thank Sian Eisenhut for field assistance, Ida Holaskova for supporting the statistical analyses, and the anonymous reviewers for improving the manuscript. We also thank the personnel at the Fernow Experimental Forest for maintaining and making available this outstanding research site and Mary Beth Adams and William Peterjohn for sharing their vast knowledge about the Fernow and the subject matter. Data on study site soil characteristics and basal area information of watersheds 10 and 13 were provided by William T. Peterjohn; these data were gathered as part of the Fernow Experimental Forest NSF LTRB awards DEB-0417678 and DEB-1019522.

Conflicts of Interest: The authors do not have any conflict of interest pertaining to this study.

References

1. Gilliam, F. The ecological significance of the herbaceous layer in temperate forest ecosystems. *Bioscience* **2007**, *57*. [CrossRef]
2. Muller, R.N. Nutrient relations of the herbaceous layer in deciduous forest ecosystems. In *The Herbaceous Layer in Forests of Eastern North America*; Gilliam, F.S., Roberts, M.R., Eds.; Oxford University Press: Oxford, UK, 2014; pp. 13–34.
3. Hooper, D.; Chapin, F.S., III; Ewel, J.J.; Hector, A.; Inchausti, P.; Lavorel, S.; Lawton, J.H.; Lodge, D.; Loreau, M.; Naeem, S.; et al. Effects of biodiversity on ecosystem functioning: A consensus of current knowledge. *Ecol. Monogr.* **2005**, *75*, 3–35. [CrossRef]

4.   Allan, E.; Weisser, W.W.; Fischer, M.; Schulze, E.-D.; Weigelt, A.; Roscher, C.; Baade, J.; Barnard, R.L.; Beßler, H.; Buchmann, N.; et al. A comparison of the strength of biodiversity effects across multiple functions. *Oecologia* **2013**, *173*, 223–237. [CrossRef] [PubMed]

5.   Tilman, D.; Knops, J.; Wedin, D.; Reich, P.; Ritchie, M.; Siemann, E. The influence of functional diversity and composition on ecosystem processes. *Science* **1997**, *277*, 1300–1302. [CrossRef]

6.   Muller, R.N.; Bormann, F.H. Role of *Erythronium americanum Ker.* in energy flow and nutrient dynamics of a northern hardwood forest ecosystem. *Science* **1976**, *193*, 1126–1128. [CrossRef] [PubMed]

7.   Melillo, J.M.; Aber, J.D.; Linkins, A.E.; Ricca, A.; Fry, B.; Nadelhoffer, K.J. Carbon and nitrogen dynamics along the decay continuum: Plant litter to soil organic matter. *Plant Soil* **1989**, *115*, 189–198. [CrossRef]

8.   Elliott, K.J.; Vose, J.; Knoepp, J.; Clinton, B.; Kloeppel, B. Functional role of the herbaceous layer in eastern deciduous forest ecosystems. *Ecosystems* **2015**, *18*, 221–236. [CrossRef]

9.   Wu, J.; Liu, Z.; Wang, X.; Zhou, L.; Lin, Y.; Fu, S. Effects of understory removal and tree girdling on soil microbial community composition and litter decomposition in two Eucalyptus plantations in South China. *Funct. Ecol.* **2011**, *25*, 921–931. [CrossRef]

10.   Maguire, D.; Forman, R.T.T. Herb cover effects on tree seedling patterns in a mature hemlock-hardwood forest. *Ecology* **1983**, *64*, 1347–1380. [CrossRef]

11.   Horsley, S.B. Role of allelopathy in hay-scented fern interference with black cherry regeneration. *J. Chem. Ecol.* **1993**, *19*, 2737–2755. [CrossRef]

12.   Gilliam, F. *The Herbaceous Layer in Forests of Eastern North America*; Oxford University Press: Oxford, UK, 2014; pp. 1–688.

13.   Bennie, J.; Huntley, B.; Wiltshire, A.; Hill, M.; Baxter, R. Slope, aspect and climate: Spatially explicit and implicit models of topographic microclimate in chalk grassland. *Ecol. Model.* **2008**, *216*, 47–59. [CrossRef]

14.   Cantlon, J.E. Vegetation and microclimates on north and south slopes of Cushetunk Mountain, New Jersey. *Ecol. Monogr.* **1953**, *23*, 241–270. [CrossRef]

15.   Måren, I.E.; Karki, S.; Prajapati, C.; Yadav, R.K.; Shrestha, B.B. Facing north or south: Does slope aspect impact forest stand characteristics and soil properties in a semiarid trans-Himalayan valley? *J. Arid Environ.* **2015**, *121*, 112–123. [CrossRef]

16.   Kang, H.; Kang, S.; Lee, D. Variations of soil enzyme activities in a temperate forest soil. *Ecol. Res.* **2009**, *24*, 1137–1143. [CrossRef]

17.   Mudrick, D.A.; Hoosein, M.; Hicks, R.R.; Townsend, E.C. Decomposition of leaf litter in an Appalachian forest: Effects of leaf species, aspect, slope position and time. *For. Ecol. Manag.* **1994**, *68*, 231–250. [CrossRef]

18.   Peterjohn, W.T.; Harlacher, M.A.; Christ, M.J.; Adams, M.B. Testing associations between tree species and nitrate availability: Do consistent patterns exist across spatial scales? *For. Ecol. Manag.* **2015**, *358*, 335–343. [CrossRef]

19.   Gilliam, F.; Roberts, M. Interactions between the Herbaceous Layer and Overstory Canopy of Eastern Forests. In *The Herbaceous Layer in Forests of Eastern North America*; Oxford University Press: Oxford, UK, 2014; pp. 233–254.

20.   Carlisle, A.; Brown, A.H.F.; White, E.J. The nutrient content of tree stem flow and ground flora litter and leachates in a sessile oak (*Quercus petraea*) woodland. *J. Ecol.* **1967**, *55*, 615–627. [CrossRef]

21.   Carl, R.C.; Ralph, E.J.B. Correlations of understory herb distribution patterns with microhabitats under different tree species in a mixed mesophytic forest. *Oecologia* **1984**, *62*, 337–343.

22.   Finzi, A.C.; Breemen, N.V.; Canham, C.D. Canopy tree-soil interactions within temperate forests: Species effects on soil carbon and nitrogen. *Ecol. Appl.* **1998**, *8*, 440–446. [CrossRef]

23.   Bigelow, S.; Canham, C. Litterfall as a niche construction process in a northern hardwood forest. *Ecosphere* **2015**, *6*, 117. [CrossRef]

24.   Cornelissen, J.; Aerts, R.; Cerabolini, B.; Werger, M.; van der Heijden, M. Carbon cycling traits of plant species are linked with mycorrhizal strategy. *Oecologia* **2001**, *129*, 611–619. [CrossRef] [PubMed]

25.   Cornelissen, J. An experimental comparison of leaf decomposition rates in a wide range of temperate plant species and types. *J. Ecol.* **1996**, *84*, 573–582. [CrossRef]

26.   Lovett, G.M.; Weathers, K.C.; Arthur, M.A. Control of nitrogen loss from forested watersheds by soil carbon:nitrogen ratio and tree species composition. *Ecosystems* **2002**, *5*, 712–718. [CrossRef]

27.   Lovett, G.M.; Mitchell, M.J. Sugar maple and nitrogen cycling in the forests of eastern North America. *Front. Ecol. Environ.* **2004**, *2*, 81–88. [CrossRef]

28.   Driscoll, C.; Whitall, D.; Aber, J.; Boyer, E.; Castro, M.; Cronan, C.; Goodale, C.; Groffman, P.; Hopkinson, C.; Lambert, K.; et al. Nitrogen pollution in the northeastern United States: Sources, effects, and management options. *Bioscience* **2003**, *53*. [CrossRef]

29.   Gilliam, F. Response of the herbaceous layer of forest ecosystems to excess nitrogen deposition. *J. Ecol.* **2006**, *94*, 1176–1191. [CrossRef]

30.   Rajaniemi, T.K. Why does fertilization reduce plant species diversity? Testing three competition-based hypotheses. *J. Ecol.* **2002**, *90*, 316–324. [CrossRef]

31.   Bobbink, R.; Hicks, K.; Galloway, J.; Spranger, T.; Alkemade, R.; Ashmore, M.; Bustamante, M.; Cinderby, S.; Davidson, E.; Dentener, F.; et al. Global assessment of nitrogen deposition effects on terrestrial plant diversity: A synthesis. *Ecol. Appl.* **2010**, *20*, 30–59. [CrossRef] [PubMed]

32.   Walter, C.A.; Raiff, D.T.; Burnham, M.B.; Gilliam, F.S.; Adams, M.B.; Peterjohn, W.T. Nitrogen fertilization interacts with light to increase Rubus spp. cover in a temperate forest. *Plant Ecol.* **2016**, *217*, 421–430. [CrossRef]

33.   Gilliam, F.; Billmyer, J.; Walter, C.; Peterjohn, W. Effects of excess nitrogen on biogeochemistry of a temperate hardwood forest: Evidence of nutrient redistribution by a forest understory species. *Atmos. Environ.* **2016**, *146*, 261–270. [CrossRef]

34. Halpern, C.B. Early successional patterns of forest species: Interactions of life history traits and disturbance. *Ecology* **1989**, *70*, 704–720. [CrossRef]
35. Kermavnar, J.; Eler, K.; Marinšek, A.; Kutnar, L. Post-harvest forest herb layer demography: General patterns are driven by pre-disturbance conditions. *For. Ecol. Manag.* **2021**, *491*, 119121. [CrossRef]
36. Götmark, F.; Paltto, H.; Nordén, B.; Götmark, E. Evaluating partial cutting in broadleaved temperate forest under strong experimental control: Short-term effects on herbaceous plants. *For. Ecol. Manag.* **2005**, *214*, 124–141. [CrossRef]
37. Roberts, M.R.; Zhu, L. Early response of the herbaceous layer to harvesting in a mixed coniferous–deciduous forest in New Brunswick, Canada. *For. Ecol. Manag.* **2002**, *155*, 17–31. [CrossRef]
38. Gilliam, F. Effects of harvesting on herbaceous layer diversity of a Central Appalachian Hardwood forest in West Virginia, USA. *For. Ecol. Manag.* **2002**, *155*, 33–43. [CrossRef]
39. Duffy, D.; Meier, A. Do Appalachian herbaceous understories ever recover from clearcutting? *Conserv. Biol.* **1992**, *6*, 196–201. [CrossRef]
40. Adams, M.B.; Edwards, P.J.; Wood, F.; Kochenderfer, J.N. Artificial watershed acidification on the Fernow Experimental Forest, USA. *J. Hydrol.* **1993**, *150*, 505–519. [CrossRef]
41. Adams, M.B.; Kochenderfer, J.N.; Wood, F.; Angradi, T.R.; Edwards, P. *Forty Years of Hydrometeorological Data from the Fernow Experiment Forest, West Virginia*; U.S. Department of Agriculture, Forest Service, Northeastern Forest Experiment Station: Radnor, PA, USA, 1994; p. 24.
42. Gilliam, F.; Turrill, N.; Aulick, S.; Evans, D.; Adams, M.B. Herbaceous layer and soil response to experimental acidification in a central Appalachian hardwood forest. *J. Environ. Qual.* **1994**, *23*, 835–844. [CrossRef]
43. Adams, M.B.; Edwards, P.J.; Ford, W.M.; Schuler, T.M.; Thomas-Van Gundy, M.; Wood, F. Fernow Experimental Forest: Research History and Opportunities. In *Experimental Forests and Ranges EFR-2*; USDA Forest Service: Washington, DC, USA, 2012.
44. Peterjohn, W.T. Fernow Experimental Forest LTREB. Soil Chemistry 2011. Available online: http://www.as.wvu.edu/fernow/data.html (accessed on 29 May 2021).
45. Adams, M.B.; Kochenderfer, J.; Edwards, P. *The Fernow Watershed Acidification Study: Ecosystem Acidification, Nitrogen Saturation and Base Cation Leaching*; Springer: Dordrecht, The Netherlands, 2007; Volume 7, pp. 267–273.
46. Trimble, G.R., Jr.; Tryon, E.H.; Smith, H.C.; Hillier, J.D. *Age and Stem Origin of Appalachian Hardwood Reproduction Following a Clearcut-Herbicide Treatment*; U.S. Department of Agriculture, Forest Service, Northeastern Forest Experiment Station: Upper Darby, PA, USA, 1986; Volume 589.
47. Kochenderfer, J.N. Fernow and the Appalachian Hardwood Region. In *The Fernow Watershed Acidification Study*; Adams, M.B., DeWalle, D.R., Hom, J.L., Eds.; Springer: Dordrecht, The Netherlands, 2006; pp. 17–19.
48. Aubertin, G.M.; Patric, J.H. Water quality after clearcutting a small watershed in West Virginia. *J. Environ. Qual.* **1974**, *3*, 243–249. [CrossRef]
49. Adams, M.B.; DeWalle, D.R.; Peterjohn, W.T.; Gilliam, F.S.; Sharpe, W.E.; Williard, K.W.J. Soil Chemical Response to Experimental Acidification Treatments. In *The Fernow Watershed Acidification Study*; Adams, M.B., DeWalle, D.R., Hom, J.L., Eds.; Springer: Dordrecht, The Netherlands, 2006; pp. 41–69.
50. Peterjohn, W.T. Fernow Experimental Forest LTREB. Tree Survey Data 2000. Available online: http://www.as.wvu.edu/fernow/data.html (accessed on 10 June 2021).
51. Walter, C.; Burnham, M.; Gilliam, F.; Peterjohn, W. A reference-based approach for estimating leaf area and cover in the forest herbaceous layer. *Environ. Monit. Assess.* **2015**, *187*, 657. [CrossRef] [PubMed]
52. Begon, M.; Harper, J.L.; Townsend, C.R. *Ecology: Individuals, Populations, and Communities*, 3rd ed.; Blackwell Publishing: Berlin, Germany, 1996.
53. Pielou, E.C. *Ecological Diversity*; Wiley: New York, NY, USA, 1975.
54. Shannon, C.E.; Weaver, W. *The Mathematical Theory of Communication*; University of Illinois Press: Champaign, IL, USA, 1949; p. 117.
55. Eisenhut, S.; Stephan, K. The role of tree species, the herb layer, and watershed characteristics on nitrogen cycling in a central Appalachian hardwood forest. 2021; (Unpublished; Manuscript in Preparation).
56. Beers, T.W.; Dress, P.E.; Wensel, L.C. Notes and observations: Aspect transformation in site productivity research. *J. For.* **1966**, *64*, 691–692.
57. Brundrett, M.; Murase, G.; Kendrick, B. Comparative anatomy of roots and mycorrhizae of common Ontario trees. *Can. J. Bot.* **1990**, *68*, 551–578. [CrossRef]
58. Wang, B.; Qiu, Y.L. Phylogenetic distribution and evolution of mycorrhizas in land plants. *Mycorrhiza* **2006**, *16*, 299–363. [CrossRef]
59. ter Braak, C.; Smilauer, P.N. *Canoco Reference Manual and User's Guide: Software for Ordination (Version 5.10)*; Microcomputer Power: Ithaca, NY, USA, 2018; p. 536.
60. McDonald, J.H. *Handbook of Biological Statistics*, 2nd ed.; Sparky House Publishing: Baltimore, MD, USA, 2014.
61. ter Braak, C.; Smilauer, P.N. *Canoco Reference Manual and CanoDraw for Windows User's Guide: Software for Canonical Community Ordination (Version 4.5)*; Microcomputer Power: Ithaca, NY, USA, 2002; p. 500.
62. Leps, J.; Smilauer, P. *Multivariate Analysis of Ecological Data Using CANOCO*; Cambridge University Press: Cambridge, UK, 2003; Volume 3, p. 282.
63. Hobbie, S.E. Effects of plant species on nutrient cycling. *Trends Ecol. Evol.* **1992**, *7*, 336–339. [CrossRef]

64. Vitousek, P.M.; Gosz, J.R.; Grier, C.C.; Melillo, J.M.; Reiners, W.A. A comparative analysis of potential nitrification and nitrate mobility in forest ecosystems. *Ecol. Monogr.* **1982**, *52*, 155–177. [CrossRef]
65. Dijkstra, F.A. Calcium mineralization in the forest floor and surface soil beneath different tree species in the northeastern US. *For. Ecol. Manag.* **2003**, *175*, 185–194. [CrossRef]
66. Boudsocq, S.; Niboyet, A.; Lata, J.-C.; Raynaud, X.; Loeuille, N.; Mathieu, J.; Blouin, M.; Abbadie, L.; Barot, S. Plant preference for ammonium versus nitrate: A neglected determinant of ecosystem functioning? *Am. Nat.* **2012**, *180*, 60–69. [CrossRef] [PubMed]
67. Peterjohn, W.T.; Adams, M.B.; Gilliam, F.S. Symptoms of nitrogen saturation in two central Appalachian hardwood forest ecosystems. *Biogeochemistry* **1996**, *35*, 507–522. [CrossRef]
68. Brundrett, M.; Kendrick, B. The roots and mycorrhizas of herbaceous woodland plants. I. Quantitative aspects of morphology. *New Phytol.* **1990**, *114*, 457–468. [CrossRef]
69. Chapman, S.K.; Langley, J.A.; Hart, S.C.; Koch, G.W. Plants actively control nitrogen cycling: Uncorking the microbial bottleneck. *New Phytol.* **2006**, *169*, 27–34. [CrossRef] [PubMed]
70. Brundrett, M.; Kendrick, B. The mycorrhizal status, root anatomy, and phenology of plants in a sugar maple forest. *Can. J. Bot.* **1988**, *66*, 1153–1173. [CrossRef]
71. Jakobsen, I. Hyphal fusion to plant species connections–Giant mycelia and community nutrient flow. *New Phytol.* **2004**, *164*, 4–7, 961–972. [CrossRef]
72. St Clair, S.B.; Lynch, J.P. Base cation stimulation of mycorrhization and photosynthesis of sugar maple on acid soils are coupled by foliar nutrient dynamics. *New Phytol.* **2005**, *165*, 581–590. [CrossRef]
73. Gilliam, F.; Adams, M.B.; Peterjohn, W. Response of soil fertility to 25 years of experimental acidification in a temperate hardwood forest. *J. Environ. Qual.* **2020**, *49*. [CrossRef]
74. Edwards, P.; Wood, F. Fernow Experimental Forest Stream Chemistry. Available online: https://doi.org/10.2737/RDS-2011-0017 (accessed on 29 May 2021).
75. Atkins, J.W.; Bond-Lamberty, B.; Fahey, R.T.; Haber, L.T.; Stuart-Haëntjens, E.; Hardiman, B.S.; LaRue, E.; McNeil, B.E.; Orwig, D.A.; Stovall, A.E.L.; et al. Application of multidimensional structural characterization to detect and describe moderate forest disturbance. *Ecosphere* **2020**, *11*, e03156. [CrossRef]
76. Bormann, F.H.; Likens, G.E. *Pattern and Process in a Forested Ecosystem: Disturbance, Development and the Steady State Based on the Hubbard Brook Ecosystem Study*; Springer: New York, NY, USA, 1979.
77. Roberts, M.; Gilliam, F. Response of the Herbaceous Layer to Disturbance in Eastern Forests. In *The Herbaceous Layer in Forests of Eastern North America*; Oxford University Press: Oxford, UK, 2014; pp. 320–339.
78. Small, C.; McCarthy, B. Spatial and temporal variation in the response of understory vegetation to disturbance in a central Appalachian oak forest. *J. Torrey Bot. Soc.* **2002**, *129*, 136. [CrossRef]
79. Melillo, J.M.; Aber, J.D.; Muratore, J.F. Nitrogen and lignin control of hardwood leaf litter decomposition dynamics. *Ecology* **1982**, *63*, 621–626. [CrossRef]
80. Gilliam, F. Response of herbaceous layer species to canopy and soil variables in a central Appalachian hardwood forest ecosystem. *Plant Ecol.* **2019**, *220*, 1131–1138. [CrossRef]
81. Fei, S.; Steiner, K. Evidence for increasing red maple abundance in the eastern United States. *For. Sci.* **2007**, *53*, 473–477.
82. Gilliam, F.; Burns, D.; Driscoll, C.; Frey, S.; Lovett, G.; Watmough, S. Decreased atmospheric nitrogen deposition in eastern North America: Predicted responses of forest ecosystems. *Environ. Pollut.* **2018**, *244*, 560–574. [CrossRef] [PubMed]

forests

Article

Long-Term Projection of Species-Specific Responses to Chronic Additions of Nitrogen, Sulfur, and Lime

Alexander Storm [1], Mary Beth Adams [2] and Jamie Schuler [3,*]

[1] Department of Arts, Sciences, and Natural Resources, Haywood Community College, Clyde, NC 28721, USA; ajstorm@haywood.edu
[2] United States Forest Service Northern Research Station (Retired), Morgantown, WV 26506, USA; MaryBeth.Adams@mail.wvu.edu
[3] Division of Forestry & Natural Resources, West Virginia University, Morgantown, WV 26506, USA
* Correspondence: Jamie.schuler@mail.wvu.edu

Abstract: Elevated acid deposition has been a concern in the central Appalachian region for decades. A long-term acidification experiment on the Fernow Experimental Forest in central West Virginia was initiated in 1996 and continues to this day. Ammonium sulfate was used to simulate elevated acid deposition. A concurrent lime treatment with an ammonium sulfate treatment was also implemented to assess the ameliorative effects of base cations to offset acidification. We show that the forest vegetation simulator growth model can be locally calibrated and used to project stand growth and development over 40 years to assess the impacts of acid deposition and liming. Modeled projections showed that pin cherry (initially) and sweet birch responded positively to nitrogen and sulfur additions, while black cherry, red maple, and cucumbertree responded positively to nitrogen, sulfur, and lime. Yellow-poplar negatively responded to both treatments. Despite these differences, our projections show a maximum of 5% difference in total stand volume among treatments after 40 years.

Keywords: nitrogen saturation; forest vegetation simulator; Appalachian hardwoods

Citation: Storm, A.; Adams, M.B.; Schuler, J. Long-Term Projection of Species-Specific Responses to Chronic Additions of Nitrogen, Sulfur, and Lime. *Forests* **2021**, *12*, 1069. https://doi.org/10.3390/f12081069

Academic Editor: Douglas Godbold

Received: 8 July 2021
Accepted: 4 August 2021
Published: 11 August 2021

Publisher's Note: MDPI stays neutral with regard to jurisdictional claims in published maps and institutional affiliations.

1. Introduction

The influence of acid rain on the health and productivity of forest ecosystems has been a topic of research interest since the late-1970s to early-1980s [1–3]. High concentrations of coal fired power plants in the Ohio River Valley have historically been a major source of nitrous oxides (NO_x) and sulfur oxides (SO_x), which are released into the atmosphere [4]. These gases are released from the combustion of fossil fuels and once in the atmosphere, both gases have a high affinity for water vapor and quickly form the most common forms of acid rain: nitric (HNO_3) and sulfuric (H_2SO_4) acids [5]. Due to the relatively short mean residence time of water in the atmosphere, these acids are deposited in higher quantities in the form of acidic precipitation across the mid-Atlantic and northeastern United States [6].

When nitric and sulfuric acids are deposited into forested areas in the Central Appalachian region, they have both direct and indirect negative influences on the associated forest soils. A large percentage of the soils of the central Appalachian hardwood forest region are at risk for base cation depletion from soil acidification and nitrogen saturation [7]. These soils are normally low in base cations such as calcium (Ca^{2+}) and magnesium (Mg^{2+}), which also limits their buffering capacity [8]. As H_2SO_4 and HNO_3 deposition increases, hydrogen ions (H^+) disassociate in soil solution, which increases the rates of mineral weathering and displaces cations from cation exchange sites [9,10]. Additionally, H^+ ions associate with aluminum (Al^{3+}) bearing compounds, increasing the amount of aluminum ions in soil solution, which negatively affects root growth [11,12]. Aluminum ions have a high affinity for cation exchange sites and displace plant essential cations such as Ca^{2+} and Mg^{2+} [11,12]. If not taken up by plants rather quickly, calcium and magnesium ions are leached from the soil, further decreasing soil fertility [4]. Increased Al^{3+} concentrations in soil can also reduce the rate of nitrate (NO_3^-) uptake by trees, allowing increased NO_3^- leaching from soils [13].

A secondary process associated with the deposition of nitric acid is nitrogen saturation. When inputs of nitrogen from atmospheric deposition, mineralization, and atmospheric sequestration become greater than the need of the organisms in the system, an ecosystem can be considered N saturated [14]. Excess nitrate is leached from the soil due to the low capacity of central Appalachian forests to retain nitrate [15]. As systems reach nitrogen saturation, the nitrification of ammonium (NH_4^+) to nitrate increases, leading to additional losses of N from the system [16]. Other negative effects of increased N include lower soil pH values [16–18], reduced base cation uptake due to Al^{3+} mobilization [18–20], and increased release of greenhouse gas emissions from soils [21,22].

The response of forests to inputs of chronic N and sulfur (S) are dependent on location and duration of the inputs. The Fork Mountain Long-Term Soil Productivity site (LTSP) was established within the U.S. Forest Service Fernow Experimental Forest (FEF) in 1996 as part of a nationwide effort to quantify the basic controls on forest soil productivity [23]. One of the goals of the Fork Mountain LTSP study was to characterize the effects of acid deposition on newly regenerating central Appalachian forest vegetation. The acid deposition was simulated by annual N + S additions at rates mimicking deposition at the time. Additionally, an ameliorative treatment in which lime was added to balance the N + S additions was also evaluated.

Research on this site and others within the FEF have highlighted some of the immediate impacts of increased nitrogen on forest vegetation. For example, commercially important species such as yellow-poplar (*Liriodendron tulipifera* L.) show decreased incremental growth after seven years of nitrogen and sulfur inputs [24]. Similarly, biomass accumulation by sweet birch (*Betula lenta* L.) and yellow-poplar decreased in areas with increased chronic N inputs [24]. Fowler et al. [25] concurrently observed a decrease in tree species diversity associated with elevated rates of N and S. However, the total plant biomass of a forest community can be stimulated by sustained increases in nitrogen inputs [26–28]. The extent and duration of these responses are variable and species-specific, but eventually, the negative effects of soil acidification are theorized to outweigh the positive effect of increased nitrogen availability [29].

Unfortunately, the long-term effects of elevated acidic deposition on forest growth have been studied to a lesser extent, mainly due to the lack of long-term/rotation length data [30]. As such, growth and yield models can offer the opportunity to examine these impacts on time scales outside the available data. The goal of our paper was to demonstrate that the Forest Vegetation Simulator (FVS), which is a distance-independent, single-tree growth model [31], can be calibrated to central Appalachian forests and used to model the impacts of elevated acidic inputs over time. FVS is a widely used growth simulator for addressing forest changes over time due to natural succession, management and natural disturbances, and proposed management. The regional variant (NE) of FVS covers the northeast region from West Virginia to Maine, which gives it broad applicability to modeling growth and stand development. However, since the model covers such a large region, model estimates tend to vary considerably, and out-of-the-box performance is often undesirable [32]. Many have noted the need to calibrate FVS in order to achieve more accurate simulations (e.g., [33,34]).

Although other studies have used FVS for similar goals (e.g., [30]), our application to the LTSP study provides a unique opportunity to model long-term stand growth for an Appalachian hardwood stand that was exposed to these conditions for over 20 years beginning at its inception, rather than modeling existing stands or simulating regeneration. This study utilizes data from the LTSP study to compare the effects of elevated acidic deposition and an ameliorative liming treatment on stand growth.

2. Materials and Methods

Data used for this study were derived from stands located in the USDA Forest Service Fernow Experimental Forest (FEF) near Parsons, West Virginia (latitude 39°04′ N, longitude 79°41′ W). The 1862-ha forest has been utilized as a research and teaching forest by the USFS since its establishment in 1934. Over this time period, key topics of long-term research have included silvicultural management for the production of high-quality hardwoods, the effects of harvesting on water quality, and ecosystem responses to acidic deposition.

The Fork Mountain Long-term Soil Productivity (LTSP) site is approximately 12.1 ha with a predominant southeast aspect and slopes between 15 and 30 percent, and the elevation ranges from approximately 792 to 853 m a.s.l. At the initiation of the study, the stand was 85 years old, and the site index (base age 50) for red oak was 24.3 m. The most prevalent species across the study site included northern red oak (*Quercus rubra* L.), sugar maple (*Acer saccharum* Marsh.), black cherry (*Prunus serotina* Ehrh.), and yellow-poplar, which comprised over 70% of the stand basal area. The soils on site are Calvin and Berks (Loamy-skeletal, mixed, active, mesic Typic Dystrudepts) and Hazelton series (Loamy-skeletal, siliceous, active, mesic Typic Dystrudepts) and are derived from sandstone colluvium, sandstone residuum, and weathered shale. Soils are well drained and loamy with the typical soil chemical characteristics for the region (Appendix A, Table A1). More specific details concerning the site and vegetation characteristics can be found in Adams et al. [35].

In the LTSP, the effects of elevated rates of acidic deposition on forest productivity were examined using four treatments replicated four times across the site. The treatments included an uncut control (CTRL), whole tree harvest (WT), whole tree harvest with the addition of ammonium sulfate fertilizer (WT + NS), and whole tree harvest + ammonium sulfate fertilizer + dolomitic lime (WT + NS + LIME). However, for the purposes of this study, we only considered the harvested plots (WT, WT + NS, and WT + NS + LIME treatments). Each treatment plot encompassed 0.4047 ha. The ammonium sulfate was added at twice the ambient nitrogen (15.0 kg N/ha/yr) and sulfur (17.0 kg S/ha/yr) deposition rates [35]. The dolomitic lime was added at twice the rate of calcium (11.2 kg Ca/ha/yr) and magnesium (5.8 kg Mg/ha/yr) export from a watershed in close proximity to the site [36]. The ammonium sulfate and dolomitic lime have been applied during March, July, and November each year since the site was harvested. The dolomitic lime addition was designed to mitigate the negative effects of soil acidification in order to better understand the effect of continual nitrogen addition on the ecosystem.

Data were collected across the study site using a total of 60 randomly selected subplots (five per treatment plot on each of the 12 treatment and block replicates) that were sampled in 2017 (age 21). Each subplot consisted of two nested circular measurement plots. Trees between 2.5–12.7 cm dbh were measured on 0.004 ha (3.59 m radius) plots, while trees greater than 12.7 cm dbh were measured on 0.04 ha (11.35 m radius) plots. Total height was measured using a clinometer on select dominant and codominant trees of the most common species on site. The number of height measurements varied by species, depending on the availability of codominant and dominant stems within the plots.

2.1. Growth Projections

Growth of the three treatments was projected using the northeast variant of FVS [37]. Datasets from 18 plots in nine calibration stands located across the FEF were used to modify the model to local conditions. These stands were all regenerated using an even age seed tree regeneration method, with initial harvest occurring between August 1960 and August 1962 [38]. Initial harvests removed most of the trees, with the remaining trees removed within three years after the initial harvest. The landscape features varied among sites (Table 1), but generally the plots ranged between 610–915 m a.s.l. in elevation. The soils are typically characterized as moderately deep, well drained residuum that were formed from the weathering of shale, sandstone, or siltstone. The Belmont series is the one exception, which was derived from mainly limestone (USDA, n.d.) and was retained to provide the range of species and stand conditions needed to calibrate the model.

Each calibration stand was quantified using two permanent 0.10 ha plots (18 total) that have been periodically measured by U.S. Forest Service personnel since the stands were approximately 20 years of age. All trees greater than 2.5 cm dbh were tagged and dbh measurements were recorded for each tree.

Thirty years of individual tree data from the 18 permanent plots on the FEF were used to calibrate FVS to local growing conditions. Initially, the base model performance was tested against this dataset to determine what, if any, modifications were necessary. Trees per ha (TPH) and basal area (m^2/ha) were used as the metrics for comparison between actual and predicted values. The evaluation of non-calibrated model projections indicated poor out-of-box performance. The two parameters of greatest concern within the model were the mortality and large tree basal area growth (trees $\geq$ 12.7 cm dbh). A workflow for the calibration and validation of the FVS NE model to local growing conditions of the FEF is provided (Figure 1).

Mortality was adjusted in the FVS based on a maximum stand density index (SDI) value. Maximum stand density index was based off calculated SDI values using the data available from the calibration stands across all time periods (ages 20–50). Within the calibration stands, the maximum observed SDI value was 786 and was used for all future model simulations. The default percentages of 55% and 85% for initiating density-dependent mortality and stand maximum density, respectively, were retained [37].

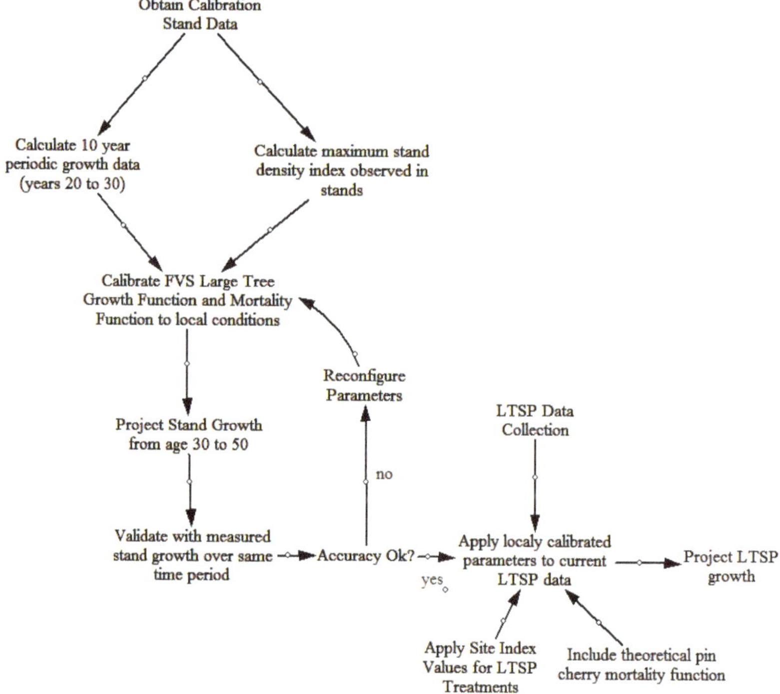

Figure 1. Workflow for calibration and validation of the northeast variant of the forest vegetation simulator for the Fernow Experimental Forest.

Table 1. Description of site features and age 20 stand parameters associated with the 18 calibration plots used for model calibration.

Stand	Soil Series	Basal Area (m²/ha)	DBH (cm)	Aspect	Elevation (m a.s.l.)	Slope (%)
32	Cateache	14.8	9.1	N	894	15
32	Cateache	17.9	10.4	NW	899	15
33	Belmont	16.0	11.7	N	781	55
33	Belmont	19.2	9.7	N	784	55
34	Calvin	16.2	10.2	NW	772	30
34	Calvin	17.1	9.9	NW	777	30
35	Calvin	15.3	9.7	SW	800	55
35	Dekalb	17.5	10.7	SW	815	55
36	Calvin	12.5	9.7	SE	600	45
36	Calvin	11.5	9.4	SE	573	45
37	Calvin	14.9	9.9	SE	583	55
37	Calvin	15.9	11.4	SE	591	55
38	Calvin	15.8	9.1	SW	774	35
38	Calvin	15.6	9.1	SW	776	35
39	Dekalb	18.7	9.9	SE	836	5
39	Dekalb	18.6	10.4	SE	836	5
43	Belmont	17.5	10.2	NW	856	20
43	Belmont	12.4	9.1	NW	863	20

The growth data from the first 10 years of periodic measurements were used to calibrate the large tree basal area growth. Each live tree (at age 30) was assigned a 10 year incremental growth value, equivalent to the observed growth from age 20–30. The "Growth" keyword was used to read the data into the model and project the growth of individuals based on observed values. The "CalbStat" keyword was used to calculate the growth of each species relative to the base model predicted growth. The model was based on a one year time step and growth modifications that were applied during every time step in the projection.

To account for differences in growth on the treatments of the LTSP plots, treatment specific site index values were included in the model. Yellow-poplar was chosen as the species for this adjustment due to its high abundance and large number of dominant and codominant stems in the stand. Site index values were estimated from site index curves for the Appalachian mountain region [38]. Site index (base age 50 years) values for the three treatments areas were calculated as 35.1 m for WT plots and 33.5 m for WT + NS and WT + NS + LIME plots. There was a consistent trend of higher site index values for the WT plots in comparison to the WT + NS and WT + NS + LIME for most species. The projections for the WT + NS and WT + NS + LIME plots assumed that the response to the treatments would remain constant across all time periods of the model projection.

A second scenario (continual decline scenario) was included in the model for the WT + NS and WT + NS + LIME treatments to describe a continued negative response of the overstory to nitrogen and sulfur inputs. On average, yellow-poplar trees in fertilized plots grew about 1.5 m less in total height (after 20 years) compared to non-fertilized trees. If this negative trend were to occur for the remainder of the projection period, the site index at the end of the projection period (age 60 years) would be reduced to 32.9 m for the fertilized treatments [39].

One final adjustment made to the final model was to account for the natural dynamics of pin cherry (*Prunus pensylvanica* L. f.). There were very few pin cherries present on the calibration plots by the time the measurements began, so it was not reasonable to think that the model would be able to predict the loss of pin cherry from the site based solely on the calibration data. A theoretical mortality function was included that would generally follow the dynamics described by [40]. In the LTSP study, pin cherry dominance had already begun to senesce based on severely declining relative importance values in 2012 and 2017 [41]. Pin cherry tree mortality was modified in the model by removing

50% of each tree record for each time step until the species was no longer present in the overstory. Mortality was initially concentrated on smaller trees and secondarily on the larger trees. This instruction to the model assumes that the smaller pin cherry trees were less competitive for light resources and therefore die sooner than the larger trees that are receiving full light [42].

2.2. Model Validation

Model validation protocol generally follows the framework developed by the USFS FVS Steering Team [43]. Variables predicted by the base model (ba/ha, TPH, quadratic mean diameter) were first verified by comparing general stand dynamic patterns. Base model performance was analyzed using mean percent error (MPE), root mean squared error (RMSE), and graphical representations of basal area and trees per ha changes over time. Locally calibrated values for maximum SDI and large tree basal area growth were included in the model. The observed vs. predicted TPH and ba/ha values for each calibration stand were again analyzed using MPE, RMSE, and graphs. Once prediction error was reduced to less than 15% MPE for both TPH and ba/ha [44], the modified model parameters were applied to the LTSP plot data.

3. Results

3.1. Model Calibration

Overall, the base model (uncalibrated) performed poorly for TPH and basal area. Uncalibrated FVS consistently predicted higher TPH and lower basal area per ha values relative to field measurements (Figure 2). Across all stands and measurement periods, FVS over-predicted TPA values by approximately 33%, with a maximum over-prediction of 90%. The mean percent error of TPH at the end of the 30 year projection period was 59% greater than measurements in the nine calibration stands. Average basal area projections were approximately 19% lower on average than the observed values for all stands and time periods.

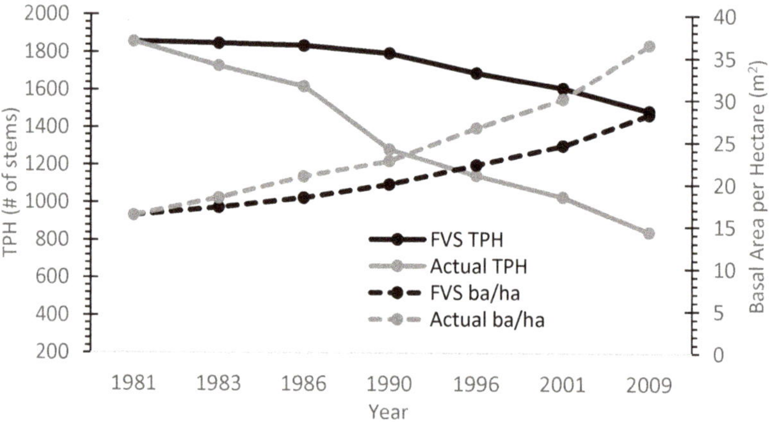

Figure 2. Base FVS model performance plotted against actual growth data for calibration stand 32. Model bias was consistent across all calibration stands.

Results from the large tree basal area growth calibration revealed several species growing at higher rates on the FEF compared to base model predictions. Increased growth rates were calculated for the following species: red maple (*Acer rubrum* L.) (13.2%), sweet birch (41.8%), yellow-poplar (14.2%), black cherry (21.5%), chestnut oak (*Quercus montana* Willd.) (10.3%), and slippery elm (*Ulmus rubra* Muhl.) (4.8%) (Appendix A, Table A2). For each species, the large tree basal area growth parameter was modified by a multiplier to increase growth at every time step.

3.2. Model Validation

After the calibration of the max SDI and large tree basal area growth functions, the model was applied to the calibration stand data for year 30 and growth was predicted over the next 20 years. The predicted values for trees/ha and basal area/ha were similar to the observed values. On average, the calibrated model reduced the systematic error in base model performance by 71% for MPE of TPH and 81% for MPE of ba/ha (Table 2). Average root mean square error values for TPH averaged 106 trees/ha for all stands. Average trees per ha MPE for all stands was ±7.7%. Basal area per ha RMSE values had an averaged value of approximately 1.8 m^2/ha. The overall average MPE for basal area per ha predictions was 3.5%. Generally, deviations from the observed values increased as the model progressed through time (Figure 3).

Table 2. Mean percent error (MPE) and root mean square error (RMSE) values for trees per hectare and basal area per hectare (m^2/ha) calculated by comparing the FVS NE base model and FEF locally calibrated model predictions to the observed values in nine calibration stands. Positive numbers represent an overestimation by the model and negative numbers represent underestimation.

Stand	Base Model (30 Years of Projections)				FEF Calibrated Model (20 Years of Projections)			
	MPE TPH	MPE ba/ha	RMSE TPH	RMSE ba/ha	MPE TPH	MPE ba/ha	RMSE TPH	RMSE ba/ha
32	40%	−15%	478	−4.7	−3%	−1%	−41	−2.1
33	42%	−13%	439	−4.1	5%	−1%	−58	−1.1
34	49%	−15%	447	−3.6	−4%	8%	−91	2.4
35	25%	−18%	305	−5.0	−12%	0%	−144	−1.1
36	16%	−30%	186	−7.8	1%	5%	29	1.4
37	48%	−23%	408	−6.3	16%	1%	119	0.7
38	21%	−14%	309	−3.8	−12%	−2%	−201	−1.1
39	50%	−19%	528	−6.8	−9%	−7%	−112	−3.5
43	16%	−18%	201	−5.7	−12%	−6%	164	2.6

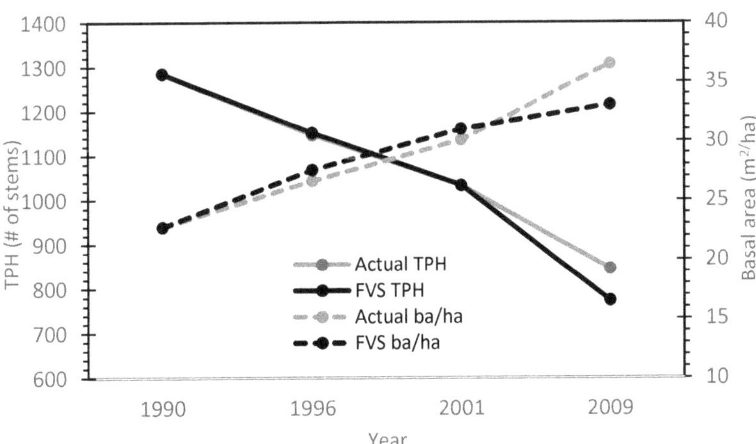

Figure 3. A visual representation of decreased model systematic error using the locally calibrated FVS model for calibration stand 32 (compared to Figure 1).

3.3. Projection of LTSP Data

The FEF calibrated model was applied to the year 21 LTSP plot data to project volumes for each treatment over 40 years. Projections indicated an initial, small decrease in volume on site for all treatments (Figure 4A). This response was due to the mortality of pin cherry as it continues its natural life cycle (Figure 5A). The reduction in total volume was less for the WT plots, which had less initial volume of pin cherry present (Table 3). As the projection continues, there is little variation in the total merchantable volume produced among treatments. Volumes at the end of the 40 year projection (61 year-old stands) for WT plots was greatest (407 m³/ha), followed by the WT + NS + LIME plots (389 m³/ha) and the WT + NS plots (386 m³/ha).

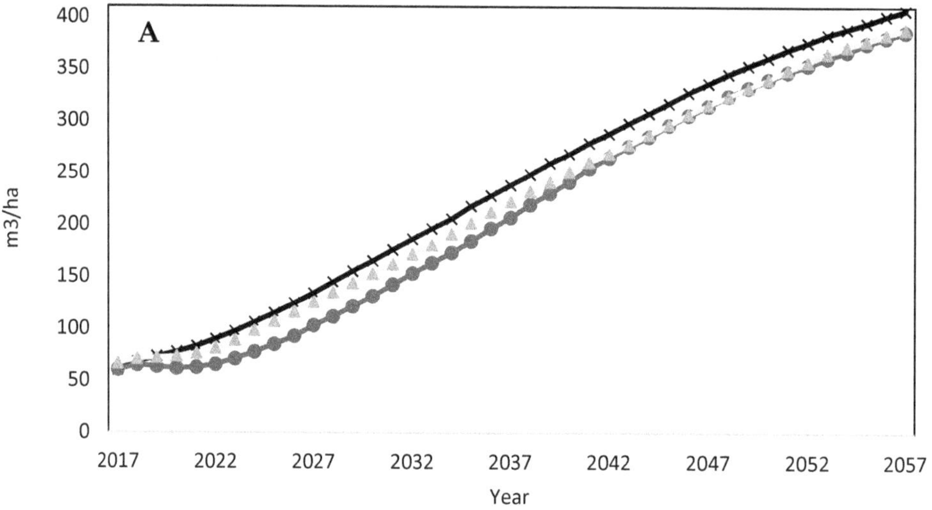

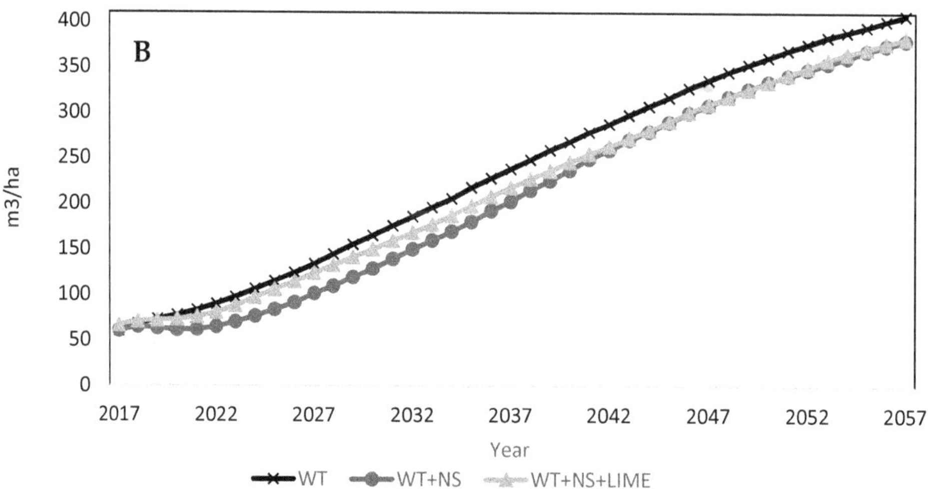

Figure 4. FVS projections for merchantable volume for all species combined by treatment. (**A**) Upper graph represents constant decline scenario. (**B**) Lower graph represents continual decline scenario.

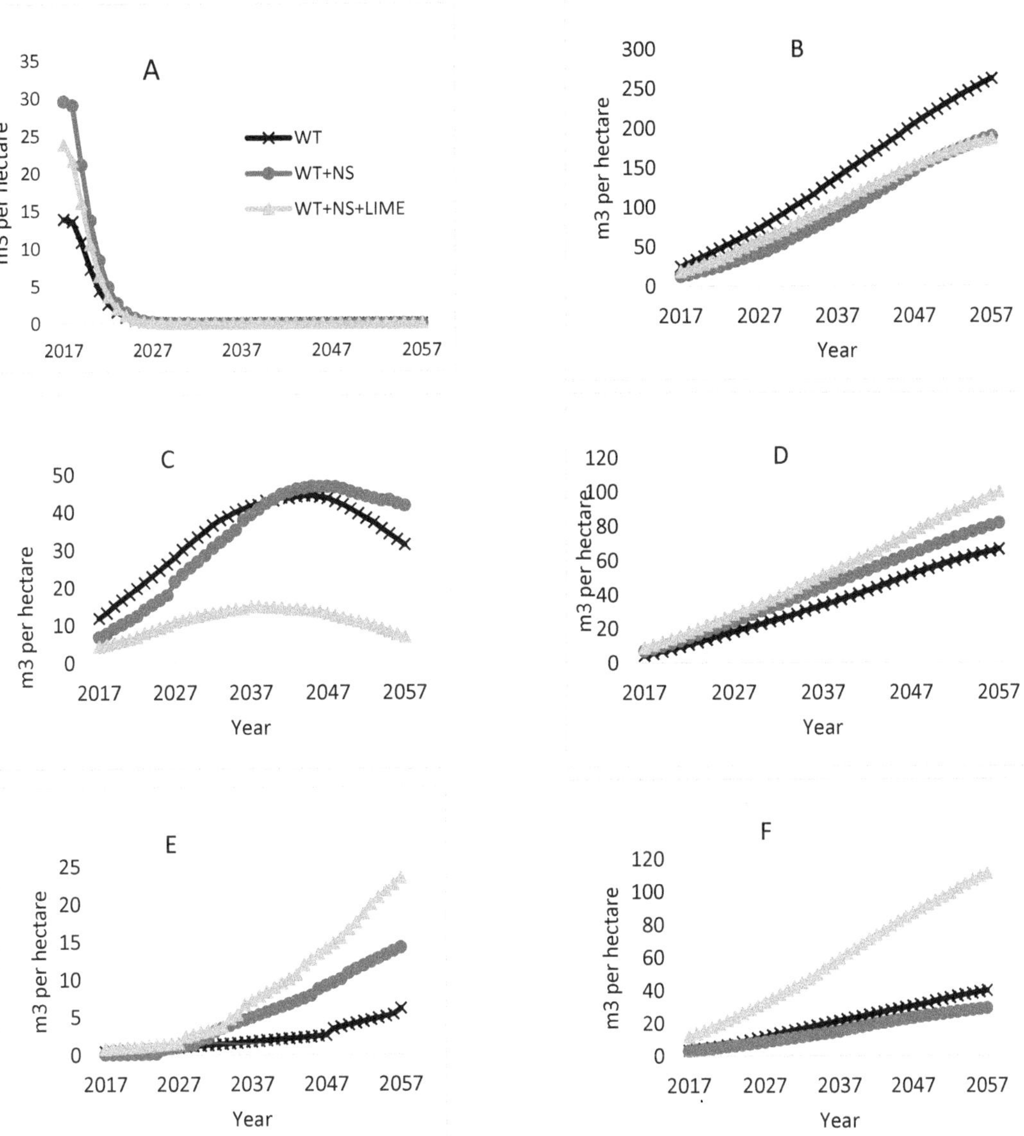

Figure 5. Species specific cumulative growth predictions for three treatments on the LTSP site. Six species (in order of highest lowest importance value at year 2017) (Storm, 2018) are presented by species: (**A**) pin cherry, (**B**) yellow-poplar, (**C**) sweet birch, (**D**) black cherry, (**E**) red maple, (**F**) cucumbertree.

For the continual decline scenario (Figure 4B), the fertilized treatments resulted in negligible differences by the end of the projection. Final volumes projected for the WT + NS and WT + NS + LIME plots were 380 m^3/ha and 382 m^3/ha, respectively. For both the declining and fixed impact projection scenarios, their respective WT + NS and WT + NS + LIME treatments had nearly identical volume estimates over time. Therefore, no further results for the continual decline scenarios are provided.

Yellow-poplar was projected to be the dominant species on the site throughout the next 40 years in terms of merchantable volume (Figure 5B) and is the driving species behind the slightly higher total volume in the WT treatment areas. By the end of the projection period, yellow-poplar made up almost 65% of the total volume on the WT plots and approximately 50% on both the WT + NS and WT + NS + LIME plots. At the start of the projection, yellow-poplar stems made up 38% of the trees on WT plots, but only 25% on the WT + NS plots and 21% on the WT + NS + LIME plots (Table 3). The greatest differences in TPH among treatments were for the smallest diameter class (Table 4). For example, yellow-poplar stems in the 5 cm diameter class were nearly three times more abundant on the WT plots than the WT + NS and WT + NS + LIME plots.

Table 3. FVS predicted stocking of mean percentage of total trees per ha for the six main species in each treatment area for 10 year projection intervals. Data for all species are included in Appendix A, Table A3.

	WT					WT + NS					WT + NS + LIME				
Species *	2017	2027	2037	2047	2057	2017	2027	2037	2047	2057	2017	2027	2037	2047	2057
black cherry	5%	6%	6%	7%	8%	5%	6%	6%	7%	9%	14%	17%	17%	16%	16%
cucumbertree	3%	3%	4%	4%	5%	3%	4%	3%	3%	4%	3%	3%	4%	6%	8%
pin cherry	11%	0%				20%	0%				16%	0%			
red maple	7%	9%	12%	16%	20%	6%	8%	10%	12%	15%	12%	16%	17%	22%	25%
sweet birch	26%	28%	25%	17%	10%	24%	30%	28%	23%	17%	11%	13%	11%	7%	2%
yellow-poplar	38%	41%	38%	39%	39%	25%	31%	31%	31%	33%	21%	24%	23%	24%	24%

* black cherry (*Prunus serotina* Ehrh.); cucumbertree (*Magnolia acuminata* L.); pin cherry (*Prunus pensylvanica* L. f.); red maple (*Acer rubrum* L.); sweet birch (*Betula lenta* L.); yellow-poplar (*Liriodendron tulipifera* L.).

Table 4. Trees per hectare for the top six species by diameter class for the Fork Mountain LTSP in 2017. Data for all species are included in Appendix A, Table A4.

	WT					WT + NS					WT + NS + LIME					
	Diameter Class (cm)															
Species	5	10	15	20	25	5	10	15	20	25	5	10	15	20	25	30
black cherry	74	37	37	7		62	12	23	22	4	259	62	27	17	4	5
cucumbertree	49	25	6	9	0	49	12	4	4	2	25	12	14	14	5	
pin cherry	62	124	148	30	2	62	148	222	65	11	37	161	170	56	9	
red maple	198	12	4			86	62	1			235	86	4		1	
sweet birch	556	185	88	19		371	148	65	7		148	111	36	7		
yellow-poplar	902	185	62	37	20	334	222	43	20	10	321	124	59	30	11	1

Final volume differences by treatment for black cherry correspond with an initial higher percent trees per ha in the WT + NS + LIME treatment (Table 3). Initially, black cherry stems were nearly three times more abundant on WT + NS + LIME plots than on the WT and WT + NS plots. In addition, there were no black cherry trees on the WT plots greater than the 20 cm diameter class (Table 4). By the end of the projection, black cherry merchantable volume for the WT + NS + LIME treatment was 34 m^3/ha more than the WT and 18 m^3/ha higher than the WT + NS (Figure 5D).

The growth of cucumbertree over time is projected to also follow a similar linear trend as black cherry and yellow-poplar. At the start of the projections, there was no difference in the initial percent of the total trees per ha in any of the treatment areas for cucumbertree (Table 3). However, there were a greater number of large individual trees that were measured on the WT + NS + LIME plots compared to trees growing in the WT and WT + NS plots (Table 4), which may indicate a positive response of this species to the WT + NS + LIME treatment during the first 21 years. The initial merchantable volume present on WT + NS + LIME plots was over three times greater than that on the other two treatments (Figure 5F). After 40 years of growth, cucumbertree is predicted to account for 100 m^3/ha on the WT + NS + LIME plots, 40 m^3/ha on the WT plots, and 29 m^3/ha on the WT + NS plots.

Red maple projections deviated from the generally linear growth trends for yellow-poplar, black cherry, and cucumbertree. Initially, there were few red maple stems in the WT plots that were greater than 5 cm, which was not the case for the WT + NS and WT + NS + LIME plots (Table 4). Red maple stems were projected to represent little merchantable volume for the first decade of the projections in all treatments (Figure 5E). Around age 33 (year 2029), red maple stems receiving additional nitrogen from the treatments are expected to begin producing merchantable volume, culminating in 24 m^3/ha on WT + NS + LIME plots and 14 m^3/ha on the WT + NS plots. The volume for red maple stems on WT plots was projected to be lower, only producing 6 m^3/ha by age 61 (year 2057).

The total volume of sweet birch is projected to increase for all treatments until the trees are 40–50 (years 2037–2047) years old, after which mortality occurs in all treatments (Figure 5C). The WT + NS + LIME plots that started out with the lowest volume of sweet birch, also reached the point of maximum volume earliest. Between the years 2038 and 2057 in the projection, the volume of sweet birch on WT + NS + LIME plots declined from 15 m^3/ha to 7 m^3/ha. Both the WT and WT + NS plots reached their respective highest points of merchantable volume between the ages of 47 and 48, respectively. The WT plots reached a maximum predicted value of 44 m^3/ha and the WT + NS plots culminated at 47 m^3/ha. After this point, the mortality induced reduction in sweet birch volume on the WT plots was projected to decrease at a rate 2.5 times faster than the trees on the WT + NS plots.

4. Discussion

The overall volume growth on the site is not projected to decline substantially over the 40-year projection for any treatment (Figure 4A). However, it does seem that there will be differences in the species that make up the final volume in the treatment areas (Table 3). This shift in species composition could have both economic and environmental implications in the future.

The 40-year volume projections followed patterns that were mostly consistent with trends observed during the first 21 years of treatment [41]. Yellow-poplar, black cherry, red maple, and cucumbertree all showed stable, long-term positive responses to treatments. With the exceptions of yellow-poplar and sweet birch, the other species had the most volume on WT + NS treated areas. For yellow-poplar, plots that received annual nitrogen and sulfur additions were projected to grow less merchantable volume than the WT (non-fertilized) areas. Sweet birch was the only species where the response to treatments changed over time. For the first 20 years of the projections, volume was highest on WT treatments, but after 20 years (41 years of treatment), volume associated with the WT+NS treatments was greatest, and then declined (Figure 5C).

As with all growth models, projections are sensitive to initial stand conditions. For example, projections for black cherry and red maple showed the greatest volume accumulation for the WT + NS + LIME treatments, while yellow-poplar volume accumulation was greatest for WT. All of these responses were associated with the treatments that also had the greatest stem densities at the beginning of the modelling period (stand age 21). However, there are indications that the acidification and liming treatments are responsible for at least some changes in stand growth and development.

In the case of yellow-poplar, which is projected to be the dominant merchantable species in all treatment areas (Figure 5B), the increased volume associated with the WT plots was at least partly attributable to higher numbers of trees per ha in 2017 (21 years-old), and possibly a reflection of early treatment impacts (Table 4). However, considering that yellow-poplar is a shade-intolerant species [45], many of these smaller stems will die as the stand continues to grow over time, which might reduce the long-term impact of the high stem counts in the smallest diameter classes (Table 3). Additionally, there appears to be an ameliorative effect of adding lime to the acidification treatment for yellow-poplar given that the total volume for WT + NS and WT + NS + LIME treatments was similar, even though the number of stems on the liming treatment were much fewer. Despite these apparent effects, the difference between treatments is marked and suggests that the acidification treatment (WT + NS) still results in reduced growth rather than just a function of starting conditions.

Growth and mortality of sweet birch on WT + NS plots differed relative to WT and WT + NS + LIME. Initially, there were fewer stems in all diameter classes on the WT + NS compared to those sampled on WT plots (Table 4), so this response cannot be attributed to additional stems present in the larger diameter classes. It is likely that the model allocated more growing space to sweet birch in the WT + NS treated areas where yellow-poplar was projected to have less volume and because of the loss of pin cherry to mortality. Since mortality in the model is based on the stand density index at any given point in time, there was less mortality assigned to sweet birch stems on the WT + NS plots compared to WT plots later in the projection due to lower number of yellow-poplar stems in the WT + NS plots.

By the end of the projection, red maple made up an increasingly larger percentage of trees in the treatment areas (Table 3). However, red maple only accounted for a small percent of total volume, suggesting that many of the red maple stems will persist in the midstory. This pattern of red maple growth dynamics is well documented [46,47]. The results suggest that chronic additions of nitrogen in the soil (either from deposition or fertilization) could lead to further increases in red maple dominance in eastern hardwood forests, which is a trend that has been noted for the last few decades [48].

The number of black cherry stems at the start of the projections was greater on the WT + NS + LIME plots (Table 4), which was also reflected in high initial stem counts shortly after the stand was regenerated in 1996 [35]. As a result, the increased volume associated with the liming treatment is difficult to separate from the impacts of having much greater stem counts in the initial stand. Additionally, although FVS predicted increased volume on the WT + NS + LIME plots, there is little evidence in the literature to support that liming increases black cherry growth. Liming studies with mature black cherry trees have shown both short- and long-term negative responses to single high rate applications (although different from the annual, low rate applications here) of dolomitic lime in areas of high historic acidic deposition [49,50]. However, young black cherry trees have been shown to increase growth and foliar nutrient concentrations after nitrogen and phosphorus fertilizer [51,52]. In the LTSP study, it seems that black cherry responded positively to the first 21 years of the ammonium sulfate fertilizer (regardless of whether dolomitic lime was added) in terms of growing larger diameter individuals (Table 4) and increased final volume estimations for both WT + NS and WT + NS + LIME treatments compared to the nonfertilized areas.

Certainly, modeling results are always subject to the inherent complexity and accuracy. Our calibration efforts significantly improved the overall model performance. The site indices for the model (YP base age 50: 35.1 m for WT plots, 30.5 m for WT + NS, and WT + NS + LIME; 32.9 m for the continual declining WT + NS and WT + NS + LIME) represent a negative response in growth from the chronic additions of N. This response is corroborated with other reports of decreased growth of yellow-poplar from chronic additions of N on the FEF [24], but does not necessarily represent the response of all species to the treatments. Since height measurements were not collected for all stems and species, the model is limited in predicting individual species response to the additions of ammonium sulfate and lime. Another limitation relates to mortality functions within FVS. Mortality is a function of tree diameter and species specific parameters [37]. The impact of treatments is not directly factored into the model, although indirectly, if a treatment negatively affects diameter growth, the mortality rate is higher. In contrast, some hardwood growth models also factor in relative size [53], so mortality is increased if a tree's relative size lags behind other stems and species.

Finally, there is a possibility that the impact of the WT + NS and N + S + LIME treatments are either inducing or exacerbating other nutrient deficiencies. There is evidence that high N enrichment, through deposition or fertilization, leads to phosphorus limitations on stand level productivity or for individual species [54–56]. Highlighting this possibility, Gress et al. [57] used root ingrowth cores and phosphatase activity in the current study area and nearby watersheds on the FEF to demonstrate a likely phosphorus deficiency and increased root growth for a prominent understory herbaceous plant, *Viola rotundifolia*, in areas associated with elevated soil, foliar, and stream nitrogen levels. Elevated nitrogen deposition can decrease species richness in certain plant communities [58–60] including forests. Whether tree species richness will decline in these central Appalachian hardwood forests because of long-term acid deposition is unknown, but our results showed only modest changes over time that do not appear specific to a particular treatment.

5. Conclusions

The goals of this paper were to provide a framework and calibration metrics for using FVS to model the impacts of acid deposition as well as supplemental liming on the growth of a central WV Appalachian hardwood stand. We showed that significant improvements (>70% improvement in error) to the base model FVS can be achieved to calibrate a locally specific model to project 40 years of additional growth and development of a forest subjected to 20+ years of experimentally elevated nitrogen, sulfur, and lime inputs. Additionally, we showed that species had different growth patterns as a result of the initial and long-term influence of acidification and liming treatments.

Over time, the dominant pin cherry will be eliminated from the stand and the more long-lived species will attain dominance. Although there does appear to be treatment impacts on growth and stand development, pinpointing the casual mechanisms (i.e., inherent stand variability vs. initial treatment effects vs. longer term treatment impacts) is challenging at this point. Likewise, continued N and S inputs may further (or begin to) alter tree growth and stand development into the future. Continued monitoring of this long-term LTSP experiment will allow us to examine the mechanisms responsible for the declining growth of the plots receiving annual nitrogen and sulfur inputs.

Author Contributions: Conceptualization, M.B.A.; Methodology, M.B.A. and J.S.; Formal analysis, A.S.; Writing—original draft preparation, A.S.; Writing—review and editing, J.S. and M.B.A.; Supervision, J.S.; Project administration, J.S. and M.B.A.; Funding acquisition, M.B.A. All authors have read and agreed to the published version of the manuscript.

Funding: This research was funded by the United States Forest Service, Northern Research Station, Timber and Watershed Lab under Joint Venture Agreement Number 15JV11242303102. This material is based upon work that is supported by the National Institute of Food and Agriculture, U.S. Department of Agriculture, McIntire Stennis project WVA00804.

Institutional Review Board Statement: Not applicable.

Informed Consent Statement: Not applicable.

Data Availability Statement: Data analyzed for this study are archived by the United States Forest Service, Northern Research Station, Timber and Watershed Lab, Parsons, WV.

Acknowledgments: The authors express their gratitude to John Juracko and Brian Simpson from the USFS for their assistance in collecting the field data.

Conflicts of Interest: The authors declare no conflict of interest.

Appendix A

Table A1. Soil chemical characteristics by soil depth at the initiation of the LTSP study (from Adams et al. 2004).

Variable	025 cm	1530 cm	3045 cm
% C	6.58	2.71	1.12
pH	4.24	4.45	4.42
Total N (%)	0.42	0.22	0.14
Ca (cmol + /kg)	0.54	0.17	0.13
Mg (cmol + /kg)	0.18	0.07	0.04
K (cmol + /kg)	0.33	0.34	0.12
Al (cmol + /kg)	3.52	3.43	3.92
Exch. acidity (cmol + /kg)	5.3	4.12	4.68
Total acidity (cmol + /kg)	27.65	18.51	13.25
Cation Exch. Capacity (cmol + /kg)	28.7	19.09	13.55
% Base Sat.	3.67	2.99	1.61

Table A2. Species growing at higher rates in calibration plots than what is normally predicted by FVS NE. Species multiplier values here were used to modify the large tree basal area growth function. Species codes are as follows: RM (red maple), SB (sweet birch), YP (yellow-poplar), BC (black cherry), CO (chestnut oak), SE (slippery elm).

Scale Factor Summary							
SPECIES	N	Min	Mean	Max	Std. Dev.	Total Tree Records	Multiplier
RM	5	1.004	1.105	1.267	0.110	244	1.132
SB	7	1.160	1.383	1.893	0.241	164	1.418
YP	6	0.669	1.160	1.543	0.330	125	1.142
BC	5	0.996	1.345	1.975	0.420	138	1.215
CO	2	1.007	1.183	1.358	0.248	67	1.103
SE	2	0.816	1.008	1.200	0.272	31	1.048

N = Number of stands that contributed scale factors; MIN = minimum initial scale factor encountered; Mean = mean initial scale factor; MAX = maximum scale factor encountered; Std. dev. = standard deviation for scale factors; Total tree records = total number of trees used to calculate scale factors; Multiplier = mean multiplier to be used to scale the growth of large trees.

Table A3. FVS predicted stocking of mean percentage of total trees per ha in each treatment area for 10 year projection intervals. Values of 0% represent less than 1% total trees per ha while dashes () represent species absence from the corresponding area.

Species *	WT					WT + NS					WT + NS + LIME				
	2017	2027	2037	2047	2057	2017	2027	2037	2047	2057	2017	2027	2037	2047	2057
black cherry	5%	6%	6%	7%	8%	5%	6%	6%	7%	9%	14%	17%	17%	16%	16%
black locust						0%	0%	0%	0%	0%	1%	1%	0%	0%	
cucumbertree	3%	3%	4%	4%	5%	3%	4%	3%	3%	4%	3%	3%	4%	6%	8%
eastern hemlock											0%	0%	0%	0%	0%
hickory spp.	0%	0%	1%	1%	1%	0%	0%	0%	0%	0%	1%	2%	2%	1%	1%
Fraser magnolia	1%	1%	1%	2%	2%	4%	6%	7%	8%	9%	2%	3%	2%	2%	2%
noncommercial						3%	3%	2%	1%	0%	2%	2%	2%	1%	0%
pin cherry	11%	0%				20%	0%				16%	0%			
red maple	7%	9%	12%	16%	20%	6%	8%	10%	12%	15%	12%	16%	17%	22%	25%
red oak	2%	2%	2%	2%	2%	3%	3%	2%	2%	2%	3%	4%	4%	4%	4%
sweet birch	26%	28%	25%	17%	10%	24%	30%	28%	23%	17%	11%	13%	11%	7%	2%
Sourwood						2%	2%	3%	4%	5%					
Serviceberry						3%	4%	4%	4%	2%	2%	2%	2%	1%	0%
sugar maple											3%	4%	4%	5%	6%
Sassafras	1%	1%	1%	0%	0%	1%	2%	2%	2%	1%					
striped maple	5%	7%	9%	12%	13%	1%	1%	2%	2%	2%	6%	7%	9%	11%	12%
white ash	1%	1%	1%	0%	0%						1%	1%	1%	0%	0%
yellow-poplar	38%	41%	38%	39%	39%	25%	31%	31%	31%	33%	21%	24%	23%	24%	24%

* black cherry (*Prunus serotina* Ehrh.); black locust (*Robinia pseudoacacia* L.); cucumbertree (*Magnolia acuminata* L.); eastern hemlock (*Tsuga canadensis* (L.) Carr.); hickory spp. (*Carya* spp.); Fraser magnolia (*Magnolia fraseri* Walt.); pin cherry (*Prunus pensylvanica* L. f.); red maple (*Acer rubrum* L.); red oak (*Quercus rubra* L.); sweet birch (*Betula lenta* L.); sourwood (*Oxydendrum arboretum* (L.) DC.); serviceberry (*Amelanchier arborea* (Michx. F.) Fernald); sugar maple (*Acer saccharum* Marsh.); sassafras (*Sassafras albidum* (Nutt.) Nees); striped maple (*Acer pensylvanicum* L.); white ash (*Fraxinus americana* L.); yellow-poplar (*Liriodendron tulipifera* L.).

Table A4. Trees per hectare for each species and diameter class for the Fork Mountain LTSP in year 2017.

	WT					WT + NS					LIME					
	Diameter Class															
Species	5	10	15	20	25	5	10	15	20	25	5	10	15	20	25	30
black cherry	74	37	37	7		62	12	23	22	4	259	62	27	17	4	5
black locust								2				12	4			
cucumbertree	49	25	6	9	0	49	12	4	4	2	25	12	14	14	5	
eastern hemlock									1				1			
hickory	12										25	12				
Fraser magnolia	12		10	4		62	37	7			12	25	16	9		
noncommercial						49	25				62					
pin cherry	62	124	148	30	2	62	148	222	65	11	37	161	170	56	9	
red maple	198	12	4			86	62	1			235	86	4		1	
red oak	62					49		1	1		49	37	4			
sweet birch	556	185	88	19		371	148	65	7		148	111	36	7		
sourwood						12	25	2								
serviceberry						74					49					
sugar maple											86					
sassafras	12	12	1			12	25									
striped maple	173					25		1			111	37				
white ash	25	12									37					
yellow-poplar	902	185	62	37	20	334	222	43	20	10	321	124	59	30	11	1
Total	2137	593	356	105	22	1248	717	374	121	27	1458	680	335	132	30	6

References

1. Johnson, A.H. Red spruce decline in the northeastern U.S.: Hypotheses regarding the role of acid rain. *J. Air Pollut. Control Assoc.* **1983**, *33*, 1049–1054. [CrossRef]
2. Johnson, A.H.; Siccama, T.G. Acid deposition and forest decline. *Environ. Sci. Technol.* **1983**, *17*, 294–305. [CrossRef]
3. Ulrich, B. Dangers for the Forest Ecosystem Due to Acid Precipitation. In *Necessary Countermeasures: Soil Liming and Exhaust Gas Purification. U.S. EPA Translation TR-82-0111*; EPA/NCSU Acid Deposition Program; North Carolina State University: Raleigh, NC, USA, 1982.
4. Driscoll, C.T.; Lawrence, G.B.; Bulger, A.J.; Butler, T.J.; Cronan, C.S.; Eagar, C.; Weathers, K.C. Acidic deposition in the northeastern United States: Sources and inputs, ecosystem effects, and management strategies. *BioScience* **2001**, *51*, 180–198. [CrossRef]
5. Galloway, J.N.; Likens, G.E.; Edgerton, E.S. Acid precipitation in the Northeastern United States: pH and acidity. *Science* **1976**, *194*, 722–724. [CrossRef] [PubMed]
6. National Atmospheric Deposition Program (NRSP-3). NADP Program Office, Wisconsin State Laboratory of Hygiene, 465 Henry Mall, Madison, WI, 53706. 2021. Available online: http://nadp.slh.wisc.edu/NADP/ (accessed on 1 July 2021).
7. Adams, M.B.; Burger, J.A.; Jenkins, A.B.; Zelazny, L. Impact of harvesting and atmospheric pollution on nutrient depletion of eastern US hardwood forests. *For. Ecol. Manag.* **2000**, *138*, 301–319. [CrossRef]
8. Spiro, T.G.; Stigliani, W.M. *Chemistry of the Environment*, 2nd ed.; Prentice Hall: Hoboken, NJ, USA, 2003.
9. Schlesinger, W.H.; Bernhardt, E.S. *Biogeochemistry: An Analysis of Global Change*, 3rd ed.; Elsevier: Amsterdam, The Netherlands, 2013.
10. Crim, P.H.; McDonald, L.M.; Cumming, J.R. Soil and Tree Nutrient Status of High Elevation Mixed Red Spruce (*Picea rubens* Sarg.) and Broadleaf Deciduous Forests. *Soil Syst.* **2019**, *3*, 80. [CrossRef]
11. Foy, C.D.; Chaney, R.L.; White, M.C. The Pysiology of Metal Toxicity in Plants. *Annu. Rev. Plant Physiol.* **1978**, *29*, 511–566. [CrossRef]
12. Foy, C.D. Physiological Effects of Hydrogen, Aluminum, and Manganese Toxicities in Acid Soil. *Soil Acidity Liming* **1984**, *12*, 57–97.
13. Burnham, M.B.; Cumming, J.R.; Adams, M.B.; Peterjohn, W.T. Soluble soil aluminum alters the relative uptake of mineral nitrogen forms by six mature temperate broadleaf tree species: Possible implications for watershed nitrate retention. *Oecologia* **2017**, *185*, 327–337. [CrossRef] [PubMed]
14. Aber, J.D.; Nadelhoffer, K.J.; Steudler, P.; Melillo, J.M. Nitrogen Saturation in Northern Forest Ecosystems. *BioScience* **1989**, *39*, 378–386. [CrossRef]
15. Peterjohn, W.T.; Adams, M.B.; Gilliam, F.S. Symptoms of Nitrogen Saturation in Two Central Appalachian Hardwood Forest Ecosystems. *Biogeochemistry* **1996**, *35*, 507–522. [CrossRef]
16. McNulty, A.S.G.; Aber, J.D.; Boone, R.D. Spatial Changes in Forest Floor and Foliar Chemistry of Spruce-Fir Forests across New England. *Biogeochemistry* **1991**, *14*, 13–29. [CrossRef]
17. Gilliam, F.S.; Adams, M.B.; Peterjohn, W.T. Response of soil fertility to 25 years of experimental acidification in a temperate hardwood forest. *J. Environ. Qual.* **2020**, *49*, 961–972. [CrossRef]
18. Van Breemen, N.; Burrough, P.A.; Velthorst, E.J.; Van Dobben, H.F.; De Wit, T.; Ridder, T.B.; Reijnders, H.F.R. Soil acidification from atmospheric ammonium sulphate in forest canopy throughfall. *Nature* **1982**, *299*, 548–550. [CrossRef]

19. Cronan, C.S.; Grigal, D.F. Use of Cacium/Aluminum Ratios as Indicators of Stress in Forest Ecosytems. *Environ. Qual.* **2004**, *24*, 209–226. [CrossRef]

20. Richter, D.D.; Johnson, D.W.; Dai, K.H. Cation exchange and Al mobilization in soils. In *Atmospheric Deposition and Forest Nutrient Cycling*; Johnson, D.W., Lindberg, S.E., Eds.; Springer: New York, NY, USA, 1991; pp. 341–377.

21. Steudler, P.A.; Bowden, R.D.; Melillo, J.M.; Aber, J.D. Influence of nitrogen fertilization on methane uptake in temperate forest soils. *Nature* **1989**, *341*, 314–316. [CrossRef]

22. Baas, P.; Knoepp, J.D.; Mohan, J.E. Well-Aerated Southern Appalachian Forest Soils Demonstrate Significant Potential for Gaseous Nitrogen Loss. *Forests* **2019**, *10*, 1155. [CrossRef]

23. Powers, R.F.; Alban, D.H.; Miller, R.E.; Tiarks, A.E.; Wells, C.G.; Avers, P.E.; Cline, R.G.; Fitzgerald, R.O.; Loftus, N.S., Jr. Sustaining site productivity in North American forests: Problems and prospects. In Proceedings of the Seventh North American Forest Soils Conference on Sustained Productivity of Forest Soils, Vancouver, BC, Canada, 24–28 July 1988; Gessel, S.P., Lacate, D.S., Weetman, G.F., Powers, R.F., Eds.; Faculty of Forestry, University of British Columbia: Vancouver, BC, Canada, 1990; pp. 49–79.

24. DeWalle, D.R.; Kochenderfer, J.N.; Adams, M.B.; Miller, G.W.; Gilliam, F.S.; Wood, F.; Sharpe, W.E. Vegetation and acidification. In *Fernow Watershed Acidif. Study*; Adams, M.B., Kochenderfer, J.N., Hom, J.L., Eds.; Springer: Dordrecht, The Netherlands, 2006; pp. 137–188.

25. Fowler, Z.K.; Adams, M.B.; Peterjohn, W.T. Will more nitrogen enhance carbon storage in young forest stands in central Appalachia? *For. Ecol. Manag.* **2015**, *337*, 144–152. [CrossRef]

26. Adams, M.B.; Kochenderfer, J.N.; Edwards, P.J. The Fernow watershed acidification study: Ecosystem acidification, nitrogen saturation and base cation leaching. *Water Air Soil Pollut.* **2007**, *7*, 267–273. [CrossRef]

27. Magill, A.H.; Aber, J.D.; Currie, W.S.; Nadelhoffer, K.J.; Martin, M.E.; McDowell, W.H.; Steudler, P. Ecosystem response to 15 years of chronic nitrogen additions at the Harvard Forest LTER, Massachusetts, USA. *For. Ecol. Manag.* **2004**, *196*, 7–28. [CrossRef]

28. Pregitzer, K.S.; Burton, A.J.; Zak, D.R.; Talhelm, A.F. Simulated chronic nitrogen deposition increases carbon storage in Northern Temperate forests. *Glob. Chang. Biol.* **2008**, *14*, 142–153. [CrossRef]

29. Aber, J.D.; McDowell, W.; Nadelhoffer, K.; Magill, A.; Berntson, G.; Kamakea, M.; Fernandez, I. Nitrogen saturation in temperate forest ecosystems—Hypotheses revisited. *Bioscience* **1998**, *48*, 921–934. [CrossRef]

30. Caputo, J.; Beier, C.M.; Sullivan, T.J.; Lawrence, G.B. Modeled effects of soil acidifcation on long-term ecological and economic outcomes for managed forests in the Adirondack region (USA). *Sci. Total Environ.* **2016**, *565*, 401–411. [CrossRef] [PubMed]

31. Dixon, G.E. *Essential FVS: A User's Guide to the Forest Vegetation Simulator*; USDA—Forest Service, Forest Management Service Center: Fort Collins, CO, USA, 2002; 212p.

32. Brooks, J.R.; Miller, G.W. An evaluation of three growth and yield simulators for even-aged hardwood forests of the mid-Appalachian region. In Proceedings of the 17th Central Hardwood Forest Conference, Lexington, KY, USA, 5–7 April 2010; Fei, S., Lhotka, J.M., Stringer, J.W., Gottschalk, K.W., Miller, G.W., Eds.; U.S. Department of Agriculture, Forest Service, Northern Research Station: Newtown Square, PA, USA, 2011.

33. Shaw, J.D.; Vacchiano, G.; DeRose, J.R.; Brough, A.; Kusbach, A.; Long, J.N. Local calibration of the Forest Vegetation Simulator (FVS) using custom inventory data. In Proceedings of the Society of American Foresters 2006 National Convention, Pittsburgh, PA, USA, 25–29 October 2006; Society of American Foresters: Bethesda, MD, USA, 2006.

34. Vandendriesche, D. FVS out of the box—Assembly required. In *Integrated Management of Carbon Sequestration and Biomass Utilization Opportunities in a Changing Climate: Proceedings of the 2009 National Silviculture Workshop*; Jain, T.B., Graham, R.T., Sandquist, J., Eds.; U.S. Department of Agriculture, Forest Service, Rocky Mountain Research Station: Fort Collins, CO, USA, 2010; pp. 289–306.

35. Adams, M.B.; Burger, J.; Zelazny, L.; Baumgras, J. *Description of the Fork Mountain Long-Term Soil Productivity Study: Site Characterization*; USDA Forest Service: Newtown Square, PA, USA, 2004; 43p.

36. Helvey, J.D.; Kunkle, S.H. *Input-Out Budgets of Selected Nutrients on an Experimental Watershed near Parsons, West Virginia*; Res. Pap. NE-584; U.S. Department of Agriculture, Forest Service, Northeastern Forest Experiment Station: Broomall, PA, USA, 1986; 7p.

37. Dixon, G.E.; Keyser, C.E. *Northeast (NE) Variant Overview*; Internal Rep.; USDA—Forest Service, Forest Management Service Center: Fort Collins, CO, USA, 2008; 56p.

38. Smith, H.C.; Rosier, R.L.; Hammack, K.P. *Reproduction 12 Years after Seed-Tree Harvest Cutting in Appalachian Hardwoods*; Res. Pap. NE-350; Northeastern Forest Experiment Station, For. Serv., US Dept Agric.: Upper Darby, PA, USA, 1976.

39. Beck, D.E. *Yellow-Poplar Site Index Curves*; Res. Note SE-180; US Department of Agriculture, Forest Service, Southeastern Experiment Station: Asheville, NC, USA, 1962; 2p.

40. Marks, P.L. The Role of Pin Cherry (*Prunus pensylvanica* L.) in the Maintenance of Stability in Northern Hardwood Ecosystems. *Ecol. Monogr.* **1974**, *44*, 73–88. [CrossRef]

41. Storm, A.J. Long-Term Effects of Chronic Additions of Nitrogen, Sulfur and Lime on the Growth and Development of a Central Appalachian Forest. Master's Thesis, West Virginia University, Morgantown, WV, USA, 2018; p. 102.

42. Oliver, C.D.; Larson, B.C. *Forest Stand Dynamics: Updated Edition*; John Wiley & Sons, Inc.: New York, NY, USA, 1996; 543p.

43. Cawrse, D.; Keyser, C.; Keyser, T.; Sanchez, A.; Smith-Mateja, E.; Van Dyck, M. *Forest Vegetation Simulator Model Validation Protocols*; USDA—Forest Service, Forest Management Service Center: Fort Collins, CO, USA, 1–10 January 2009.

44. Russell, M.B.; Weiskittel, A.R.; Kershaw, J.A., Jr. Benchmarking and Calibration of Forest Vegetation Simulator Individual Tree Attribute Predictions across the Northeastern United States. *North. J. Appl. For.* **2013**, *30*, 75–84. [CrossRef]

45. Beck, D.E. Yellow-poplar. In *Silvics of North America: 2. Hardwoods*; Burns, R.M., Barbara, H.H., Eds.; Agriculture Handbook 654; U.S. Department of Agriculture, Forest Service: Washington, DC, USA, 1990; 877p.

46. Lorimer, C.G. Development of the red maple understory in northeastern oak forests. *For. Sci.* **1984**, *30*, 3–22.

47. Oliver, C.D. The development of northern red oak in mixed stands in central New England. *Yale Sch. For. Environ. Stud. Bull.* **1978**, *91*, 63.

48. Abrams, M.D. The red maple paradox. *BioScience* **1998**, *48*, 355–364. [CrossRef]

49. Long, R.P.; Horsley, S.B.; Hall, T.J. Long-term impact of liming on growth and vigor of northern hardwoods. *Can. J. For. Res.* **2011**, *41*, 1295–1307. [CrossRef]

50. Long, R.P.; Horsley, S.B.; Lilja, P.R. Impact of forest liming on growth and crown vigor of sugar maple and associated hardwoods. *Can. J. For. Res.* **1997**, *27*, 1560–1573. [CrossRef]

51. Auchmoody, L.R. Response of young black cherry stands to fertilization. *Can. J. For. Res.* **1992**, *12*, 319–325. [CrossRef]

52. Stanturf, J.A. Effects of added nitrogen on growth of hardwood trees in southern New York. *Can. J. For. Res.* **1989**, *19*, 279–284. [CrossRef]

53. Marquis, D.A.; Ernst, R.L.; Stout, S.L. *Prescribing Silvicultural Treatments in Hardwood Stands of the Alleghenies, (Revised)*; US Department of Agriculture, Forest Service, Northeastern Forest Experiment Station: Broomall, PA, USA, 1992; 101p.

54. Mohren, G.M.J.; Van Den Burg, J.; Burger, F.W. Phosphorus deficiency induced by nitrogen input in Douglas fir in The Netherlands. *Plant Soil* **1986**, *95*, 191–200. [CrossRef]

55. Vitousek, P.M.; Porder, S.; Houlton, B.Z.; Oliver, C.A. Terrestrial phosphorus limitation: Mechanisms, implications, and nitrogen–phosphorus interactions. *Ecol. Appl.* **2010**, *20*, 5–15. [CrossRef]

56. Goswami, S.; Fisk, M.C.; Vadeboncoeur, M.A.; Garrison-Johnston, M.; Yanai, R.D.; Fahey, T.J. Phosphorus limitation of aboveground production in northern hardwood forests. *Ecology* **2018**, *99*, 438–449. [CrossRef] [PubMed]

57. Gress, S.E.; Nichols, T.D.; Northcraft, C.C.; Peterjohn, W.T. Nutrient limitation in soils exhibiting differing nitrogen availabilities:what lies beyond nitrogen saturation. *Ecology* **2007**, *88*, 119–130. [CrossRef]

58. Soons, M.B.; Hefting, M.M.; Dorland, E.; Lamers, L.P.M.; Versteeg, C.; Bobbink, R. Nitrogen effects on plant species richness in herbaceous communities are more widespread and stronger than those of phosphorus. *Biol. Conserv.* **2017**, *212*, 390–397. [CrossRef]

59. Walter, C.A.; MBAdams FSGilliam, W.T. Peterjohn. Non-random species loss in a forest herbaceous layer following nitrogen addition. *Ecology* **2017**, *98*, 2322–2332. [CrossRef]

60. Gilliam, F.S. Excess nitrogen in temperate forest ecosystems decreases herbaceous layer diversity and shifts control from soil to canopy structure. *Forests* **2019**, *10*, 66. [CrossRef]

forests

MDPI

Article

Assessing the Linkages between Tree Species Composition and Stream Water Nitrate in a Reference Watershed in Central Appalachia

Mark B. Burnham [1,*], Martin J. Christ [2], Mary Beth Adams [3] and William T. Peterjohn [4]

[1] Center for Advanced Bioenergy and Bioproducts Innovation, University of Illinois at Urbana-Champaign, Urbana, IL 61801, USA
[2] West Virginia Department of Environmental Protection, Fairmont, WV 26554, USA; Martin.J.Christ@wv.gov
[3] USDA Forest Service Northern Research Station, Morgantown, WV 26505, USA; mbadams@fs.fed.us
[4] Department of Biology, West Virginia University, Morgantown, WV 26505, USA; William.Peterjohn@mail.wvu.edu
* Correspondence: mburnham@illinois.edu

Abstract: Many factors govern the flow of deposited nitrogen (N) through forest ecosystems and into stream water. At the Fernow Experimental Forest in WV, stream water nitrate (NO_3^-) export from a long-term reference watershed (WS 4) increased in approximately 1980 and has remained elevated despite more recent reductions in chronic N deposition. Long-term changes in species composition may have altered forest N demand and the retention of deposited N. In particular, the abundance and importance value of *Acer saccharum* have increased since the 1950s, and this species is thought to have a low affinity for NO_3^-. We measured the relative uptake of NO_3^- and ammonium (NH_4^+) by six important temperate broadleaf tree species and estimated stand uptake of total N, NO_3^-, and NH_4^+. We then used records of stream water NO_3^- and stand composition to evaluate the potential impact of changes in species composition on NO_3^- export. Surprisingly, the tree species we examined all used both mineral N forms approximately equally. Overall, the total N taken up by the stand into aboveground tissues increased from 1959 through 2001 (30.9 to 35.2 kg N ha^{-1} yr^{-1}). However, changes in species composition may have altered the net supply of NO_3^- in the soil since *A. saccharum* is associated with high nitrification rates. Increases in *A. saccharum* importance value could result in an increase of 3.9 kg NO_3^--N ha^{-1} yr^{-1} produced via nitrification. Thus, shifting forest species composition resulted in partially offsetting changes in NO_3^- supply and demand, with a small net increase of 1.2 kg N ha^{-1} yr^{-1} in NO_3^- available for leaching. Given the persistence of high stream water NO_3^- export and relatively abrupt (~9 year) change in stream water NO_3^- concentration circa 1980, patterns of NO_3^- export appear to be driven by long-term deposition with a lag in the recovery of stream water NO_3^- after more recent declines in atmospheric N input.

Keywords: watershed biogeochemistry; nitrogen cycle; nitrification; nitrogen uptake; nitrate export; *Acer saccharum*

Citation: Burnham, M.B.; Christ, M.J.; Adams, M.B.; Peterjohn, W.T. Assessing the Linkages between Tree Species Composition and Stream Water Nitrate in a Reference Watershed in Central Appalachia. *Forests* **2021**, *12*, 1116. https://doi.org/10.3390/f12081116

Academic Editor: Thomas H. DeLuca

Received: 3 June 2021
Accepted: 17 August 2021
Published: 20 August 2021

Publisher's Note: MDPI stays neutral with regard to jurisdictional claims in published maps and institutional affiliations.

1. Introduction

The northeastern United States experienced relatively high atmospheric N deposition during the latter half of the 20th century [1,2], increasing N supply into some forested ecosystems enough that the availability of N exceeded stand N demand—a situation that can cause significant nitrate (NO_3^-) leaching [3]. Substantial loss of NO_3^- contributes to an associated leaching of base cations, such as calcium and magnesium, which are important to plant growth [4–6], and may also have negative effects downstream [1]. Since the passage and subsequent amendment of the Clean Air Act, national emissions of NO_x and atmospheric N deposition have steadily declined; however, the response of forested catchments is variable. Some have lower N export following national emission and deposition trends, while the levels of N export in others remain high and result in

137

declining inorganic N retention [7–9]. Given the ecological implications of N export into stream water, it is important to understand what controls watershed responses to changes in N deposition through time.

Many factors (both belowground and aboveground) can affect the retention and export of N deposited into forests [10]. Below ground, soil organic matter is the largest pool of N in temperate forests and is a major sink for added N [11]. Microbial immobilization, plant uptake, mineralization, and nitrification control mineral N availability in the soil [12], and net nitrification has a large impact on N export due to the mobility of NO_3^- in soils. Above ground, stand age has a large impact on N retention, as young, aggrading stands usually retain more N due to greater N demand [10]. Even between stands of similar age, differences in species composition can lead to differences in N retention and loss [3,13–16]. As a result, gradual changes in species composition through time could also impact watershed N retention but are more challenging to study due to the need for long-term records.

Fortunately, there are long-term records of changes in both stream-water NO_3^- (since 1970) and the composition of tree species (since 1959) in a reference watershed (WS 4) at the Fernow Experimental Forest (FEF) in the central Appalachian Mountains of West Virginia. From 1975 to 1984, there was a 435% increase (1.3 to 6.9 kg N ha^{-1} yr^{-1}) in stream water NO_3^- export, and one assessment of 24 watersheds in the eastern United States found that WS 4 at the FEF had the lowest retention of inorganic N among those examined [17]. This relatively abrupt increase in stream water NO_3^- export along with other changes in stream water chemistry were likely symptoms of nitrogen saturation caused by long-term N deposition [18]. In addition, nearby measurements show a significant increase in the importance of *A. saccharum* through time [19], which is a species associated with high rates of NO_3^- production. The maintenance of high NO_3^- export from WS 4 despite a reduction in N deposition suggests that long-term changes within the watershed may be responsible, and that these changes may not be quickly reversed. Thus, long-term data sets for WS 4 afford the unique opportunity to assess the potential impact of changes in stand species composition on stream water NO_3^- loss and its potentially long-lasting effect on inorganic N retention.

Tree species composition could impact N retention due to interspecific differences in rate of total N uptake, and interspecific differences in their reliance on different forms of mineral N. Relatively slow-growing *Fagus* species, as well as coniferous species, tend to have lower rates of total N uptake, while other species, including *A. saccharum* and European *Fraxinus* and *Tilia* species, have higher rates of N uptake [20–23]. Therefore, should species with different N uptake requirements change in relative abundance, the overall stand demand for N could shift and alter watershed N retention.

Similarly, differences among species with respect to the mineral forms of N they prefer could also affect watershed N retention if the composition of tree species is altered. The relative uptake of different forms of N varies from species that rely mostly on NO_3^- [24], to species that prefer NH_4^+ [25–28], to species that change their preference to match the form that is most available [29,30]. More specifically, *A. saccharum* trees, which are often abundant in northeastern and Appalachian deciduous forests, may have a strong preference for NH_4^+ [21,31–34]. While many other trees also preferentially take up NH_4^+, some acquire most of their N as NO_3^- [22]. Indeed, seedlings of several species found in central Appalachian forests (*Fagus grandifolia*, *Tsuga canadensis*, *Quercus rubra*, and *Betula lenta*) either take up more NO_3^- than NH_4^+ [21], or grow better under NO_3^- additions [35]. Thus, both the total uptake of N and the variability in relative uptake of different mineral N forms by overstory trees could impact NO_3^- losses following shifts in stand species composition.

Given the variation between species in both total N uptake and relative utilization of different mineral forms, it is interesting that the importance of *A. saccharum* in the FEF has increased substantially over the past century [19]. Since this species appears to strongly prefer NH_4^+, a shift towards a greater influence of *A. saccharum* on the overall community could partially explain the maintenance of stream water NO_3^- exhibited in

FEF WS 4 despite recent reductions in N deposition, particularly if the species it replaces preferentially utilizes NO_3^-. In addition, *A. saccharum* in the FEF is associated with soils having higher NO_3^- production rates and higher soil water NO_3^- concentrations at the scale of individual trees, plots, and entire watersheds [15]. Thus, an increase in the relative importance of this species may not only diminish the demand for NO_3^- but also increase its supply. These combined effects indicate that shifts in species composition and stand NO_3^- utilization may contribute to the temporal trends observed in stream NO_3^- export from WS 4.

To assess whether changing tree species composition in WS 4 could reduce long-term watershed N retention, we took advantage of the relatively unique stand inventory and stream water chemistry data at the FEF by coupling these data with in situ measurements of NO_3^- versus NH_4^+ preference for the dominant, overstory tree species found at this location. This combination of data was then used to estimate total N uptake and temporal changes in stand composition in order to evaluate the hypothesis that changes in species composition at this site have contributed to elevated NO_3^- export in stream water.

2. Materials and Methods

2.1. Study Site

The focus of this study was a long-term reference watershed and a nearby untreated stand at the FEF. The reference watershed (WS 4) is 39 ha at an average elevation of 792 m and has a southeastern aspect. The predominant soil type is a Calvin channery silt loam (loamy-skeletal, mixed, mesic Typic Dystrochrept), and the average annual precipitation is ~145 cm [36]. The forest in WS 4—and the entire FEF—was heavily cut in approximately 1905–1910, and since that time the forest in WS 4 has been left uncut and untreated. WS 4 is dominated by temperate broadleaf trees, with *Quercus* spp., *Acer* spp., *Liriodendron tulipifera*, and *Prunus serotina* making up >75% of the tree stems. In this watershed, the forest canopy is closed along the drainage and there is no clear delineation separating the riparian zone from surrounding areas and no discernable difference in riparian vegetation compared to that of the surrounding slopes.

Continuous stream flow measurements for WS 4 began in 1951 [37], and weekly or bi-weekly stream water samples have been analyzed for their NO_3^- concentration since 1970 [36]. All precipitation and stream water chemistry variables were measured using EPA-approved protocols by the USDA Forest Service's Timber and Watershed Laboratory in Parsons, WV. The analyses and quality control measures are detailed by Edwards and Wood, 1993 [38]. From 1975 through 1984, NO_3^- export in stream water increased by 5.6 kg N ha^{-1} yr^{-1} (~435%); since that time, NO_3^- levels have remained elevated, with fairly regular ~ 5–10 year oscillations (Figure 1). Stream water NH_4^+ concentrations average ~0.05% of NO_3^- concentrations, and although dissolved organic N is not regularly measured in stream water at this site, one year of measurements in the 1995 show that ~87% of N export is as NO_3^-; thus, we focused on stream water NO_3^- export. Historically, the area has received high rates of N deposition (Figure 1), with total (wet + dry) deposition estimated to be ~10 kg N ha^{-1} year^{-1} from 1986 to 2002 [15].

2.2. Species Composition and Stand N Uptake

Complete inventories of all trees in WS 4, including the total number of live trees of all species in 2 inch diameter at breast height (DBH) categories, were completed by the US Forest Service in 1959, 1964, 1972, 1984, and 2001 [39]. To investigate changes in species composition, we calculated relative importance value (RIV) for each species in each inventory year as the average of its relative abundance (RA, the number of stems of that species divided by the total number of tree stems) and its relative basal area (RBA, the basal area of that species divided by the total tree basal area).

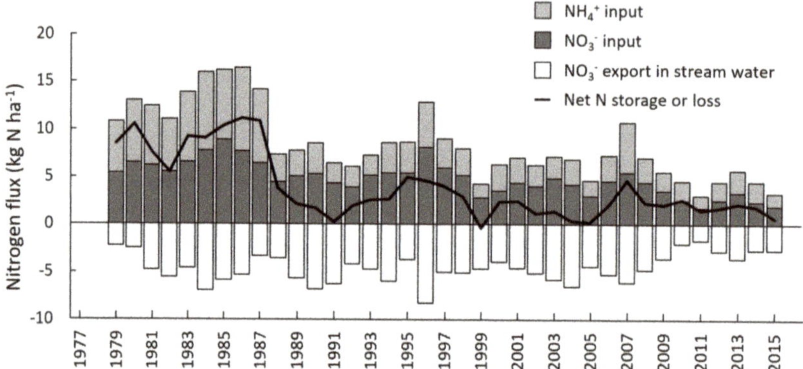

Figure 1. Annual NH$_4^+$ and NO$_3^-$ inputs into and stream NO$_3^-$ export from FEF WS 4, and the net N storage or loss from the catchment. Export of NH$_4^+$ in stream water is negligible (~0.05% of NO$_3^-$ export).

We estimated the total N uptake by the trees in WS 4 as the sum of annual N storage in aboveground woody biomass and annual N return to the soil via litterfall. Complete forest inventory data (1959–2001) were used to estimate annual woody N storage, and since these were 100% live-tree inventories, tree death is accounted for in these measurements, and in our estimates. To determine the N concentration in aboveground woody tissue, trees greater than 8 cm in DBH were cored in 16 plots (10 m radius) spread evenly throughout WS 4 in the summer of 1998 (Christ and others 2002). Using these cores, the width of the last 5 growth rings was measured, and the wood within 1 cm of the bark was ground and analyzed for N concentration by Dumas combustion [40] using a Carlo Erba 1500 CNS elemental analyzer. The total aboveground woody biomass of each tree was estimated with FEF-specific allometric equations [41], and annual N storage was then calculated as the product of annual biomass increment and woody tissue N concentration. Using the DBH and annual N storage, a regression equation was built to estimate the annual woody N storage based on the DBH of any tree in the watershed ($R^2 = 0.790$):

$$\log(\text{annual woody N storage}) = -2.256 + 2.182 \log(\text{DBH}) + a$$

where a is a species-specific constant (Table A1) based on the average residual for each species (Christ and Peterjohn, unpublished data).

Total autumnal litter fall mass (~September through December) was collected annually beginning in 1988 by the US Forest Service using 25 litter traps throughout the watershed (0.7679 m^2 wooden frames with bottoms of ~0.625 × 0.625 cm-opening metal mesh). A relationship between autumnal litter fall and total stand basal area was created using the total basal area measured at 13 long-term growth plots in WS 4, and the total litter fall measured in 1989, 1994, 1999, and 2009 ($R^2 = 0.887$). Using this relationship, we estimated total litter fall for the years of stand inventories prior to the start of the collection of litterfall data (1959, 1964, 1972, and 1984). We then estimated each species' litter N returns for all inventory years using the relationships between a tree species' RBA and the species-specific litterfall N contents at 16 plots in 1998 [13].

2.3. ^{15}N Labeling

To avoid affecting the δ^{15}N of materials in the long-term reference watershed, we used a "test area" located in a nearby untreated area of the FEF (<1 km from WS 4) to measure the relative uptake of NO$_3^-$ versus NH$_4^+$. This area has a similar elevation, slope, and tree composition to WS 4, and an east-northeasterly aspect. Unlike WS 4, small (0.2-ha) plots in this portion of the FEF were harvested to selected basal areas in the 1980s. However, for this study we selected trees within an area showing no signs of harvest, and the trees selected were of similar size to those in WS 4.

At our "test area" in early July 2014, we conducted a ^{15}N-labeling experiment similar to one by performed by McKane et al. [42] to determine the relative uptake of NH_4^+ vs. NO_3^- for 6 major tree species at the FEF: *A. rubrum, A. saccharum, B. lenta, L. tulipifera, Q. rubra,* and *P. serotina.* We used the holes in pieces of commercial peg board (625 cm^2 each, with 10 rows $\times$ 10 columns of holes spaced 2.54 cm apart) to evenly space injections of 3.5 mM ^{15}N as K^{15}NO$_3$ in one area (1 mL per hole), and 3.5 mM ^{15}N as ^{15}NH$_4$Cl in another area under the canopy (within ~3 m of the trunk) of five mature trees of each species. The solutions were injected midday at approximately the boundary between organic and mineral soil horizons—a depth of ~3 cm—using a syringe needle with four side ports. Based on the soil NH_4^+ and NO_3^- concentrations, we estimate that this injection increased background N concentrations by 10% and 5%, respectively. After three hours, we harvested fine roots (<2 mm diameter) from a depth of ~3 cm at each injection site, and roots from one unlabeled area under each tree to measure the natural ^{15}N abundance of root tissue. The sampled roots were traced as far as possible towards the target canopy tree, and we compared the morphology of the collected roots to the fine roots of nearby seedlings of the same species. All species had distinct root characteristics except the two *Acer* species. Thus, we selected *A. saccharum* and *A. rubrum* trees that had no nearby *Acer* spp. within ~15 m.

All harvested roots were placed on ice and transported to the lab, where they were soaked in 1 M CaSO$_4$ for 1 min to remove unassimilated N from the Donnan free space [43]. They were then dried at 65 °C for 48 h and ground to a fine powder in a dental amalgamator (Henry Schein, Inc., Melville, NY, USA). Approximately 5 mg of each sample was wrapped in tin capsules and analyzed for δ^{15}N via isotope ratio gas chromatography–mass spectrometry at the Central Appalachian Stable Isotope Facility that is part of the University of Maryland Center for Environmental Science Appalachian Laboratory (Frostburg, MD, USA).

We calculated root uptake of ^{15}N from the labeled N pool as described in Burnham and others [44]. We first converted δ^{15}N values to the fraction of the heavy isotope in the sample (*F*) using the ^{15}N/^{14}N ratio in each sample (R_{sample}) [45]:

$$R_{sample} = \left(\left(\frac{\delta^{15}N}{1000} \right) * R_{std} \right) + R_{std}$$

$$F = \frac{R_{sample}}{1 + R_{sample}}$$

where R_{std} = ^{15}N/^{14}N ratio in atmospheric N$_2$ (0.0036764). Using the root tissue *N* content and *F*, we calculated the μmol ^{15}N g^{-1} root, and then estimated the rate of ^{15}N uptake from the ^{15}N-labeled pools by dividing the ^{15}N excess (^{15}N content of labeled—unlabeled roots from the same tree) by the exposure time (3 h). Finally, we calculated total uptake of ^{15}N label (^{15}NH$_4^+$ + ^{15}NO$_3^-$) and the percent that was taken up as NH_4^+ and NO_3^-.

2.4. Data Analysis

Our overall ^{15}N label study design included six species, and five trees per species, with a measurement of NO_3^- vs. NH_4^+ uptake associated with each tree. We used a nested ANOVA with Tukey's HSD post hoc test ($\alpha = 0.05$) to determine if the percent of total N taken up as NO_3^- varied by species. The model included the effect of tree nested within species. We then performed one-tailed *t*-tests to determine if the contribution of NO_3^- to total uptake of N from the labeled pool was greater than 50%, which would indicate a significant preference of NO_3^- over NH_4^+.

We used the error terms in our plot-level RBA vs. leaf litter N return and BA vs. woody N storage models to run a Monte Carlo simulation to estimate the uncertainty in our total stand N uptake calculations. For this simulation, we assumed errors were normally distributed and randomly sampled 100 times from the error distribution, and we report uncertainty estimates in woody N storage, litter N return, and total N uptake are reported as 95% confidence intervals.

3. Results

From 1959 to 2001, total stand density in WS 4 decreased 18% (from 372 to 305 trees ha^{-1}) and total stand basal area increased 45% (from 24.3 to 35.2 m^2 ha^{-1}). In 2001, eight species accounted for ~85% of the stand composition (84.6% of stems and 85.8% of basal area): *Quercus rubra, Q. prinus, Acer saccharum, A. rubrum, Liridendron tulipifera, Prunus serotina, Betula lenta,* and *Fagus grandifolia.* Over the study period, five of these species increased in RIV, and three decreased (Figure 2). The RIVs of *A. saccharum* and *A. rubrum* increased 5.8 and 8.5%, respectively, the most of any species. While the RIV of *A. saccharum* increased, its relative basal area decreased slightly (1.4%) and the number of stems increased substantially (from 8.9% to 21.9%) throughout the period examined. The RIV of *Q. rubra* increased to a more modest degree (2.9%), with its relative basal area increasing from 22.6% to 32.3% and its relative abundance decreasing from 20.4% to 16.7% throughout the study period. The RIV of *Q. prinus, B. lenta,* and *F. grandifolia* all declined through the study period (Figure 2). The RIV of *Q. prinus* fell from 6.8% to 5.6%, and the RIV of *B. lenta* fell from 6.9% to 3.8%. While there was only a slight decline in the RIV of *F. grandifolia,* from 4.1% to 3.7%, its relative basal area fell from 5.4% of the stand to 3.4%, but its relative abundance increased from 2.8% to 4.0%.

Aboveground woody N storage increased from 6.4 (6.1–6.7 95% CI) to 9.8 (9.2–10.4) kg N ha^{-1} yr^{-1} (+53.5%) and litter N return increased from 24.5 (19.3–29.7) to 25.4 (21.8–29.0) kg N ha^{-1} yr^{-1} (+3.5%) over this period. In total, stand N uptake increased from 30.9 (25.7–36.1) kg N ha^{-1} yr^{-1} in 1959 to 35.2 (31.7–38.7) kg N ha^{-1} yr^{-1} in 2001 (+13.8%). The percent of mineral N uptake as NO$_3^-$ ranged from 52.7% (*L. tulipifera*) to 75.3% (*A. rubrum*) but was not significantly different between species (Table 1). When these rates of NO$_3^-$ vs. NH$_4^+$ uptake were applied to the estimates of total N uptake within the watershed, NO$_3^-$ uptake increased from 18.7 to 21.4 kg N ha^{-1} yr^{-1} (14.5%), and NH$_4^+$ uptake increased from 12.2 to 13.8 kg N ha^{-1} yr^{-1} (12.8%) from 1959 to 2001. The percent of total stand uptake of N taken up as NO$_3^-$ thus increased only 0.4%.

Table 1. The percent of total uptake of mineral N as NO$_3^-$ for six major overstory trees at the FEF, measured in situ using ^{15}N-labeled NO$_3^-$ and NH$_4^+$.

Species	Percent of N Uptake as NO$_3^-$ ($\pm$SE)
A. rubrum	75.3 ($\pm$12.5)
A. saccharum	53.6 ($\pm$16.0)
B. lenta	54.7 ($\pm$11.5)
L. tulipifera	52.7 ($\pm$13.0)
P. serotina	61.6 ($\pm$11.3)
Q. rubra	56.4 ($\pm$11.5)

Prior studies, using other methods and some using more sampling dates, found much lower rates of N uptake as NO$_3^-$ by *A. saccharum* (average of 15.8%, vs. 53.6% in this study) (Table 2). Given the range of values reported for the affinity of *A. saccharum* for NO$_3^-$, we assessed the potential impact that changes in this particular species might have on stand uptake of NO$_3^-$ by considering two scenarios. First, we used the average relative contribution of NO$_3^-$ to tree uptake of N (15.8%) reported in previous studies. Second, we used the average of all available estimates of NO$_3^-$ uptake by *A. saccharum*, which raised the average to 23.4%. In both scenarios, to estimate stand uptake of NO$_3^-$ we used the average of our measured values of NO$_3^-$ uptake for all unmeasured species. For the first scenario, when values from previous studies were applied to the estimates of total N uptake within WS 4 at the FEF, NO$_3^-$ uptake increased from 17.5 to 20.3 kg N ha^{-1} yr^{-1} (2.8%) and NH$_4^+$ uptake increased from 13.4 to 14.9 kg N ha^{-1} yr^{-1} (1.5%) from 1959 to 2001. Under this scenario, the percent of total stand uptake of N as NO$_3^-$ increased slightly, from 56.7% to 57.7% (Figure 3). For the second scenario, using all available estimates of NO_3^- uptake, the stand uptake of NO_3^- increased from 17.6 to 20.5 kg N ha^{-1} yr^{-1} (2.6%) and uptake of NH_4^+ increased from 13.1 to 14.7 kg N ha^{-1} yr^{-1} (1.5%) from 1959 to 2001. In addition, the percent of total stand uptake of N as NO_3^- increased slightly 57.4% to 58.3% (Figure 3). Thus, in neither of the two scenarios did the observed change in the importance of *A. saccharum* reduce the absolute amount NO_3^- uptake, and in only one scenario was the relative amount of NO_3^- uptake reduced—but this apparent reduction was extremely small.

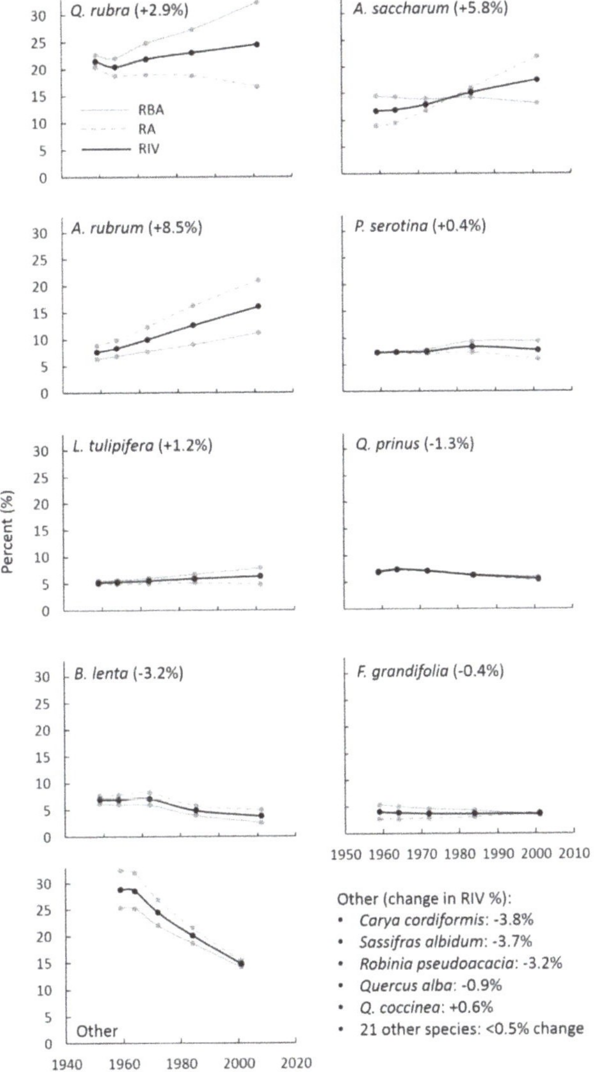

Figure 2. Tree species' relative importance, abundance, and basal area (%) in FEF WS 4 from 1959 to 2001. The percent changes for species listed under "other" are changes in RIV. Data from the USDA Forest Service Northern Research Station [39].

Table 2. All available estimates of the percent of total uptake of mineral N as NO_3^- and estimated N uptake rates (μmol NO_3^--N g dry $root^{-1}$ hr^{-1}) for *A. saccharum*. Measurement methods and parameters varied by study.

Study	Method	*A. saccharum* N Uptake as NO_3^- (%)	Estimated Uptake Rate (μmol N g^{-1} h^{-1})
BassiriRad et al. (1999)	In situ N depletion, excavated intact roots, V_{max}	31	9
Eddy et al. (2008)	Excised root ^{15}N uptake, V_{max}	11.2	0.63
Rothstein et al. (1996)	Excised root ^{15}N uptake, V_{max}	3	1.0
Templer and Dawson (2004)	^{15}N addition to seedlings, greenhouse, roots in native soil	18	1.0 [1]
This study	In situ ^{15}N addition to mature trees, roots left in native soil	53.6	11.6 [2]

[1] Estimated using the reported values of root biomass, total plant biomass, and N uptake per total plant biomass. [2] Estimated assuming that the soil ^{15}N atom percent after labeling was similar to that of the root after 3 h of uptake.

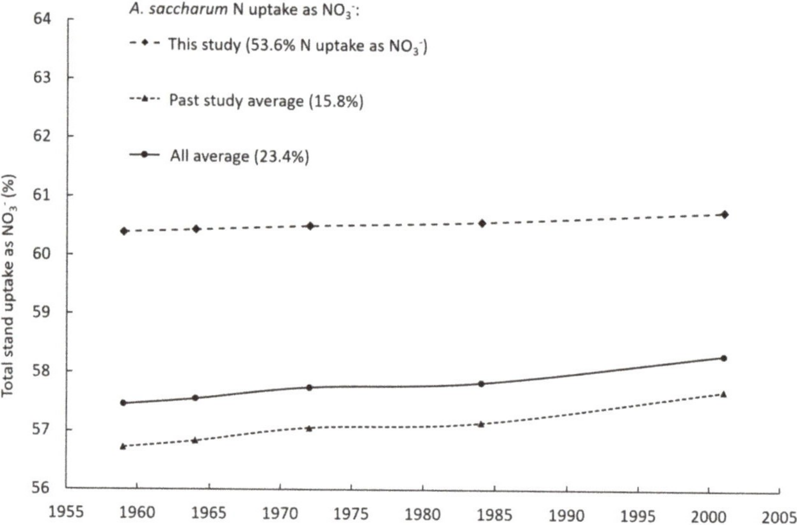

Figure 3. The contribution of NO_3^- to total stand uptake of N from 1959 to 2001. Different lines represent different estimates of uptake of N as NO_3^- for *A. saccharum*, based on prior studies, this study, and the average of all available rates.

4. Discussion

Unexpectedly, the tree species we considered did not differ in their relative uptake of NH_4^+ and NO_3^- and utilized significant amounts of both forms in their mineral N nutrition. This is surprising because prior studies found large differences in the relative uptake of N as NH_4^+ vs. NO_3^- for temperate forest species [21,22]. Notably, in past studies, *A. saccharum* trees took up substantially less NO_3^- than we found using an in situ ^{15}N-labeling technique (Table 2) [21,32–34], and it seems likely that methodological differences could account for the higher relative NO_3^- uptake in this study [23]. Most of the prior research on the form of mineral N uptake utilized seedlings [21], hydroponic techniques [27,46], or N depletion in a simulated soil solution [20,27]—techniques that do not account for some aspects of in situ soil N dynamics. Perhaps most importantly, the differential diffusional resistances of NH_4^+ and NO_3^- in soils [47] are not represented in hydroponic and simulated soil solution techniques. It is possible that tree preferences for NH_4^+ vs. NO_3^- are dynamic through time, particularly as the rate of N deposition changes. However, the relative contributions of NH_4^+ and NO_3^- to total N deposition have not changed substantially (Figure 1), and we therefore believe that large changes in tree N form preference due to changing relative availability of the two mineral N forms is unlikely. Thus, assuming that our ^{15}N-labeling experiment is representative of the long-term mineral N form preference of these tree species, NO_3^- may contribute more to N nutrition of trees than previously thought due to the greater rates of transfer of NO_3^- to roots in the soil.

Since the species examined did not differ in their relative contribution of NO_3^- to total N uptake, it seems unlikely that changes in stand composition contributed to the relatively rapid increase in NO_3^- export or to the long-term persistence of low N retention via a reduction in the demand by trees for NO_3^-. Furthermore, since the stand N demand may have increased over the second half of the last century, it may have contributed to the gradual and slight decrease in soil and stream water NO_3^- since the early 1980s [48]. Although a forest inventory has not conducted after 2001, there have been no major changes in the stand or significant disturbances in this time. We speculate that the reduction in stream water NO_3^- concentration circa 2010 resulted from decreasing N deposition with a significant lag after this decline in deposition started in the early-1990s. Thus, it appears that the large increase observed in NO_3^- export from WS 4 in approximately 1980 resulted from an enhanced supply of available NO_3^- via deposition, and the long-term trend in stream water NO_3^- is controlled primarily by atmospheric N inputs with a lag in recovery as inputs decline.

Although changes in stand NO_3^- demand do not seem to account for the increase in NO_3^- export in stream water, shifts in stand composition could still affect NO_3^- production in the soil and thus contribute to a lag in the recovery of stream water NO_3^- export after deposition declines. At several locations in the eastern U.S., *A. saccharum* trees are associated with high rates of soil net nitrification and low soil C:N ratios [16,31,32,49,50], including WS 4 and other locations in the FEF [13,15], and nitrification rates are positively associated with stream NO_3^- export [51]. The relationship between *A. saccharum* abundance and nitrification is driven, in part, by relatively labile litter and low N residence time [15,52]. To make an initial assessment of the potential impact of species shifts on soil NO_3^- production and stream water NO_3^- export, we used previous plot-level measurements of net nitrification potential and the relative importance and relative basal area of tree species in WS 4. We estimated that net nitrification potential increases 0.02 kg ha^{-1} day^{-1} for every 1% increase in *A. saccharum* importance value ($R^2 = 0.45$) and decreases 0.017 kg ha^{-1} day^{-1} for every 1% increase in *A. rubrum* importance value ($R^2 = 0.13$) [14]. Similarly, net nitrification potential increases 0.017 kg ha^{-1} day^{-1} for every 1% increase in *A. saccharum* relative basal area ($R^2 = 0.20$) and decreases 0.016 kg ha^{-1} day^{-1} for every 1% increase in *A. rubrum* relative basal area ($R^2 = 0.12$) When analyzed in the same manner, no other species was associated with significant changes in net nitrification potential. Since *A. saccharum* and *A. rubrum* had large changes in relative importance value and basal area from 1959 through 2001, and have opposite associations with net nitrification potential, we assessed their potential impact on soil NO_3^- supply and NO_3^- loss to stream water. To arrive at an annual estimate, we assumed that: (1) the estimated daily rate of change in net nitrification potential applied during the months of May through August; (2) only 50% of the estimated daily rate occurred during March, April, and September through November, when the rate of nitrification is lower [53]; and (3) the species change had no effect on net nitrification potential during the months of December through February, when very little nitrification takes place.

The decline in *A. saccharum* and increase in *A. rubrum* relative basal area in WS4 suggest that nitrate production via nitrification was 19.4 kg N ha^{-1} yr^{-1} lower in 2001 than in 1959. However, our plot-level data show a stronger relationship between relative importance value and nitrification potential. Furthermore, past studies have detected a strong relationship between soil NO_3^- concentration and *A. saccharum* abundance [31,54]. Thus, using the relationship between these species' relative importance values and nitrification potential, our initial approximation suggests that the effects of *A. saccharum* and *A. rubrum* on soil NO_3^- production from 1959 to 2001 mostly offset each other, with the negative effect of *A. rubrum* on nitrification causing a net decrease in the rate of NO_3^- production of 2.6 kg NO_3^--N ha^{-1} yr^{-1} within WS 4. However, the majority of the increase observed in the importance of *A. rubrum* occurred in a silvicultural compartment of the watershed (compartment WS 4c) that produces very little NO_3^- in the soil, and that has very low NO_3^- concentrations in soil water collected by tension-free lysimeters [55]. Thus, it is unlikely that this region of the WS 4 contributed to the observed patterns in stream NO_3^- export. Additionally, this subcompartment contains no *A. saccharum* trees, so the increased importance of this species only occurred in the portions of the watershed where nitrification and soil solution NO_3^- levels are currently much higher [13,55]. Although it is unclear why *A. saccharum* has increased in importance at this site, we believe that this is a long-term successional change due to the decline of other subcanopy species.

Considering these known spatial patterns in NO_3^- availability, we refined our initial assessment to ~86% of WS 4 by excluding compartment WS 4c where NO_3^- availability is very low. Taking this approach, we estimate that the net effect of changes in the importance of *A. saccharum* and *A. rubrum* was to increase soil NO_3^- production by 3.9 kg NO_3^--N ha^{-1} yr^{-1} from 1959 through 2001. The long-term change in species composition resulted in a 2.7 kg N ha^{-1} yr^{-1} increase in NO_3^- demand, which mostly offsets the estimated increase in soil NO_3^- production. Thus, we estimate that a net increase of 1.2 kg NO_3^--N ha^{-1} yr^{-1} was available for leaching into stream water. Consequently, it seems that patterns of NO_3^- export were primarily driven by long-term changes in N deposition, but changes in tree species composition may have contributed an increase in soil NO_3^- production and thus to a lag in the recovery of stream water NO_3^- export, which remained 3.5 kg N ha^{-1} yr^{-1} higher from 1992 to 2001 (~5.0 kg NO_3^--N ha^{-1} yr^{-1}) (Figure 1) than the export that occurred from 1970 to 1979 (~ 1.5 kg NO_3^--N ha^{-1} yr^{-1}).

This first-order estimate illustrates that understanding the effect of N deposition on the temporal dynamics of stream water NO_3^- loss requires a relatively complete understanding of how changes in forest species composition can influence the balance between nutrient supply and demand. Moreover, the spatial patterning of N supply and demand within a watershed and connectivity to stream discharge and N export may also be important. We suggest that the recent reductions in atmospheric inputs of N in the eastern US may result in a delayed return of stream water NO_3^- losses to "baseline" levels in situations where a long-lasting shift in the composition of tree species changes the inherent rates of soil NO_3^- production and biotic NO_3^- demand.

Author Contributions: M.B.B. and W.T.P. designed this study; M.B.B., M.J.C. and M.B.A. performed the research and analyzed data; M.B.B. and W.T.P. wrote this paper. All authors have read and agreed to the published version of the manuscript.

Funding: This research was funded by the Long-Term Research in Environmental Biology (LTREB) program at the National Science Foundation (Grant Nos. DEB-0417678, DEB-1019522, and DEB-1455785).

Data Availability Statement: Relevant data can be accessed in a GitHub repository at https://github.com/markbburnham/Forests-Species-Nitrate-Appalachia (accessed on 3 June 2021).

Acknowledgments: We thank Thomas Schuler and the Fernow Experimental Forest personnel for providing and maintaining the long-term data sets for WS 4, and for access to the field site. We also thank Christopher Walter, Rachel Arrick, Jessica Graham, and Hoff Lindberg for their help with the stable isotope-labeling field experiment.

Conflicts of Interest: The authors declare no conflict of interest.

Appendix A

Table A1. Species-specific constants (α) used in woody N storage estimation.

Species	A
Acer rubrum	0.149
Acer saccharum	−0.097
Fagus grandifolia	0.301
Liriodendron tulipifera	0.338
Magnolia acuminata	0.276
Nyssa sylvatica	−0.220
Oxydendrum arboreum	0.187
Prunus serotina	0.222
Quercus alba	−0.041
Quercus coccinea	0.304
Quercus prinus	0.440
Quercus rubra	0.327
Tilia americana	0.000
Other species	0.168

References

1. Driscoll, C.T.; Lawrence, G.B.; Bulger, A.J.; Butler, T.J.; Cronan, C.S.; Eagar, C.; Lambert, K.F.; Likens, G.E.; Stoddard, J.L.; Weathers, K.C. Acidic deposition in the northeastern United States: Sources and inputs, ecosystems effects, and management strategies. *Bioscience* **2001**, *51*, 180–198. [CrossRef]
2. Galloway, J.; Dentener, F.; Capone, D.; Boyer, E.; Howarth, R.; Seitzinger, S.; Asner, G.; Cleveland, C.; Green, P.; Holland, E.; et al. Nitrogen cycles: Past, present, and future. *Biogeochemistry* **2004**, *70*, 153–226. [CrossRef]
3. Aber, J.; McDowell, W.; Nadelhoffer, K.; Magill, A.; Bernston, G.; Kamakea, M.; McNulty, S.; Currie, W.; Rustad, L.; Fernandez, I. Nitrogen saturation in temperate forest ecosystems. *Bioscience* **1998**, *48*, 921–934. [CrossRef]
4. Edwards, P.J.; Williard, K.W.J.; Wood, F.; Sharpe, W.E. Soil water and stream water chemical responses. In *The Fernow Watershed Acidification Study*; Adams, M., DeWalle, D., Hom, J., Eds.; Springer: Dordrecht, The Netherlands, 2006; pp. 71–136.
5. Adams, M.B.; Angradi, T.R.; Kochenderfer, J.N. Stream water and soil solution responses to 5 years of nitrogen and sulfur additions at the Fernow Experimental Forest, West Virginia. *For. Ecol. Manag.* **1997**, *95*, 79–91. [CrossRef]
6. Boggs, J.L.; Mcnulty, S.G.; Gavazzi, M.J.; Myers, J.M. Tree growth, foliar chemistry, and nitrogen cycling across a nitrogen deposition gradient in southern Appalachian deciduous forests. *Can. J. For. Res.* **2005**, *35*, 1901–1913. [CrossRef]
7. Likens, G.E.; Buso, D.C. Dilution and the elusive baseline. *Environ. Sci. Technol.* **2012**, *46*, 4382–4387. [CrossRef] [PubMed]
8. Argerich, A.; Johnson, S.L.; Sebestyen, S.D.; Rhoades, C.C.; Greathouse, E.; Knoepp, J.D.; Adams, M.B.; Likens, G.E.; Campbell, J.L.; McDowell, W.H.; et al. Trends in stream nitrogen concentrations for forested reference catchments across the USA. *Environ. Res. Lett.* **2013**, *8*, 1–8. [CrossRef]
9. Skjelkvåle, B.L.; Stoddard, J.L.; Jeffries, D.S.; Tørseth, K.; Høgåsen, T.; Bowman, J.; Mannio, J.; Monteith, D.T.; Mosello, R.; Rogora, M.; et al. Regional scale evidence for improvements in surface water chemistry 1990-2001. *Environ. Pollut.* **2005**, *137*, 165–176. [CrossRef]
10. Fenn, M.E.; Poth, M.A.; Aber, J.D.; Baron, J.S.; Bormann, B.T.; Johnson, D.W.; Lemly, A.D.; Mcnulty, S.G.; Ryan, D.F.; Stottlemyer, R. Nitrogen excess in North American ecosystems: Predisposing factors, ecosystem responses, and management strategies. *Ecol. Appl.* **1998**, *8*, 706–733. [CrossRef]
11. Nadelhoffer, K.; Downs, M.; Fry, B.; Magill, A.; Aber, J. Controls on N retention and exports in a forested watershed. *Environ. Monit. Assess.* **1999**, *55*, 187–210. [CrossRef]
12. Goodale, C.L.; Fredriksen, G.; Weiss, M.S.; McCalley, C.K.; Sparks, J.P.; Thomas, S.A. Soil processes drive seasonal variation in retention of 15N tracers in a deciduous forest catchment. *Ecology* **2015**, *96*, 2653–2668. [CrossRef] [PubMed]
13. Christ, M.; Peterjohn, W.; Cumming, J.; Adams, M. Nitrification potentials and landscape, soil and vegetation characteristics in two Central Appalachian watersheds differing in NO_3^- export. *For. Ecol. Manag.* **2002**, *159*, 145–158. [CrossRef]
14. Lovett, G.M.; Weathers, K.C.; Arthur, M.A.; Schultz, J.C. Nitrogen cycling in a northern hardwood forest: Do species matter? *Biogeochemistry* **2004**, *67*, 289–308. [CrossRef]
15. Peterjohn, W.T.; Harlacher, M.A.; Christ, M.J.; Adams, M.B. Testing associations between tree species and nitrate availability: Do consistent patterns exist across spatial scales? *For. Ecol. Manag.* **2015**, *358*, 335–343. [CrossRef]
16. Lovett, G.M.; Weathers, K.C.; Arthur, M.A. Control of nitrogen loss from forested watersheds by soil carbon: Nitrogen ratio and tree species composition. *Ecosystems* **2002**, *5*, 712–718. [CrossRef]
17. Campbell, J.L.; Hornbeck, J.W.; Mitchell, M.J.; Adams, M.B.; Castro, M.S.; Driscoll, C.T.; Kahl, J.S.; Kochenderfer, J.N.; Likens, G.E.; Lynch, J.A.; et al. Input-output budgets of inorganic nitrogen for 24 forest watersheds in the northeastern United States: A review. *Water Air Soil Pollut.* **2004**, *151*, 373–396. [CrossRef]

18. Peterjohn, W.T.; Adams, M.B.; Gilliam, F.S. Symptoms of nitrogen saturation in two central Appalachian hardwood forest ecosystems. *Biogeochemistry* **1996**, *35*, 507–522. [CrossRef]
19. Schuler, T.; Gillespie, A. Temporal patterns of woody species diversity in a central Appalachian forest from 1856 to 1997. *J. Torrey Bot. Soc.* **2000**, *127*, 149–161. [CrossRef]
20. McFarlane, K.J.; Yanai, R.D. Measuring nitrogen and phosphorus uptake by intact roots of mature Acer saccharum Marsh., *Pinus resinosa* Ait., and *Picea abies* (L.) Karst. *Plant Soil* **2006**, *279*, 163–172. [CrossRef]
21. Templer, P.H.; Dawson, T.E. Nitrogen uptake by four tree species of the Catskill Mountains, New York: Implications for forest N dynamics. *Plant Soil* **2004**, *262*, 251–261. [CrossRef]
22. Schulz, H.; Härtling, S.; Stange, C.F. Species-specific differences in nitrogen uptake and utilization by six European tree species. *J. Plant Nutr. Soil Sci.* **2011**, *174*, 28–37. [CrossRef]
23. Jacob, A.; Leuschner, C. Complementarity in the use of nitrogen forms in a temperate broad-leaved mixed forest. *Plant Ecol. Divers.* **2014**, 1–16. [CrossRef]
24. Rennenberg, H.; Wildhagen, H.; Ehlting, B. Nitrogen nutrition of poplar trees. *Plant Biol.* **2010**, *12*, 275–291. [CrossRef] [PubMed]
25. DesRochers, A.; van den Driessche, R.; Thomas, B.R. The interaction between nitrogen source, soil pH, and drought in the growth and physiology of three poplar clones. *Can. J. Bot.* **2007**, *85*, 1046–1057. [CrossRef]
26. Buchmann, N.; Schultz, E.; Gebauer, G. 15N-ammonium and 15N-nitrate uptake of a 15-Year-Old Picea abies plantation. *Oecologia* **1995**, *102*, 361–370. [CrossRef]
27. Gessler, A.; Schneider, S.; Von Sengbusch, D.; Weber, P.; Haneman, U.; Huber, C.; Rothe, A.; Kreutzer, K.; Rennenberg, H. Field and laboratory experiments on net uptake of nitrate and ammonium by the roots of spruce (*Picea abies*) and beech (*Fagus sylvatica*) trees. *New Phytol.* **1998**, *138*, 275–285. [CrossRef]
28. Gessler, A.; Jung, K.; Gasche, R.; Papen, H.; Heidenfelder, A.; Börner, E.; Metzler, B.; Augustin, S.; Hildebrand, E.; Rennenberg, H. Climate and forest management influence nitrogen balance of European beech forests: Microbial N transformations and inorganic N net uptake capacity of mycorrhizal roots. *Eur. J. For. Res.* **2005**, *124*, 95–111. [CrossRef]
29. Malagoli, M.; Canal, A.D.; Quaggiotti, S.; Pegoraro, P.; Bottacin, A. Differences in nitrate and ammonium uptake between Scots pine and European larch. *Plant Soil* **2000**, *221*, 1–3. [CrossRef]
30. Gallet-Budynek, A.; Brzostek, E.; Rodgers, V.L.; Talbot, J.M.; Hyzy, S.; Finzi, A.C. Intact amino acid uptake by northern hardwood and conifer trees. *Oecologia* **2009**, *160*, 129–138. [CrossRef] [PubMed]
31. Lovett, G.; Mitchell, M. Sugar maple and nitrogen cycling in the forests of Eastern North America. *Front. Ecol. Environ.* **2004**, *2*, 81–88. [CrossRef]
32. Rothstein, D.E.; Zak, D.R.; Pregitzer, K.S.; Url, S.; Zak, R. Nitrate deposition in northern hardwood forests and the nitrogen metabolism of Acer saccharum marsh. *Oecologia* **1996**, *108*, 338–344. [CrossRef] [PubMed]
33. BassiriRad, H.; Prior, S.; Norby, R.; Rogers, H. A field method of determining NH4+ and NO3- uptake kinetics in intact roots: Effects of CO2 enrichment on trees and crop species. *Plant Soil* **1999**, *217*, 195–204. [CrossRef]
34. Eddy, W.C.; Zak, D.R.; Holmes, W.E.; Pregitzer, K.S. Chronic Atmospheric NO_3^- Deposition Does Not Induce NO_3^- Use by Acer saccharum Marsh. *Ecosystems* **2008**, *11*, 469–477. [CrossRef]
35. Crabtree, R.; Bazzaz, F. Seedling response of four birch species to simulated nitrogen deposition: Ammonium vs. nitrate. *Ecol. Appl.* **1993**, *3*, 315–321. [CrossRef]
36. Kochenderfer, J.N. Fernow and the Appalachian Hardwood Region. In *The Fernow Watershed Acidification Study*; Adams, M., DeWalle, D., Hom, J., Eds.; Springer: Dordrecht, The Netherlands, 2006; pp. 17–39.
37. Reinhart, K.; Eschner, A.; Trimble, G. *US Forest Service Research Paper NE-1. Effect on Streamflow of Four Forest Practices in the Mountains of West*; Upper Darby: Virginia, PA, USA, 1963.
38. Edwards, P.J.; Wood, F. *USDA Forest Service Northern Forest Experimental Station General Technical Report NE-177. Field and Laboratory Quality Assurance/Quality Control Protocols and Accomplishments for the Fernow Experimental Forest Watershed Acidification Study*; Forest Service: Radnor, PA, USA, 1993.
39. Schuler, T.M.; Wood, F. *Fernow Experimental Forest watershed 4 overstory tree data, 1959–2001*; U.S. Department of Agriculture Forest Service, Northern Research Station: Parsons, WV, USA, 2015.
40. Bremner, J.M.; Mulvaney, C.S. N—Total. In *Methods of Soil Analysis. Part 2*; Page, A.L., Ed.; American Society of Agronomy and Soil Science Society of America: Madison, WI, USA, 1982; pp. 595–624.
41. Brenneman, B.; Frederick, D.; Gardner, W.; Schoenhofen, L.; Marsh, P. Biomass of species and stands of West Virginia hardwoods. In Proceedings of the 2nd Central Hardwood Forest Conference, West Lafayette, IN, USA, 14–16 November 1978; pp. 159–178.
42. McKane, R.B.; Johnson, L.C.; Shaver, G.R.; Nadelhoffer, K.J.; Rastetter, E.B.; Fry, B.; Giblin, A.E.; Kielland, K.; Kwiatkowski, B.L.; Laundre, J.A.; et al. Resource-based niches provide a basis for plant species diversity and dominance in arctic tundra. *Nature* **2002**, *415*, 68–71. [CrossRef]
43. Thornton, B.; Osborne, S.M.; Paterson, E.; Cash, P. A proteomic and targeted metabolomic approach to investigate change in Lolium perenne roots when challenged with glycine. *J. Exp. Bot.* **2007**, *58*, 1581–1590. [CrossRef]
44. Burnham, M.B.; Cumming, J.R.; Adams, M.B.; Peterjohn, W.T. Soluble soil aluminum alters the relative uptake of mineral nitrogen forms by six mature temperate broadleaf tree species: Possible implications for watershed nitrate retention. *Oecologia* **2017**. [CrossRef]
45. Fry, B. *Stable Isotope Ecology*, 1st ed.; Springer: New York, NY, USA, 2006; ISBN 0-387-33745-8.

46. Kronzucker, H.J.; Siddiqi, M.Y.; Glass, A.D.M. Conifer root discrimination against soil nitrate and the ecology of forest succession. *Nature* **1997**, *385*, 59–61. [CrossRef]
47. Chapman, N.; Miller, A.J.; Lindsey, K.; Whalley, W.R. Roots, water, and nutrient acquisition: Let's get physical. *Trends Plant Sci.* **2012**, *17*, 701–710. [CrossRef]
48. Gilliam, F.S.; Adams, M.B. Effects of nitrogen on temporal and spatial patterns of nitrate in streams and soil solution of a central hardwood forest. *ISRN Ecol.* **2011**, *2011*, 1–9. [CrossRef]
49. Pastor, J.; Aber, J.D.; Mcclaugherty, C.A.; Melillo, J.M. Aboveground production and N and P cycling along a nitrogen mineralization gradient on Blackhawk Island, Wisconsin. *Ecology* **1984**, *65*, 256–268. [CrossRef]
50. Zak, D.R.; Pregitzer, K.S. Spatial and temporal variability of nitrogen cycling in northern Lower Michigan. *For. Sci.* **1990**, *36*, 367–380.
51. Ross, D.S.; Shanley, J.B.; Campbell, J.L.; Lawrence, G.B.; Bailey, S.W.; Likens, G.E.; Wemple, B.C.; Fredriksen, G.; Jamison, A.E. Spatial patterns of soil nitrification and nitrate export from forested headwaters in the northeastern United States. *J. Geophys. Res. Biogeosci.* **2012**, *117*, 1–14. [CrossRef]
52. Pregitzer, K.S.; Zak, D.R.; Talhelm, A.F.; Burton, A.J.; Eikenberry, J.R. Nitrogen turnover in the leaf litter and fine roots of sugar maple. *Ecology* **2010**, *91*, 3456–3462. [CrossRef] [PubMed]
53. Gilliam, F.S.; Yurish, B.M.; Adams, M.B. Temporal and spatial variation of nitrogen transformations in nitrogen-saturated soils of a central Appalachian hardwood forest. *Can. J. For. Res.* **2001**, *31*, 1768–1785. [CrossRef]
54. Mitchell, M.J. Nitrate dynamics of forested watersheds: Spatial and temporal patterns in North America, Europe and Japan. *J. For. Res.* **2011**, *16*, 333–340. [CrossRef]
55. Peterjohn, W.T.; Foster, C.J.; Christ, M.J.; Adams, M.B. Patterns of nitrogen availability within a forested watershed exhibiting symptoms of nitrogen saturation. *For. Ecol. Manag.* **1999**, *119*, 247–257. [CrossRef]

Review

Response of Temperate Forest Ecosystems under Decreased Nitrogen Deposition: Research Challenges and Opportunities

Frank S. Gilliam

Department of Biology, University of West Florida, Pensacola, FL 32514, USA; fgilliam@uwf.edu

Abstract: Although past increases in emissions and atmospheric deposition of reactive nitrogen (N_r) provided the impetus for extensive research investigating the effects of excess N in terrestrial and aquatic ecosystems, the Clean Air Act and associated rules have led to decreases in emissions and deposition of oxidized N, especially in the eastern U.S., but also in other regions of the world. Thus, research in the near future must address the mechanisms and processes of recovery for impacted forests as they experience chronically less N in atmospheric depositions. Recently, a hysteretic model was proposed to predict this recovery. By definition, hysteresis is any phenomenon in which the state of a property depends on its history and lags behind changes in the effect causing it. Long-term whole-watershed additions of N at the Fernow Experimental Forest allow for tests of the ascending limb of the hysteretic model and provide an opportunity to assess the projected changes following cessation of these additions. A review of 10 studies published in the peer-reviewed literature indicate there was a lag time of 3–6 years before responses to N treatments became apparent. Consistent with the model, I predict significant lag times for recovery of this temperate hardwood ecosystem following decreases in N deposition.

Keywords: decreased N deposition; forest recovery; hysteresis; temperate hardwood forests; Fernow Experimental Forest

Citation: Gilliam, F.S. Response of Temperate Forest Ecosystems under Decreased Nitrogen Deposition: Research Challenges and Opportunities. *Forests* **2021**, *12*, 509. https://doi.org/10.3390/f12040509

Academic Editor: Christopher Gough

Received: 19 March 2021
Accepted: 16 April 2021
Published: 19 April 2021

Publisher's Note: MDPI stays neutral with regard to jurisdictional claims in published maps and institutional affiliations.

1. Introduction

Historic increases in atmospheric deposition of reactive nitrogen (N_r, primarily NH_4^+ and NO_3^-, but including numerous other reactive species [1]), and modeled projections for future increases on a global scale, have led to a proliferation of studies on the effects of excess N on aquatic and terrestrial ecosystems over the past several decades. Galloway et al. [2] estimated that the total global atmospheric deposition of NH_4^+ and NO_3^- in terrestrial ecosystems increased from 17 Tg N yr^{-1} in 1860 to 64 Tg N yr^{-1} in the early 1990s. They also projected further increases to 125 Tg N yr^{-1} by 2050, a >7-fold increase during this time period. Bobbink et al. [3] predicted similar increases in N deposition by 2030. Although most terrestrial ecosystems studied were initially herb- and grass-dominated [4,5], recent decades have witnessed an expansion of studies in forest ecosystems [6–13]. Seminal papers on then-new perspectives of N biogeochemistry, e.g., [14–16], changed our views of N in forest ecosystems from a predominantly growth-limiting nutrient to one whose excess could threaten the structure and function of forests, including net primary productivity and biodiversity, a phenomenon called N saturation [17].

Understanding the general responses of temperate forests to changes in N biogeochemistry is of particular importance because of their (1) global distribution (Figure 1), (2) high biodiversity, and (3) close proximity to high densities of human populations [18]. Early predictions for temperate forests in response to increasing N deposition included increased nitrification leading to leaching of NO_3^- and the associated loss of nutrient cations (Ca^{2+}, Mg^{2+}, K^+) [16,17]. Although most of these predictions were initially—and continue to be—tested and refined via field N manipulation studies, e.g., [19], enhanced NO_3^- and cation leaching was also observed on the long-term untreated reference watershed (WS4) at Fernow Experimental Forest (FEF), West Virginia, based on stream chemistry data from 1969 to 1990 [20]. Peterjohn et al. [20] further identified seven symptoms of N saturation at FEF, including high relative rates of net nitrification, relatively high concentrations of NO_3^- in soil solution, low seasonal variability in stream NO_3^- concentrations, high loss of NO_3^- from an aggrading forest, a rapid increase in NO_3^- loss following fertilization from an aggrading forest, and low retention of inorganic N compared to other forested sites. They concluded that untreated watersheds at FEF were perhaps the best example of a temperate forest that had undergone N saturation via ambient deposition of N.

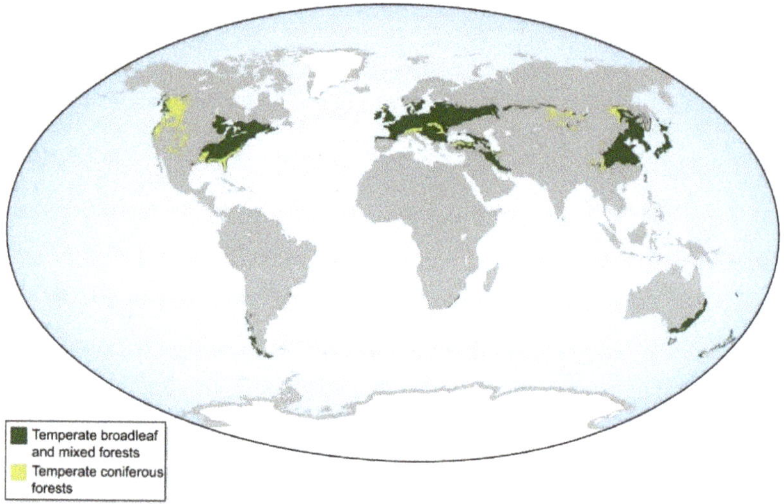

Figure 1. The distribution of temperate forests. Reprinted with permission from ref. [18]. Copyright 2016 Gilliam.

Other research in temperate forests has examined the effects of excess N on plant biodiversity, microbial communities, and forest health. In general, chronic increases in N inputs have decreased the diversity of the herbaceous layer [3,11,21], altered the structure and composition of soil microbial communities [22], and the threatened the growth and vitality of some temperate forests [23]. How these ecosystem properties will respond in the future, however, remains an open, and important, question.

The Clean Air Act (CAA) of 1970 in the U.S. and similar environmental regulations worldwide have been exemplars of environmental legislation, leading to global-scale improvements in many facets of environmental health, including lower human mortality rates in the U.S. [24]. Initially targeting anthropogenic emissions of sulfur, subsequent rules addressed emissions of N, leading to reductions of >50% from vehicles and power plants in the U.S. [25]. On the other hand, the focus of the CAA has been on N oxides, rather than reduced forms of N, such that temporal trends of N deposition vary between oxidized versus reduced N (Figure 2). Thus, although total deposition of NO_3^- peaked in the mid-1990s and is currently declining throughout most of the U.S. [25,26], total deposition of NH_4^+ has either increased of remained stable [27,28].

This recent trend in atmospheric deposition of N, at least for NO_3^-, challenges some of the earlier models predicting future increases in N deposition, e.g., [2,3], and alters the way N biogeochemists must think about N-impacted ecosystems in the future. Recent papers have hypothesized several future scenarios for changes in terrestrial ecosystems in response to reduced levels of atmospheric N input [29–33]. Gilliam et al. [28] proposed a hysteretic model to predict future changes in forests of eastern North America, given current patterns of declines in atmospheric deposition of N (Figure 3). Hysteresis is a phenomenon in which the state of a property depends on its history and lags behind changes in the effect causing it. Thus, it is a system property wherein output is not a strict function of corresponding input, resulting in a response to decreasing inputs that exhibits a different trajectory than the response to increasing inputs.

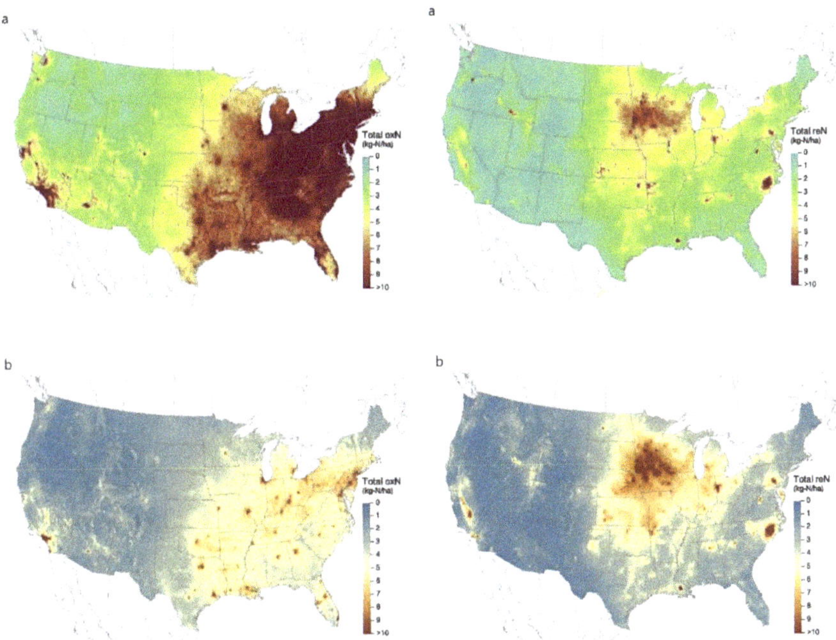

Figure 2. Total oxidized N and total reduced N deposition. Figures marked (**a**) are from 2000; figures marked (**b**) are from 2014. Reprinted with permission from ref. [28]. Copyright 2019 Gilliam et al.

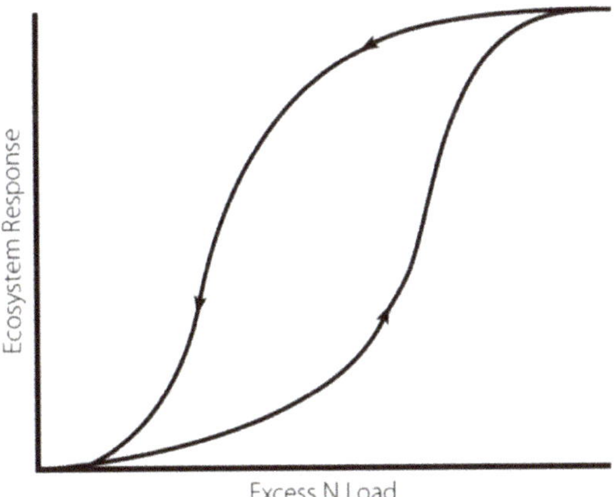

Figure 3. Hysteretic model for responses of terrestrial ecosystems to atmospheric deposition of N. Note lack of units and variables, as this is a conceptual model, with variables and units varying by ecosystem process and N status of the ecosystem. Reprinted with permission from ref. [28]. Copyright 2019 Gilliam et al.

With this in mind, the curvilinear nature of the hysteretic model when applied to important ecosystem responses at FEF (e.g., plant diversity and soil nutrient availability) suggests that initial effects may have been buffered and slow to occur after experimental increases in N deposition. Following declines in atmospheric N inputs, the hysteretic model predicts varying lag times in the recovery of virtually all components of temperate forest ecosystems, including soil nutrients and microbial communities, plant biodiversity, and surface water chemistry toward conditions prior to N saturation as deposition of N continues to decline.

The purpose of this paper is to use results from a long-term experiment at FEF to assess the hysteretic model of ecosystem response for a forest during a time of elevated N inputs, and to consider its implications for the recovery of forest ecosystem processes during a period of declining N inputs. First, I describe FEF as a study site, along with a general overview of field design and sampling associated with numerous investigations over the past three decades on the effects of whole-watershed additions of N. Next, I summarize the findings of those studies in the context of the hysteretic model. Finally, I describe future directions for work at FEF and elsewhere as N biogeochemists address this relatively new paradigm of decreasing N deposition, especially in the context of climate change.

2. Research at Fernow Experimental Forest

2.1. Background

Forest ecological research at FEF has a long, productive history [34], with many efforts initially focused on silvicultural practices [35]. The collection of hydrochemical data for the long-term reference watershed (WS4) began in 1951. Investigations into the effects of acid deposition were initiated via the Fernow Watershed Acidification Study (WAS), beginning as a now-terminated pilot study in 1987 on a watershed adjacent to FEF. The study was established on FEF watersheds in 1989 and remains on-going. A notable characteristic of the WAS is that it employs a whole-watershed application of simulated acid deposition via three aerial additions of $(NH_4)_2SO_4$ per year as a solid powder by fixed- and variable-wing aircraft, representing a total N addition of 35 kg N ha^{-1} yr^{-1}, and it was one of few studies utilizing watershed-scale simulations of atmospheric deposition. After three decades, the $(NH_4)_2SO_4$ additions were discontinued in 2019, creating the unique opportunity to examine experimentally the recovery of a temperate hardwood forest ecosystem following a significant decrease in N inputs [34].

2.2. Site Description

FEF comprises approximately 1900 ha of the Allegheny Mountain section of the unglaciated Allegheny Plateau near Parsons, West Virginia (39°03'16″ N, 79°41'0″ W). Mean annual precipitation is approximately 1430 mm yr^{-1}. Concentrations of acid ions in wet deposition (i.e., H^+, $SO_4^=$, and NO_3^-) were once among the highest in North America. Soils of the experimental watersheds are coarse-textured Inceptisols (loamy-skeletal, mixed mesic Typic Dystrochrept) of the Berks and Calvin series sandy loams largely derived from sandstone [34].

Forest stands of FEF watersheds are composed of mixed hardwood species generally varying with age. Young watersheds support early-successional species, such as black birch (*Betula lenta* L.), black cherry (*Prunus serotina* Ehrh.), and yellow-poplar (*Liriodendron tulipifera* L.), whereas older watersheds support late-successional species, such as sugar maple (*Acer saccharum* Marshall) and northern red oak (*Q. rubra* L.) [34]. Herbaceous layer communities initially exhibited less age-related variation, and included wood-nettle (*Laportea canadensis* (L.) Wedd.), violets (*Viola* spp.), and several ferns [36].

2.3. Field Design

Although FEF has numerous experimental watersheds that have been, and continue to be, studied in a variety of contexts, the WAS focused on three watersheds (Figure 4). WS4 supports >100 yr-old even-aged stands and has served as the long-time reference watershed at FEF, whereas WS7 supports >40 yr-old even-age stands, and both are used as untreated watersheds of contrasting stand ages. WS3 supports an approximately 40 yr-old even-age stand and serves as the treatment watershed, receiving three aerial applications of $(NH_4)_2SO_4$ yr^{-1}, beginning in 1989 and extending to 2019. March (or sometimes April) and November applications were approximately 7.1 kg N ha^{-1}; July applications were approximately 21.2 kg N ha^{-1} [34].

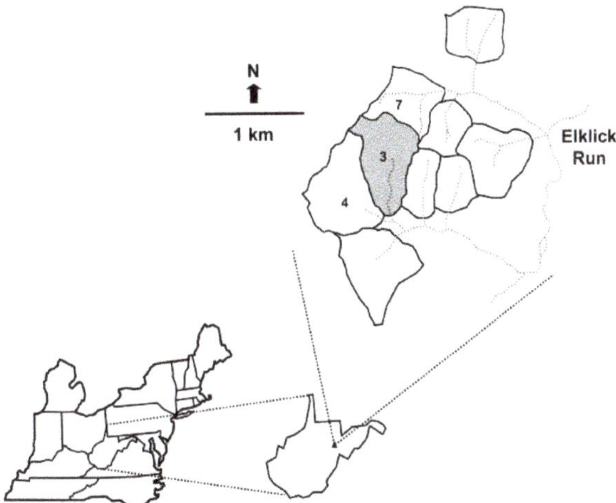

Figure 4. Map depicting watersheds of the Fernow Experimental Forest, West Virginia, USA. WS4 is the long-term reference watershed, WS3 is the treated watershed as part of the Watershed Acidification Study, and WS7 is an additional reference watershed of stand age similar to that of WS3.

2.4. Findings

A notable number of investigations have been carried out within the design of the WAS, including several Master's theses and Ph.D. dissertations from multiple institutions, such as Marshall University, the Pennsylvania State University, the University of Pittsburgh, and West Virginia University. To date, these have resulted in ~130 publications, mostly articles in the peer-reviewed ecological literature, but also books, e.g., [34,37,38], symposia proceedings, book and proceedings chapters, and USDA Forest Service research publications. To assess the findings of the WAS in the context of the hysteretic model, I have summarized 10 studies from among these publications using three main criteria. First, each must have been published in the peer-reviewed literature, thus excluding books, proceedings, chapters, and USDA Forest Service publications. Second, each must represent a specific study with a particular sampling within WAS watersheds, thus excluding review articles (with the exception of Peterjohn et al. [20], a historically important synthesis placing WAS watersheds in the context of N saturation). A good example of an excluded peer-reviewed paper would be Gilliam [21], which used extensive results from the WAS, but did so only in the context of general responses of forest herb communities to excess N deposition, including studies from throughout Europe and North America. Third, studies should represent largely continuous measurements of variables (e.g., herb communities, soil fertility, stream chemistry) over time.

These studies are summarized in Table 1 and placed in order according to the cumulative number of years that $(NH_4)_2SO_4$ had been added to WS3 when the study was conducted. Since several studies used multiple sampling periods (e.g., stream chemistry), the number of years listed represents the latest sampling included in the paper. Although many, perhaps most, studies were multifaceted in their scope, for brevity only the main focus and findings are summarized. All told, these studies comprise a sample period that extends from three years [39] to 30 years of treatment [12].

Table 1. A brief summary of 10 studies published in the peer-reviewed literature. Yr indicates the number of years of treatment on WS3 represented by latest year of sampling of a given study. Many studies were multi-focused with multiple findings. For brevity, only the principle areas of focus and findings are summarized herein.

Study	Yr	Focus	Findings
Gilliam et al. (1994) [39]	3	herb layer/soil nutrients	no significant differences between WS3, WS4, and WS7
Gilliam et al. (1996) [40]	4	N mineralization, foliar nutrients	no differences for N mineralization
Peterjohn et al. (1996) [20]	4	symptoms of N saturation	clear evidence of N saturation on untreated WS4
Gilliam et al. (2006) [36]	5	herb layer communities	no differences in composition and diversity
Edwards et al. (2002) [41]	8	soil solution chemistry	higher NO_3^- and cation concentrations on WS3
Edwards et al. (2002) [42]	8	stream chemistry	increases in NO_3^-, cations, and acidity, following ~2 yr lag
Gilliam et al. (2016) [11]	25	herb layer composition/diversity	N alters herb layer composition, decreases species diversity
Gilliam et al. (2018) [43]	25	soil N mineralization/nitrification	no watershed differences, increased homogeneity on WS3
Gilliam et al. (2020) [44]	25	soil fertility	significant decreases in base cations and soil pH
Eastman et al. (2021) [12]	30	long-term carbon and N budgets	added N leads to greater C storage in vegetation and soil

The timeline generated by these summarized results generally follows the trajectory depicted in the hysteretic model (Figure 3). Early on, most studies reported minimal effects of added N to WS3. After 3 yr of treatment, there were no treatment-related differences in either soil or herb foliar nutrients [39], but 25 yr of treatment yielded numerous effects [44]. Following even as many as five years of treatment, Gilliam et al. [36] reported no significant differences between WS3 and WS4 in species composition and diversity in forest herb communities, yet 25 yr of treatment greatly altered composition and decreased diversity [11]. By 8–10 years of treatment, Edwards et al. [41,42] reported higher concentrations of NO_3^- and base cations in both streamflow and soil solution, but also demonstrated a time lag of 2–3 years before responses became apparent, also consistent with the model and confirmed more recently by Gilliam et al. [44]. Essentially all studies after this period reported significant treatment effects on WS3.

It should be noted that many studies within the WAS were not carried out in ways that allow direct assessment with the hysteretic model. On the other hand, the complete body of published work merits summarization of the effects of excess N at FEF.

Following a lag period of 3–5 years, excess N elicited several responses in a temperate hardwood forest ecosystem. It decreased rates of decomposition of organic matter [45], likely through the alteration of microbial communities and extracellular enzymes [46]. Excess N increased NO_3^- mobility, accompanied by the leaching of base cations [41,42] and leading to decreases in soil fertility [44], increased Al mobility [47], increased NO and NO_2 emissions [48,49], and limitation by P [50]. Several of these responses are also evidenced in both tree and herb tissues [19,51]. Although there was an initial effect of fertilizer added on tree growth, decreased soil fertility has led to slower growth of prominent tree species [52–54]. There was, however, a high degree of interspecific variability in such a response. Despite slower tree growth, excess N has led to greater storage of C in soil and vegetation [12]. Finally, there has been a pronounced shift in herb layer composition that arose from increases in a nitrophilic species, blackberry (*Rubus allegheniensis* Porter) (hereafter, *Rubus*) [55], that competitively eliminated numerous N-efficient herbaceous species, resulting in a loss of plant diversity [11]. The observed increase in cover of *Rubus* at FEF in response to increasing deposition of N, as both a function of time and N loading, is consistent with the ascending limb of the hysteretic model (Figure 5).

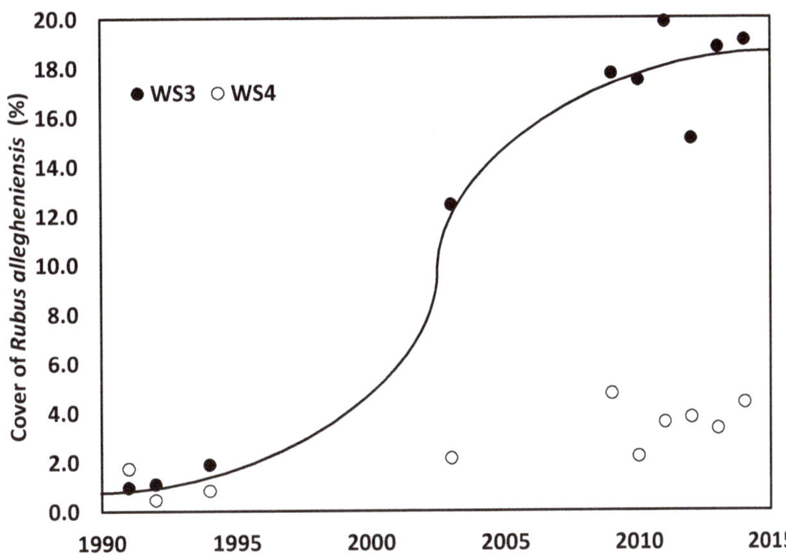

Figure 5. Changes in cover of *Rubus allegheniensis* in response to a quarter century of N additions at Fernow Experimental Forest, West Virginia. Data taken from Gilliam et al. [11].

3. Response of Terrestrial Ecosystems to Decreased N Deposition

Relative to temporal patterns of increases in Nr on a global scale, decreased N deposition is a somewhat novel phenomenon. As a result, few studies have directly assessed how terrestrial ecosystems respond to such decreases. Using whole-watershed additions of $(NH_4)_2SO_4$ (same as the WAS at FEF) from 1989 to 2016 at Bear Brook Watershed in Maine, Patel et al. [13] found relatively rapid recovery (within one year) of stream NO_3^- outputs following cessation of treatment. They concluded, however, that the pattern for base cation loss distinctly followed the predictions of the hysteretic model [13].

Earlier studies on potential recovery for forest herb communities following decreased N deposition were all carried out in Europe, including the NITREX roof experiments [56] and forest fertilization studies, e.g., [57]. In the 1990s, NITREX comprised a network of sites throughout Europe that, among several integrated investigations, used roofs to experimentally decrease high ambient N deposition. They found that nitrophilous plant species, all of which had increased under high N deposition, declined over a 5-yr period in plots under the roofs, suggesting a relatively rapid recovery [56]. By contrast, Strengbom et al. [57] addressed what they called 'legacy effects' of excess N in forests of northern Sweden by examining forest herbs 9 yr after cessation of 20 yr of N fertilization. They found that N-mediated declines in ericaceous species and increases in a nitrophilous grass, wavy hairgrass—*Deschampsia flexuosa* [(L.) Trin.]—persisted after this period of nearly a decade, concluding that the effects of excess N can be long-lived for forest herbs. As with biogeochemical responses, these sharply contrasting results highlight both the research challenges and opportunities for the temperate forest sites in the eastern U.S.

Stevens [31] reviewed evidence from long-term experiments in grasslands, forests, heathlands, and wetlands experiencing lower N inputs, primarily from cessation of experimental additions of N. She concluded that plant species composition and soil microbial communities may be slow in recovery, whereas soil N dynamics may respond more rapidly. Based primarily on long-term research at Hubbard Brook Experimental Forest, New Hampshire, USA, Groffman et al. [32] observed declines in the atmospheric deposition of N over a half century from 1964–2014 and referred to ecosystem responses to this as *N oligotrophication*. They suggested that this phenomenon is driven via increased C flux from the atmosphere to soils in ways that stimulate microbial immobilization of N, decreasing the plant's available N pool, a response further exacerbated by climate change, including lengthening of the growing season. Similar to patterns reported for North America [26], Schmitz et al. [33] reported declines in N deposition throughout Europe since the 1990s. They reviewed both observational (generally using deposition gradients) and experimental studies for changes in soil acidification and eutrophication, understory vegetation, tree foliar nutrients, and tree growth and vitality. In contrast to trajectories for the U.S., they predicted that further declines in N deposition will be minimal, but also affirmed the hysteretic model for forests of Europe.

4. Future Directions

With the cessation of whole-watershed additions of $(NH_4)_2SO_4$ at FEF in 2019 following 30 years of continuous treatment, the WAS at FEF is uniquely positioned to provide empirical documentation of the recovery of a temperate hardwood forest following decreases in N deposition. Some of these investigations are already in place and are currently of a monitoring nature. Plots to assess the response of the forest herb layer community were established in 1991 and were sampled at various time intervals until 2009, after which they have been sampled annually in early July [11]. Beginning in 1993, in situ incubations of mineral soil were made monthly during the growing seasons of several years until 2007, after which they have received monthly growing season sampling every year [43]. In addition, personnel of the Timber and Watershed Laboratory, Parsons, West Virginia— which manages the FEF—take weekly grab samples of stream water at calibrated weirs for complete chemical analysis. As previously suggested, temporal patterns of stream chemistry for treatment WS3 are consistent with the ascending curve of the hysteretic model, exhibiting a lag time for the response of pH, NO_3^-, and $SO_4^=$ of ~3 yr (Figure 6). The extension of all such investigations will provide a continuous timeline for potential changes under conditions of decreased N deposition.

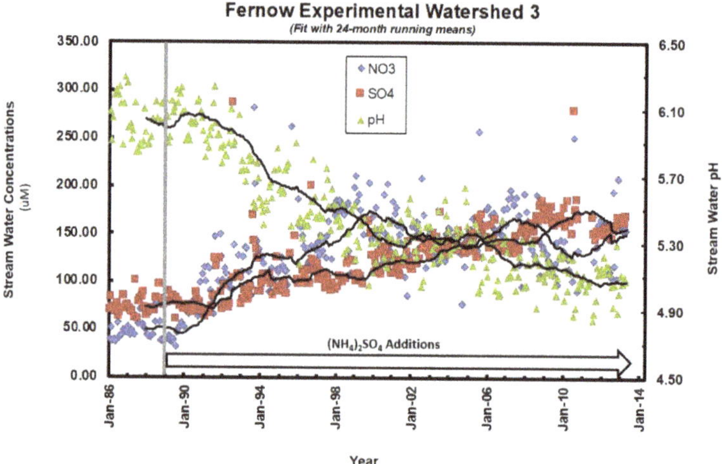

Figure 6. Monthly mean stream water pH and concentrations of NO_3^- and $SO_4^=$ for treated WS3 from 1986 to 2014.

It is indeed likely that the temporal patterns of recovery will be as varied and nonlinear as the responses to 30 years of treatment. It is reasonable to hypothesize that most ecosystem processes will display lag times as predicted by the hysteretic model (Figure 3), but that the length of this period will vary considerably among processes. Once again, Patel et al. [13] reported an immediate decline in stream NO_3^- output at Bear Brook Watershed, Maine, following cessation of whole-watershed additions identical to those at FEF. Base cations, however, exhibited the hysteretic pattern. It is likely that the NO_3^- response at FEF will be less immediate than in Maine because of the higher conifer component of stands at Bear Brook Watershed, Maine [58], especially at higher elevations [13].

Of particular interest will be the response of forest herb communities on WS3. Once again, other studies have exhibited conflicting results. A reasonable hypothesis for FEF in particular is that there will be a very pronounced lag time in the response of herb community composition and diversity. The major change in response to N additions on WS3 has been the increased dominance of a single nitrophilous species that was once only a minor component—*Rubus* (Figure 5). Among the many effects this increase has had on the biotic and abiotic environment of the forest understory is that it is has effectively redistributed high, possibly phytotoxic, levels of Mn from deep soil to surface horizons [50]. Because *Rubus* accumulates Mn in pre-senescent leaves and further concentrates Mn in post-senescent leaves [59], this effect may persistent for several years.

Finally, regarding future directions, perhaps the greatest challenge is the 'moving target' represented by climate change. Consider that total inorganic deposition in the U.S. declined by nearly 25% from a maximum in 1995 to 2012, at which time it equaled the deposition in 1985 [26]. Mean annual temperature in the U.S., however, was 10.7 C in 1985, increasing to 13.2 C in 2012. More importantly, CO_2 concentrations for those two years in the Northern Hemisphere were 346 and 395 ppm, respectively, an increase of >1% over that 30-yr period. Accordingly, whereas it is certain that N deposition will decline to previous levels, it is equally certain that ambient temperature and CO_2 will not.

5. Conclusions

In conclusion, the Clean Air Act in the U.S. and similar environmental regulations worldwide have exhibited a high degree of efficacy with respect to decreasing emissions of pollutants, including and especially N_r. As a result, atmospheric deposition of N has declined over the past few decades on a global scale, including North America [59], Europe [33], and China [60], representing a challenge to the way N biogeochemists have thought and carried out research over this period of time. Moving forward, such challenges lead to new research opportunities and directions.

The hysteretic model proposed by Gilliam et al. (2019) has received support in the literature in predicting a non-linear response of forest ecosystems to declines in atmospheric deposition of N_r, coupled with time lags of varying duration, depending on specific conditions for a particular site [13,61,62]. Studies over the past three decades at the Fernow Experimental Forest, West Virginia, generally support the ascending limb of the hysteretic model during which time aerial applications were added to an entire watershed. Cessation of this treatment in 2019 now allows for further empirical testing of the model via ongoing studies of several components of this hardwood forest ecosystem, including biogeochemical cycling and dynamics of the forest herbaceous layer.

Author Contributions: F.S.G. wrote the manuscript. The author has read and agreed to the published version of the manuscript.

Funding: This research was funded in part through the USDA Forest Service, Fernow Experimental Forest, Timber and Watershed Laboratory, Parsons, W.V., under USDA Forest Service Cooperative Grants 23-165, 23-590, and 23-842 to Marshall University. Additional funding was provided by USDA National Research Initiative Competitive Grants (Grant NRICGP #2006-35101-17097) to Marshall University and the Long Term Research in Environmental Biology program at the National Science Foundation (Grant Nos. DEB0417678 and DEB-1019522) to West Virginia University.

Institutional Review Board Statement: Not applicable.

Informed Consent Statement: Not applicable.

Data Availability Statement: Not applicable.

Acknowledgments: I am indebted to numerous undergraduate and graduate students at Marshall University and West Virginia University for field assistance. I especially thank Bill Peterjohn for numerous constructive comments on a previous draft, and for providing Figure 6. I also thank Mary Beth Adams (USDA Forest Service, ret.) for field and logistical support. The long-term support of the USDA Forest Service in establishing and maintaining the research watersheds is acknowledged.

Conflicts of Interest: The author declares no conflict of interest.

References

1. Horii, C.V.; Munger, J.W.; Wofsy, S.C.; Zahniser, M.; Nelson, D.; McManus, J.B. Atmospheric reactive nitrogen concentration and flux budgets at a Northeastern U.S. forest site. *Agric. For. Meteorol.* **2005**, *133*, 210–225. [CrossRef]
2. Galloway, J.N.; Dentener, F.J.; Capone, D.G.; Boyer, E.W.; Howarth, R.W.; Seitzinger, S.P.; Asner, G.P.; Cleveland, C.C.; Green, P.A.; Holland, E.A.; et al. Nitrogen cycles: Past, present, and future. *Biogeochemistry* **2004**, *70*, 153–226. [CrossRef]
3. Bobbink, R.; Hicks, K.; Galloway, J.; Spranger, T.; Alkemade, R.; Ashmore, M.; Bustamante, M.; Cinderby, S.; Davidson, E.; Dentener, F.; et al. Global assessment of nitrogen deposition effects on terrestrial plant diversity effects of terrestrial ecosystems: A synthesis. *Ecol. Appl.* **2010**, *20*, 30–59. [CrossRef] [PubMed]
4. Silvertown, J.; Poulton, P.; Johnson, A.E.; Edwards, G.; Heard, M.; Biss, P.M. The Park Grass Experiment 1865-2006: Its contribution to ecology. *J. Ecol.* **2006**, *94*, 801–814. [CrossRef]
5. De Schrijver, A.; de Frenne, P.; Ampoorter, E.; van Nevel, L.; Demey, A.; Wuyts, K.; Verheyen, K. Cumulative nitrogen inputs drives species loss in terrestrial ecosystems. *Glob. Ecol. Biogeogr.* **2011**, *20*, 803–816. [CrossRef]
6. Bernhardt-Römermann, M.; Römermann, C.; Pillar, V.D.; Kudernatsch, T.; Fischer, A. High functional diversity is related to high nitrogen availability in a deciduous forest—Evidence from a functional trait approach. *Folia Geobot.* **2010**, *45*, 111–124. [CrossRef]
7. Bernhardt-Römermann, M.; Baeten, L.; Craven, D.; De Frenne, P.; Hédl, R.; Lenoir, J.; Bert, D.; Brunet, J.; Chudomelová, M.; Decocq, G.; et al. Drivers of temporal changes in temperate forest plant diversity vary across spatial scales. *Glob. Chang. Biol.* **2015**, *21*, 3726–3737. [CrossRef]
8. Verheyen, K.; Baeten, L.; De Frenne, P.; Bernhardt-Römermann, M.; Brunet, J.; Cornelis, J.; Decocq, G.; Dierschke, H.; Eriksson, O.; Hédl, R.; et al. Driving factors behind the eutrophication signal in understorey plant communities of deciduous temperate forests. *J. Ecol.* **2012**, *100*, 352–365. [CrossRef]
9. Dirnböck, T.; Grandin, U.; Bernhardt-Römermann, M.; Beudert, B.; Canullo, R.; Forsius, M.; Grabner, M.-T.; Holmberg, M.; Kleemola, S.; Lundin, L.; et al. Forest floor vegetation response to nitrogen deposition in Europe. *Glob. Chang. Biol.* **2014**, *20*, 429–440. [CrossRef]
10. Ferretti, M.; Marchetto, A.; Arisci, S.; Bussotti, F.; Calderisi, M.; Carnicelli, S.; Cecchini, G.; Fabbio, G.; Bertini, G.; Matteucci, G.; et al. On the tracks of Nitrogen deposition effects on temperate forests at their southern European range—an observational study from Italy. *Glob. Chang. Biol.* **2014**, *20*, 3423–3438. [CrossRef] [PubMed]
11. Gilliam, F.S.; Welch, N.T.; Phillips, A.H.; Billmyer, J.H.; Peterjohn, W.T.; Fowler, Z.K.; Walter, C.A.; Burnham, M.B.; May, J.D.; Adams, M.B. Twenty-five year response of the herbaceous layer of a temperate hardwood forest to elevated nitrogen deposition. *Ecosphere* **2016**, *7*, e01250. [CrossRef]
12. Eastman, B.A.; Adams, M.B.; Brzotek, E.R.; Burnham, M.B.; Carrara, J.E.; Kelly, C.; McNeil, B.E.; Walter, C.A.; Peterjohn, W.T. Altered plant carbon partitioning enhanced forest ecosystem carbon storage after 25 years of nitrogen additions. *New Phytol.* **2021**, *230*, 1435–1448. [CrossRef]
13. Patel, K.F.; Fernandez, I.J.; Nelson, S.J.; Malcom, J.; Norton, S.A. Contrasting stream nitrate and sulfate response to recovery from experimental watershed acidification. *Biogeochemistry* **2020**, *151*, 127–138. [CrossRef]
14. Ågren, G.I.; Bosatta, E. Nitrogen saturation of terrestrial ecosystems. *Environ. Pollut.* **1988**, *54*, 185–197. [CrossRef]
15. Skeffington, R.A.; Wilson, E.J. Excess nitrogen deposition: Issues for consideration. *Environ. Pollut.* **1988**, *54*, 159–184. [CrossRef]
16. Aber, J.D.; McDowell, W.H.; Nadelhoffer, K.J.; Magill, A.; Berntson, G.; Kamakea, M.; McNulty, S.G.; Currie, W.; Rustad, L.; Fernandez, I. Nitrogen saturation in temperate forest ecosystems: Hypotheses revisited. *BioScience* **1998**, *48*, 921–934. [CrossRef]
17. Aber, J.D.; Nadelhoffer, K.J.; Steudler, P.; Melillo, J.M. Nitrogen saturation in northern forest ecosystems—Hypotheses and implications. *BioScience* **1998**, *39*, 378–386. [CrossRef]
18. Gilliam, F.S. Forest ecosystems of temperate climatic regions: From ancient use to climate change. *New Phytol.* **2016**, *212*, 871–887. [CrossRef]
19. Lovett, G.M.; Goodale, C.L. A new conceptual model of nitrogen saturation based on experimental nitrogen addition to an oak forest. *Ecosystems* **2011**, *14*, 615–631. [CrossRef]
20. Peterjohn, W.T.; Adams, M.B.; Gilliam, F.S. Symptoms of nitrogen saturation in two central Appalachian hardwood forests. *Biogeochemistry* **1996**, *35*, 507–522. [CrossRef]
21. Gilliam, F.S. Response of the herbaceous layer of forest ecosystems to excess nitrogen deposition. *J. Ecol.* **2006**, *94*, 1176–1191. [CrossRef]
22. Moore, J.A.M.; Anthony, M.A.; Pec, G.J.; Trocha, L.K.; Trzebny, A.; Geyer, K.M.; van Diepen, L.T.A.; Frey, S.D. Fungal community structure and function shifts with atmospheric nitrogen deposition. *Glob. Chang. Biol.* **2021**, *27*, 1349–1364. [CrossRef]
23. Thomas, R.Q.; Canham, C.D.; Weathers, K.C.; Goodale, C.L. Increased tree carbon storage in response to nitrogen deposition in the US. *Nat. Geosci.* **2010**, *3*, 13–17. [CrossRef]
24. Peel, J.L.; Haeuber, R.; Garcia, V.; Russell, A.G.; Neas, L. Impact of nitrogen and climate change interactions on ambient air pollution and human health. *Biogeochemistry* **2013**, *114*, 121–134. [CrossRef]

25. Lloret, J.; Valiela, I. Unprecedented decrease in deposition of nitrogen oxides over North America: The relative effects of emission controls and prevailing air-mass trajectories. *Biogeochemistry* **2016**, *129*, 165–180. [CrossRef]
26. Du, E. Rise and fall of nitrogen deposition in the United States. *Proc. Natl. Acad. Sci. USA* **2016**, *113*, E3594–E3595. [CrossRef]
27. Warner, J.X.; Dickerson, R.R.; Wei, Z.; Strow, L.L.; Wang, Y.; Liang, Q. Increased atmospheric ammonia over the world's major agricultural areas detected from space. *Geophys. Res. Lett.* **2017**, *44*, 2875–2884. [CrossRef] [PubMed]
28. Gilliam, F.S.; Burns, D.A.; Driscoll, C.T.; Frey, S.D.; Lovett, G.M.; Watmough, S.A. Decreased atmospheric nitrogen deposition in eastern North America: Predicted responses of forest ecosystems. *Environ. Pollut.* **2019**, *244*, 560–574. [CrossRef]
29. Högberg, P.; Johannisson, C.; Yarwood, S.; Callesen, I.; Näsholm, T.; Myrold, D.D.; Högberg, M.N. Recovery of ectomycorrhiza after 'nitrogen saturation' of a conifer forest. *New Phytol.* **2011**, *189*, 515–525. [CrossRef]
30. Rowe, E.C.; Jones, L.; Dise, N.B.; Evans, C.D.; Mills, G.; Hall, J.; Stevens, C.J.; Mitchell, R.J.; Field, C.; Caporn, S.J.M.; et al. Metrics for evaluation the ecological benefits of decreased nitrogen deposition. *Biol. Conserv.* **2014**, *212*, 454–463. [CrossRef]
31. Stevens, C.J. How long do ecosystems take to recover from atmospheric nitrogen deposition? *Biol. Conserv.* **2016**, *200*, 160–167. [CrossRef]
32. Groffman, P.M.; Driscoll, C.T.; Durán, J.; Campbell, J.L.; Christenson, L.M.; Fahey, T.J.; Fisk, M.C.; Fuss, C.; Likens, G.E.; Lovett, G.; et al. Nitrogen oligotrophication in northern hardwood forests. *Biogeochemistry* **2018**, *141*, 523–539. [CrossRef]
33. Schmitz, A.; Sanders, T.G.M.; Bolte, A.; Bussotti, F.; Dirnböck, T.; Johnson, J.; Peñuelas, J.; Pollastrini, M.; Prescher, A.-K.; Sardans, J.; et al. Responses of forest ecosystems in Europe to decreasing nitrogen deposition. *Environ. Pollut.* **2019**, *244*, 980–994. [CrossRef]
34. Adams, M.B.; DeWalle, D.R.; Hom, J. *The Fernow Watershed Acidification Study*; Environmental Pollution Series 11; Springer: New York, NY, USA, 2006.
35. Weitzman, S. *Five Years of Research on the Fernow Experimental Forest*; Station Paper NE-61; U.S. Department of Agriculture, Forest Service, Northeastern Forest Experiment Station: Upper Darby, PA, USA, 1953.
36. Gilliam, F.S.; Hockenberry, A.W.; Adams, M.B. Effects of atmospheric nitrogen deposition on the herbaceous layer of a central Appalachian hardwood forest. *J. Torrey Bot. Soc.* **2006**, *133*, 240–254. [CrossRef]
37. Gilliam, F.S.; Roberts, M.R. *The Herbaceous Layer in Forests of Eastern North America*; Oxford University Press, Inc.: New York, NY, USA, 2003.
38. Gilliam, F.S. *The Herbaceous Layer in Forests of Eastern North America*, 2nd ed.; Oxford University Press, Inc.: New York, NY, USA, 2014.
39. Gilliam, F.S.; Turrill, N.L.; Aulick, S.D.; Evans, D.K.; Adams, M.B. Herbaceous layer and soil response to experimental acidification in a Central Appalachian hardwood forest. *J. Environ. Qual.* **1994**, *23*, 835–844. [CrossRef]
40. Gilliam, F.S.; Adams, M.B.; Yurish, B.M. Ecosystem nutrient response to chronic nitrogen inputs at Fernow Experimental Forest, West Virginia. *Can. J. For. Res.* **1996**, *26*, 196–205. [CrossRef]
41. Edwards, P.J.; Kochenderfer, J.N.; Coble, D.W.; Adams, M.B. Soil leachate responses during 10 years of induced whole-watershed acidification. *Water Air Soil Pollut.* **2002**, *140*, 99–118. [CrossRef]
42. Edwards, P.J.; Wood, F.; Kochenderfer, J.N. Baseflow and peakflow chemical responses to experimental applications of ammonium sulphate to forested watersheds in north-central West Virginia, USA. *Hydrol. Process.* **2002**, *16*, 2287–2310. [CrossRef]
43. Gilliam, F.S.; Walter, C.A.; Adams, M.B.; Peterjohn, W.T. Nitrogen (N) dynamics in the mineral soil of a Central Appalachian hardwood forest during a quarter century of whole-watershed N additions. *Ecosystems* **2018**, *21*, 1489–1504. [CrossRef]
44. Gilliam, F.S.; Adams, M.B.; Peterjohn, W.T. Response of soil fertility to 25 years of experimental acidification in a temperate hardwood forest. *J. Environ. Qual.* **2020**, *49*, 961–972. [CrossRef]
45. Adams, M.B.; Angradi, T.R. Decomposition and nutrient dynamics of hardwood leaf litter in the Fernow Whole-Watershed Acidification Experiment. *For. Ecol. Manag.* **1996**, *83*, 61–69. [CrossRef]
46. Carrara, J.E.; Walter, C.A.; Hawkins, J.S.; Peterjohn, W.T.; Averill, C.; Brzostek, E.R. Interactions among plants, bacteria, and fungi reduce extracellular enzyme activities under long-term N fertilization. *Glob. Chang. Biol.* **2017**, *24*, 2721–2734. [CrossRef] [PubMed]
47. Burnham, M.B.; Cumming, J.R.; Adams, M.B.; Peterjohn, W.T. Soluble soil aluminum alters the relative uptake of mineral nitrogen forms by six mature temperate broadleaf tree species: Possible implications for watershed nitrate retention. *Oecologia* **2017**, *185*, 327–337. [CrossRef] [PubMed]
48. Venterea, R.T.; Groffman, P.M.; Castro, M.S.; Verchot, L.V.; Fernandez, I.J.; Adams, M.B. Soil emissions of nitric oxide in two forest watersheds subjected to elevated N inputs. *For. Ecol. Manag.* **2004**, *196*, 335–349. [CrossRef]
49. Peterjohn, W.T.; McGervey, R.J.; Sexstone, A.J.; Christ, M.J.; Foster, C.J.; Adams, M.B. Nitrous oxide production in two forested watersheds exhibiting symptoms of nitrogen saturation. *Can. J. For. Res.* **1998**, *28*, 1723–1732. [CrossRef]
50. Gress, S.E.; Nichols, T.D.; Northcraft, C.C.; Peterjohn, W.T. Nutrient limitation in soils exhibiting differing nitrogen availabilities: What lies beyond nitrogen saturation. *Ecology* **2007**, *88*, 119–130. [CrossRef]
51. Gilliam, F.S.; Billmyer, J.H.; Walter, C.A.; Peterjohn, W.T. Effects of excess nitrogen on biogeochemistry of a temperate hardwood forest: Evidence of nutrient redistribution by a forest understory species. *Atmos. Environ.* **2016**, *146*, 261–270. [CrossRef]
52. DeWalle, D.R.; Tepp, J.S.; Swistock, B.R.; Sharpe, W.E.; Edwards, P.J. Tree-ring cation response to experimental watershed acidification in West Virginia and Maine. *J. Environ. Qual.* **1999**, *28*, 299–309. [CrossRef]

53. May, J.D.; Burdette, S.B.; Gilliam, F.S.; Adams, M.B. Interspecific divergence in foliar nutrient dynamics and stem growth in a temperate forest in response to chronic nitrogen inputs. *Can. J. For. Res.* **2005**, *35*, 1023–1030. [CrossRef]
54. Malcomb, J.D.; Scanlon, T.M.; Epstein, H.E.; Druckenbrod, D.L.; Vadeboncoeur, M.A.; Lanning, M.; Adams, M.B.; Wang, L. Assessing temperate forest growth and climate sensitivity in response to a long-term whole-watershed acidification experiment. *J. Geophys. Res. Biogeosci.* **2020**, *125*, e2019JG005560. [CrossRef]
55. Walter, C.A.; Raiff, D.T.; Burnham, M.B.; Gilliam, F.S.; Adams, M.B.; Peterjohn, W.T. Nitrogen fertilization interacts with light to increase *Rubus* spp. cover in a temperate forest. *Plant Ecol.* **2016**, *217*, 421–430. [CrossRef]
56. Boxman, A.W.; van der Ven, P.J.M.; Roelofs, J.G.M. Ecosystem recovery after a decrease in nitrogen input to a Scots pine stand at Ysselsteyn, the Netherlands. *For. Ecol. Manag.* **1998**, *101*, 155–163. [CrossRef]
57. Strengbom, J.; Nordin, A.; Nasholm, T.; Ericson, L. Slow recovery of boreal forest ecosystem following decreased nitrogen input. *Funct. Ecol.* **2001**, *15*, 451–457. [CrossRef]
58. Booth, M.S.; Stark, J.M.; Rastetter, E. Controls on nitrogen cycling in terrestrial ecosystems: A synthetic analysis of literature data. *Ecol. Monogr.* **2005**, *75*, 139–157. [CrossRef]
59. Gilliam, F.S.; May, J.D.; Adams, M.B. Response of foliar nutrients of Rubus allegheniensis to nutrient amendments in a central Appalachian hardwood forest. *For. Ecol. Mang.* **2018**, *411*, 101–107. [CrossRef]
60. Wen, Z.; Xu, W.; Li, Q.; Han, M.; Tang, A.; Zhang, Y.; Luo, X.; Shen, J.; Wang, W.; Li, K.; et al. Changes of nitrogen deposition in China from 1980 to 2018. *Environ. Int.* **2020**, *144*, 106022. [CrossRef]
61. Nieland, M.A.; Moley, P.; Hanschu, J.; Zeglin, L.H. Differential resilience of soil microbes and ecosystem functions following cessation of long-term fertilization. *Ecosystems* **2021**. [CrossRef]
62. Newcomer, M.E.; Bouskill, N.J.; Wainwright, H.; Maavara, T.; Arora, B.; Siirila-Woodburn, E.R.; Dwivedi, D.; Williams, K.H.; Steefel, C.; Hubbard, S.S. Hysteresis patterns of watershed nitrogen retention and loss over the past 50 years in United States hydrological basins. *Glob. Biogeochem. Cycles* **2021**, *35*, e2020GB006777. [CrossRef]

MDPI
St. Alban-Anlage 66
4052 Basel
Switzerland
Tel. +41 61 683 77 34
Fax +41 61 302 89 18
www.mdpi.com

Forests Editorial Office
E-mail: forests@mdpi.com
www.mdpi.com/journal/forests